Advances in Evaporation and Evaporative Demand

Advances in Evaporation and Evaporative Demand

Editors

Aristoteles Tegos
Nikolaos Malamos

MDPI • Basel • Beijing • Wuhan • Barcelona • Belgrade • Manchester • Tokyo • Cluj • Tianjin

Editors
Aristoteles Tegos
School of Civil Engineering
National Technical University
Athens
Greece

Nikolaos Malamos
Department of Agriculture
University of Patras
Amaliada
Greece

Editorial Office
MDPI
St. Alban-Anlage 66
4052 Basel, Switzerland

This is a reprint of articles from the Special Issue published online in the open access journal *Hydrology* (ISSN 2306-5338) (available at: www.mdpi.com/journal/hydrology/special_issues/evaporation).

For citation purposes, cite each article independently as indicated on the article page online and as indicated below:

LastName, A.A.; LastName, B.B.; LastName, C.C. Article Title. *Journal Name* **Year**, *Volume Number*, Page Range.

ISBN 978-3-0365-4254-6 (Hbk)
ISBN 978-3-0365-4253-9 (PDF)

Contents

About the Editors

Aristoteles Tegos

Aristoteles Tegos (Dr.) is a Senior Engineer at Ryan Hanley Ltd, Ireland and a Researcher at the National Technical University of Athens- School of Civil Engineering. He has a PhD in the fields of Hydrology, and his primary interests are related to evapotranspiration, flooding, and environmental flow modelling. His research activity focuses on the definition of data-driven methods with applications in multiple fields, including fluvial flood risk assessment, hydrological models and river monitoring.

Nikolaos Malamos

Nikolaos Malamos is Associate Professor of Irrigation and Drainage, Modelling of Soil-Plant-Atmosphere interactions, at the Department of Agriculture, of the University of Patras He graduated from the Agricultural University of Athens, he received his M.Sc. from the School of Civil Engineering at the National Technical University of Athens and his Doctorate in Agricultural Sciences from the Department of Natural Resources and Agricultural Engineering. He received a scholarship from the State Scholarships Foundation, for the attainment of his doctoral thesis in the field of Agricultural Hydraulics –Irrigation. He worked in the private and public sector as an irrigation expert. He participated in several research projects on the field of irrigation, hydroinformatics and geographic information systems. His activity focuses on Water Resources Management (mainly on agricultural uses), in Spatial Analysis and Hydroinformatics. He is a member of the Editorial Board (Associate Editor) of the Hydrological Sciences Journal, which is the official Journal of the International Association of Hydrological Sciences –IAHS.

hydrology MDPI

Editorial

Advances in Evaporation and Evaporative Demand

Nikolaos Malamos [1,*] and **Aristoteles Tegos** [2]

[1] Department of Agriculture, University of Patras, GR-27200 Amaliáda, Greece
[2] Department of Water Resources and Environmental Engineering, National Technical University of Athens, Heroon Polytechneiou 5, GR-15780 Athens, Greece; tegosaris@itia.ntua.gr
* Correspondence: nmalamos@upatras.gr

Citation: Malamos, N.; Tegos, A. Advances in Evaporation and Evaporative Demand. *Hydrology* 2022, 9, 78. https://doi.org/10.3390/hydrology9050078

Received: 26 April 2022
Accepted: 5 May 2022
Published: 6 May 2022

Publisher's Note: MDPI stays neutral with regard to jurisdictional claims in published maps and institutional affiliations.

1. Introduction

The importance of evapotranspiration is well-established in various disciplines such as hydrology, agronomy, climatology, and other geosciences. Reliable estimates of evapotranspiration are also vital to develop criteria for in-season irrigation management, water resource allocation, long-term estimates of water supply, demand and use, design and management of water resources infrastructure, and evaluation of the effect of land use and management changes on the water balance.

The objective of the Special Issue "Advances in Evaporation and Evaporative Demand" was to define and discuss several related terms, including potential, reference, and actual evapotranspiration, and to present a wide spectrum of innovative research papers and case studies.

In this Special Issue there were eleven contributions that tackled the aforementioned goals.

2. Contributed Papers

The articles in this Special Issue address a wide variety of topics reflecting the challenges mentioned above ranging from urban hydrology to global evapotranspiration modelling:

The paper **"Determinants of Evapotranspiration in Urban Rain Gardens: A Case Study with Lysimeters under Temperate Climate"** [1] by Ahmeda Assann Ouédraogo, Emmanuel Berthier, Brigitte Durand and Marie-Christine Gromaire explores ET in urban rain gardens, a topic receiving more and more attention from both rain garden designers for a better consideration of ET in their designs and hydrology researchers for a more accurate description of the flux in the urban context. The city of Paris has instrumented eight rain garden lysimeters to obtain a better understanding and prediction of their hydrological behavior. In order to extrapolate on real situations, experimental rain gardens of reduced size and well-known structures were designed. Monitoring was carried out with lysimeters, i.e., mechanisms that enable the water balance components (exfiltration, water storage, etc.) to be observed, with measurements made by weighing variations in water content of the lysimeter. The aim was also to test different vegetation configurations and internal storage options, and to implement replicas in order to test the validity of the measurements. Thehe purpose of this study consists of three main points: estimating the actual evapotranspiration (ET) of these rain gardens at daily steps; assessing the impact of different configurations on ET fluxes; and comparing the actual ETs obtained from the lysimeters with reference to ET values, such as evaporation, from a pan evaporimeter and some models taken from the literature. The seasonal dynamics and the relative significance of each determinant of ET in the rain gardens were highlighted and the results could be used to investigate the modelling of hydrological processes in urban rain gardens.

The paper **"Stochastic Analysis of Hourly to Monthly Potential Evapotranspiration with a Focus on the Long-Range Dependence and Application with Reanalysis and Ground-Station Data"** [2] by Panayiotis Dimitriadis, Aristoteles Tegos and Demetris

Koutsoyiannis, explores the stochastic structure of the potential evapotranspiration process, ranging from hourly to climatic scales, in terms of Hurst–Kolmogorov (HK) dynamics, which describes all the processes exhibiting the Hurst phenomenon (i.e., with a power-law autocorrelation function at large scales). They focused on the marginal structure of the PE process as fitted through the Pareto–Burr–Feller distribution. Both marginal and second-order dependence structures of the HK dynamics were estimated and compared to the ones identified from global-scale analyses in other key hydrometeorological processes that form the hydrological-cycle path driven by atmospheric turbulence, such as temperature, wind, solar radiation, and relative humidity. It was found that both the marginal probability distributions of PEV and PET are lighttailed when estimated through the Pareto–Burr–Feller distribution function. Additionally, the long-range dependence of both the PEV and PET was found to be of moderate strength, quantified through a Hurst parameter of 0.64 and 0.69, respectively. Both PET and PEV can be placed between the stochastic structures of temperature, relative humidity, solar radiation, and wind speed (i.e., strong LRD and light-to medium-tail) and the precipitation's structures (i.e., weak LRD and heavy tail).

The paper "**Precipitation and Potential Evapotranspiration Temporal Variability and Their Relationship in Two Forest Ecosystems in Greece**" [3] by Stefanos Stefanidis and Vasileios Alexandridis, aimed to investigate temporal variability and detect trends in drought conditions in two different types of forest ecosystems using long–term timeseries meteorological data from mountainous meteorological stations. For this purpose, the ratio of precipitation to potential evapotranspiration was used as a proxy indicator for the evaluation of drought conditions at different timescales (annual/seasonal). The Mann–Kendal and Sen's slope methods were applied in order to evaluate the significance and magnitude of the tendency, and to identify the time of abrupt changes. The results indicated that humid conditions prevail in both forest areas and that dry conditions occur in summer. The examined parameters present significant variability between seasons, following the Mediterranean climate pattern. The trend analysis showed that the reported upward and downward trends in Aridity Index are, in general, statistically insignificant, and the magnitude of the trend is considered negligible.

The paper "**Evaluation of Evaporation from Water Reservoirs in Local Conditions at Czech Republic**" [4] by Eva Melišová, Adam Vizina, Martin Hanel, Petr Pavlík and Petra Šuhájková, aimed to explore the relationships for the calculation of evaporation from water surface in the Czech Republic using reanalyzed climate data and the constructed linear models (LM) and random forest models (RFM) for the calculation of evaporation. The main objective of the evaporation estimation from the water surface was to derive a universal relationship for the whole territory of the Czech Republic. The derivation of the relationship for evaporation was based on the multiple linear regression method, where the values of the dependent variable (evaporation) were sought, based on two or more variables (predictors: air temperature, surface temperature, wind speed, surface net solar radiation, dew point, surface pressure, dew point, altitude, latitude, longitude and calculated humidity). The construction of the models was performed (i) manually, where the evaluation was perfomed using the AIC parameter and the quantile–quantile was used for visual diagnostics, (ii) using stepwise regression, where the predictorswere entered sequentially and models from one to X-selected variables were generated. Random forest regression was used to account for non-linear relationships. Linear and random forest regression models were cross-validated and evaluated using criterion functions (R^2, RMSE, MAE and RERR). Finally, 3(+1) LM models and 3 RF models were selected. It turned out that geomorphological information (elevation and location) appeared more in the manually derived models than in the to models constructed using the stepwise regression method. In the comparison between linear models (LM) and random forest models (RFM), LM was found to have much more variability in the outcome compared to the RFM. Among the best models that were evaluated by linear regression, models LM1 from the manual linear regression group and LM12 from the stepwise regression group were used. Model LM1 was selected as the best model among the six predictors. The LM1 model can be replaced by an alternative model LM12 which also performed satisfactorily in terms of four predictors.

The paper "**Integrating Drone Technology into an Innovative Agrometeorological Methodology for the Precise and Real-Time Estimation of Crop Water Requirements**" [5] by Stavros Alexandris, Emmanouil Psomiadis, Nikolaos Proutsos, Panos Philippopoulos, Ioannis Charalampopoulos, George Kakaletris, Eleni-Magda Papoutsi, Stylianos Vassilakis and Antonios Paraskevopoulos, aimed to present a new methodology and the equipment used in the assessment of crop water stress by spatial measurements of canopy temperature, air temperature, and relative humidity from sensors incorporated into an unmanned aerial vehicle (UAV) from a pilot implementation in a potato cultivation field. The functionality of the proposed system was certified (accuracy of the UAV path and flight altitude, reliability of the aerial data acquisition system, communication stability between UAVs and ground base). Their findings indicated that the canopy temperatures derived from the ground meteorological station, the onboard aerial micrometeorological system, and the portable IRT radiometers produced a suitable thermal image from the surface of the crop. The subsystems can be useful for supporting applications that are significant for irrigation water management and programming, such as irrigation alerting and scheduling, crop surveillance, and irrigation water management. However, the authors state that more efforts are necessary to make these technologies more user-friendly and available for all end users, covering different advantages for a precise crop water stress evaluation.

The paper "**Estimation of Daily Potential Evapotranspiration in Real-Time from GK2A/AMI Data Using Artificial Neural Network for the Korean Peninsula**" [6] by Jae-Cheol Jang, Eun-Ha Sohn, Ki-Hong Park and Soobong Lee, developed a model that estimates the daily PET based on ANN using the GEOstationary Korea Multi-Purpose SATellite 2A (GEO-KOMPSAT 2A, GK2A). The objective was to retrieve real-time daily ET with a spatial resolution of 1 km for hydrological resource monitoring on the Korean Peninsula. To reflect the complex relationships and nonlinearity between the GK2A-derived data and ET, the precipitation and digital elevation data were used as input for the ANN. Daily PET from KMA were used as reference data for the ANN model training. The accuracy of the model was verified by comparing the modeled data with the ET from in-situ measurements of the KMA and National Institute of Forest Science (NIFoS). In comparison with the station-derived PM-ET, the ANN-based derived PET showed high accuracy, while validating the spatial distribution and the ANN model-estimated daily PET showed high accuracy at all KMA stations. Additionally, the derived PET models performed particularly better than the Terra/MODIS PET product for the eastern coastal region of the Korean Peninsula, where the elevation changes dramatically.

The paper "**Simplified Interception/Evaporation Model**" [7] by Giorgio Baiamonte, explored the rainfall partitioning in net rainfall and evaporation losses by the canopy, using a very simplified sketch of the interception process, which combines a modified exponential equation from the literature (Merrian model), accounting for the antecedent volume stored on the canopy, and a simple power-law equation to compute the evaporation by the wet canopy. Even though the considered approach is far from the physically based approaches, the latter may require many parameters that are not easy to determine. It is shown that the simplified parsimonious approach may lead to a reasonable quantification of this important component of the hydrologic cycle, which can be useful when a rough estimate is required, in the absence of a detailed characterization of the canopy and of the climate conditions. It is also shown that the Merrian model can be derived by considering a simple linear storage model. The application of the suggested procedure was performed for faba bean cover crop, which was described according to the general lengths of four distinct growth stages considered in FAO56, whereas LAI and interception capacity were obtained from the literature. Since few are parameters required, this simple approach could be applied at large scale when a rough estimate of evaporation loss by wet canopy is necessary, in the absence of a detailed characterization of canopy and climate.

The paper "**Sensitivity of the Evapotranspiration Deficit Index to Its Parameters and Different Temporal Scales**" [8] by Frank Joseph Wambura, investigated the sensitivity of the evapotranspiration deficit index (ETDI) to its parameters, and to data at different temporal scales. The parameter sensitivity test revealed that ETDI is less sensitive when the (a, b)-parameters range from (0.1, 1.8) to (0.5, 1.0) inclusive, and more sensitive when they approach (0.9, 0.2). Since the ETDI is sensitive to different parameter combinations, the selection of an optimal parameter combination might rely on information from specific locations. Moreover, an optimal parameter combination can also be obtained when ETDI is calibrated against other drought indices or durations of historically severe drought events. The temporal scale sensitivity test at the twelve points in the river basin showed that the number of drought events, the total drought durations, and durations per event decrease as the temporal scale increases. Therefore, small temporal scale ET data are highly recommended to increase the accuracy of ETDI-based drought characteristics.

The paper "**Estimation of Reference Evapotranspiration Using Spatial and Temporal Machine Learning Approaches**" [9] by Ali Rashid Niaghi, Oveis Hassanijalilian and Jalal Shiri, investigated the effect of different input combinations of meteorological data on the accuracy of daily ETo estimation in subhumid climate using gene expression programming (GEP), support vector machine (SVM), multiple linear regression (LR), and random forest (RF) methods. They compared the spatial and local prediction capabilities of the different machine learning (ML) techniques in ETo estimation and evaluated the performance of the models based on the various study years and meteorological stations. The comparison of the performance accuracy of the applied models revealed that the RF model was, in general, the best for all combinations among the four defined models, in general. The LR, GEP, and SVM models were improved when a local approach was used, except for the RF model, which was less accurate with a local approach. The radiation-based combination was the most accurate predictor among all models tested. The results showed that due to the flat topography of the study area with high wind speeds during the growing season, the inclusion of the wind used as a parameter to build the model architecture and estimate the ETo could increase the accuracy of the prediction. In addition, it might be more practical to apply the spatial RF model for stations with missing meteorological data without the need for local training. The recommended application of spatial RF using radiation combination allows for a more reliable estimate of ETo to fill the missing values for more precise water management purposes.

The paper "**Evapotranspiration Trends and Interactions in Light of the Anthropogenic Footprint and the Climate Crisis: A Review**" [10] by Stavroula Dimitriadou and Konstantinos G. Nikolakopoulos, reviewed emerging ET trends over the latest decades in areas with different environmental conditions in the context of the ongoing climate change. Additionally, they focused on critical components such as the anthropogenic impact on ET and, the mechanisms in which ET participates in forest land-cover and wildfires, croplands (irrigation and cultivation practices), groundwater (quantity and quality), and ambient air. Five broad conclusions were deducted: First, Mediterranean climate regions (MCRs) appear to be vulnerable to the impacts of the ongoing increase in ET, especially during summertime, due to the ongoing precipitation shifting in winter and the air temperature warming (especially the rise in the minimum air temperature values) which is expected to be more severe in MCRs such as Southern Europe, in the summertime. Air temperature is considered a proxy of the energy state of the system. In water-limited areas, evaporative fractions can serve as a water-stress indicator. Second, the ET in tropical forests plays a rather beneficial role since it moderates the flooding risk during the wet season resulting in a net cooling effect. Third, in semi-arid to arid areas, an increase in ET and especially of evaporation constitutes an important problem due to sustained baseflow recessions which exacerbate the limited water availability. In these drought-prone areas, ET exacerbates soil salinization. Fourth, the relationship between ET and wildfires is of major importance. The impacts are site-specific and climate, and fire-severity-dependent. The hydrological processes may be altered if a critical amount of canopy loss occurs (e.g., 20% for semi-arid regions, 45% for tropical forests) occurs. Concurrently, the Reference evapotranspiration could serve as a fuel aridity measure to assess forest fire risk. Fifth, along with climate change, the consequences of human activity such as air pollution (aerosols, CO_2 emissions), land use/land cover shifting to agricultural uses with intensive productivity practices, large reforestation implementation, and large constructions (e.g., dams, dense and high urban buildings) have substantially changed the actual evapotranspiration rates during recent decades. Via the human footprint, the interpretation of the evaporation paradox has been made plausible.

Finally, Aristoteles Tegos, Nikolaos Malamos and Demetris Koutsoyiannis in their paper "**RASPOTION—A New Global PET Dataset by Means of Remote Monthly Temperature Data and Parametric Modelling**" [11], introduced a new monthly global PET dataset, named RASPOTION, by implementing the Parametric model with remote sensing data of mean air temperature, provided by a recent remote mean temperature dataset from 2003 to 2016. The dataset was validated with in situ samples (USA, Germany, Spain, Ireland, Greece, Australia, China) and by using spatial Penman–Monteith estimates in England. Overall, for the majority of the Earth's surface, RASPOTION constitutes a reliable monthly PET dataset, freely available to scientists across different research disciplines in order to assist scientific studies into the global hydrological cycle and decisions for both short- and long-term hydro-climatic policy actions.

3. Conclusions

Since we have been conducting research in the field of evaporation assessment for more than a decade and considering the remaining challenges within evaporation assessment research, this SI was a great opportunity to discover and promote new trends in evaporation analysis.

The state-of- the art review study research presented by MchMahon et al. [12] identifies six areas for further research which are: (i) hard-wired potential evaporation estimates; (ii) estimating evaporation without wind data; (iii) estimating evaporation without at-site data; (iv) dealing with an environment undergoing climate: increasing annual air temperature but decreasing pan evaporation rates; (v) daily meteorological data averaging over 24 h or day-light hours only; and (vi) finally, uncertainty in evaporation estimates. In addition to the above key research topics advanced remote sensing techniques can be further support the water engineering and scientific community and some featured papers with modern views have been presented in our SI along with the key outstanding issues presented above.

As Guest Editors, we are sharing our enthusiasm with the successful completion of the SI and we trust that the selected research papers will be a valuable contribution to the domain of geosciences in the years to come.

Author Contributions: Writing—original draft preparation, N.M.; writing—review and editing, A.T. All authors have read and agreed to the published version of the manuscript.

Funding: This research received no external funding.

Acknowledgments: We would like to acknowledge the efforts of all authors who contributed to this Special Issue.

Conflicts of Interest: The authors declare no conflict of interest.

References

1. Ouédraogo, A.A.; Berthier, E.; Durand, B.; Gromaire, M.-C. Determinants of Evapotranspiration in Urban Rain Gardens: A Case Study with Lysimeters under Temperate Climate. *Hydrology* **2022**, *9*, 42. [CrossRef]
2. Dimitriadis, P.; Tegos, A.; Koutsoyiannis, D. Stochastic analysis of hourly to monthly potential evapotranspiration with a focus on the long-range dependence and application with reanalysis and ground-station data. *Hydrology* **2021**, *8*, 177. [CrossRef]
3. Stefanidis, S.; Alexandridis, V. Precipitation and potential evapotranspiration temporal variability and their relationship in two forest ecosystems in greece. *Hydrology* **2021**, *8*, 160. [CrossRef]
4. Melišová, E.; Vizina, A.; Hanel, M.; Pavlík, P.; Šuhájková, P. Evaluation of evaporation from water reservoirs in local conditions at czech republic. *Hydrology* **2021**, *8*, 153. [CrossRef]

5.   Alexandris, S.; Psomiadis, E.; Proutsos, N.; Philippopoulos, P.; Charalampopoulos, I.; Kakaletris, G.; Papoutsi, E.-M.; Vassilakis, S.; Paraskevopoulos, A. Integrating Drone Technology into an Innovative Agrometeorological Methodology for the Precise and Real-Time Estimation of Crop Water Requirements. *Hydrology* **2021**, *8*, 131. [CrossRef]
6.   Jang, J.C.; Sohn, E.H.; Park, K.H.; Lee, S. Estimation of daily potential evapotranspiration in real-time from gk2a/ami data using artificial neural network for the korean peninsula. *Hydrology* **2021**, *8*, 129. [CrossRef]
7.   Baiamonte, G. Simplified Interception/Evaporation Model. *Hydrology* **2021**, *8*, 99. [CrossRef]
8.   Wambura, F.J. Sensitivity of the evapotranspiration deficit index to its parameters and different temporal scales. *Hydrology* **2021**, *8*, 26. [CrossRef]
9.   Niaghi, A.R.; Hassanijalilian, O.; Shiri, J. Estimation of reference evapotranspiration using spatial and temporal machine learning approaches. *Hydrology* **2021**, *8*, 25. [CrossRef]
10.   Dimitriadou, S.; Nikolakopoulos, K.G. Evapotranspiration trends and interactions in light of the anthropogenic footprint and the climate crisis: A review. *Hydrology* **2021**, *8*, 163. [CrossRef]
11.   Tegos, A.; Malamos, N.; Koutsoyiannis, D. RASPOTION—A New Global PET Dataset by Means of Remote Monthly Temperature Data and Parametric Modelling. *Hydrology* **2022**, *9*, 32. [CrossRef]
12.   McMahon, T.A.; Peel, M.C.; Lowe, L.; Srikanthan, R.; McVicar, T.R. Estimating actual, potential, reference crop and pan evaporation using standard meteorological data: A pragmatic synthesis. *Hydrol. Earth Syst. Sci.* **2013**, *17*, 1331–1363. [CrossRef]

Article

Determinants of Evapotranspiration in Urban Rain Gardens: A Case Study with Lysimeters under Temperate Climate

Ahmeda Assann Ouédraogo [1,*], Emmanuel Berthier [1,*], Brigitte Durand [2] and Marie-Christine Gromaire [3]

1 Equipe TEAM, Centre d'Etudes et d'Expertise sur les Risques, l'Environnement, la Mobilité et l'Aménagement (Cerema), 12 rue Teisserenc de Bort, F 78190 Trappes, France
2 Division Etudes et Ingénierie, Direction de la Propreté et de l'Eau, Service Technique de l'Eau et de l'Assainissement (DPE-STEA), 27 rue du Commandeur, F 75014 Paris, France; brigitte.durand1@paris.fr
3 Leesu, Ecole des Ponts, Université Paris Est Creteil, F 77455 Marne-la-Vallee, France; marie-christine.gromaire@enpc.fr
* Correspondence: ahmeda.ouedraogo@cerema.fr (A.A.O.); emmanuel.berthier@cerema.fr (E.B.)

Abstract: Accurate evaluation of evapotranspiration (ET) flux is an important issue in sustainable urban drainage systems that target not only flow rate limitations, but also aim at the restoration of natural water balances. This is especially true in context where infiltration possibilities are limited. However, its assessment suffers from insufficient understanding. In this study, ET in 1 m^3 pilot rain gardens were studied from eight lysimeters monitored for three years in Paris (France). Daily ET was calculated for each lysimeter based on a mass balance approach and the related uncertainties were assessed at ±0.42 to 0.58 mm. Results showed that for these lysimeters, ET is the major term in water budget (61 to 90% of the precipitations) with maximum values reaching 8–12 mm. Furthermore, the major determinants of ET are the existence or not of an internal water storage and the atmospheric factors. The vegetation type is a secondary determinant, with little difference between herbaceous and shrub configurations, maximum ET for spontaneous vegetation, and minimal values when vegetation was regularly removed. Shading of lysimeters by surroundings buildings is also important, leading to lower values. Finally, ET of lysimeters is higher than tested reference values (evaporimeter, FAO-56, and local Météo-France equations).

Keywords: evapotranspiration estimation; urban rain gardens; lysimeters; evapotranspiration models

Citation: Ouédraogo, A.A.; Berthier, E.; Durand, B.; Gromaire, M.-C. Determinants of Evapotranspiration in Urban Rain Gardens: A Case Study with Lysimeters under Temperate Climate. *Hydrology* **2022**, *9*, 42. https://doi.org/10.3390/hydrology9030042

Academic Editor: Aristoteles Tegos

Received: 31 January 2022
Accepted: 21 February 2022
Published: 23 February 2022

Publisher's Note: MDPI stays neutral with regard to jurisdictional claims in published maps and institutional affiliations.

1. Introduction

Urbanization has a great impact on cities' hydrological cycle: runoff is increased to the detriment of infiltration and evapotranspiration (ET), leading to an increase in risks linked to flooding and deterioration of the receiving environments. Urban stormwater management policies have been developed in recent years that favour runoff management in green infrastructure systems (GIS) in order to store the water before to infiltrate, evaporate and transpire it. These sustainable urban drainage systems (SUDS) are considered as a viable mechanism that can substitute or complete the traditional sewerage system (canalisation, underground basins, pipes, etc.) and also provide environmental benefits apart from hydraulic services [1–4]. SUDS uses a set of GIS, such as green roofs, rain gardens, infiltration basins, rain trees, etc.

Rain gardens are recognised as one of the best stormwater management practices in countries such as Northern Europe, the United States, Canada, Japan, and Australia, since in addition to reducing the runoff, they also allow for water treatment and promote biodiversity in the urban environment [5,6]. Rain gardens are, by definition, a local structure with a shallow depression that receives rainwater from upstream can infiltrate, evaporate, transpire, or treat this water [7,8]. Significant hydrological processes in a rain garden include the exfiltration to the underlying soil or by drainage system, the evapotranspiration and the interception from vegetation. These processes "should work together" for being able to control large flows and reduce the total volume of small storms [9].

Jennings et al. [10], in a study on the efficiency of residential rain gardens in terms of runoff reduction, in Ohio, temperate climate in the USA, attribute a major role to the exfiltration process and a minor role to the evapotranspiration as regards their contributions of 85% and 0.32% respectively, in reducing runoff volumes. However, more recent experimental studies [11–16] have shown a greater importance of ET in GIS. In rain gardens, daily ET rates are generally low, around 1–5 mm per day, a rate that is sufficient to restore the retention capacity of the structure between two rain events [17]. Studies estimated the ET between 43 and 70% [8] and sometimes up to 78% of the collected rainfall [13].

The ET is known as a dynamic process and it depends on meteorological factors (e.g., precipitation characteristics, air relative humidity and temperature and wind speed), GIS properties (e.g., drainage system, soil, etc.), and vegetation [15]. While the ET process has been investigated widely in agriculture, it remains relatively unknown in urban areas, and particularly in SUDS. Even though progress has been made in the study of ET in urban areas, in particular with the development of approaches based on remote sensing, the current models are still imprecise and do not always account for all the specificities (spatial heterogeneity, microclimatic variability, etc.) associated with the urban environment [18–21].

In the hydrological modelling aspect of these GIS structures, the representation of the water transfer processes in the soil (infiltration of water in the soil, exfiltration, etc.) have been prioritized in the preliminary studies. A review of 11 urban hydrological models used for modelling in SUDS, including rain gardens, by Kaykhosravi et al. [19] also noted that despite recent improvements in existing models, their ability to model multi-layered soil systems, trees or vegetation processes (interception, absorption, and evapotranspiration), snowmelt, and runoff at different spatial scales is limited and further research are needed. In these hydrological models, the ET is usually estimated and represented by predictive equations based on physical approaches that require significant input data (Penman–Monteith [22] model is a reference and its variants of Fao-56 [23] or ASCE [24] methods) or other more conceptual approaches that use less data (Hargreaves and Allen [25], Priesley–Taylor [26]). These predictive equations have been evaluated with the estimated ET in pilot rain gardens lysimeters in the literature [12–14,27]. The findings of these research show that the classical equations for ET are not always satisfactory with either underestimations or overestimations of the observed ET data. Another method proposed by Hess et al. [17], and based on water content measurements at different soil depths seems to be less expensive in terms of input data, and provides comparable results to the classical assessment methods of Penman–Monteith [22] and Hargreaves and Allen [25]. The main limitation of using water content profiles can be their non-representativeness of the spatial variation in water content in gardens due to its important heterogeneity.

For urban rain gardens, recent research has shown the significance of ET, but there are not enough case studies estimating the flux and the factors involved. Note that this lack is particularly related to the difficulty in measuring the flux on the one hand and, on the other hand, the fact that some preliminary studies have minimised its importance [15]. Thus, to the challenges of stormwater management and also urban heat islands, ET in rain gardens is a topic receiving more and more attention from both rain garden designers for a better consideration of ET in the design and hydrology researchers for a more accurate description of the flux in the urban context.

In some countries, such as Australia and the United States, legislation is already taking form to include ET in the design of rain gardens [17]. In France, the Paris Council with its "ParisPluie" plan seeks to develop the rain garden method [28,29]. The city has instrumented eight rain garden lysimeters for a better understanding and prediction of their hydrological behaviour. In order to extrapolate on real situations, experimental rain gardens of reduced size and well-known structures were designed. Monitoring was carried out with lysimeters, i.e., mechanisms that enable the water balance components (exfiltration, water storage, etc.) to be observed, with measurements by weighing the variations in water content of the lysimeter. The aim was also to test different vegetation configurations and internal storage options, and to implement replicas in order to test the validity of the measurements. In this study, the purpose consists of three main points: estimate the actual evapotranspiration (ET) of these rain gardens at daily steps; assess the impact of different configurations on ET fluxes; and compare the actual ETs obtained from the lysimeters with reference to ET values, such as evaporation, from a pan evaporimeter and some models taken from the literature.

2. Materials and Methods

2.1. General Context of the Study Area

The site is located at 43 rue Buffon in Paris, France, within the Museum National d'Histoire Naturelle (MNHN) (Figure 1). In the Paris region, there is no strong topographical contrast and the agglomeration of Paris is very dense, with an estimated population of nearly two million people and 9 million in the 1500 cities and villages that constitute its suburbs [29]. Paris has a fairly temperate climate, with moderately warm summers (average temperature of 19 °C in July) and moderately cold winters (average temperature of 3 °C in January), with rare snow. The urban dominance leads to urban heat islands (UHIs), characterised by night-time temperatures that are about 2.5 °C higher (annual average) compared to rural areas [30]. The average annual rainfall of 650 mm is evenly distributed over the year and the annual potential ET is in average around 850 mm with higher values in summer and limited values in winter (data from Météo-France, The French Meteorological Service).

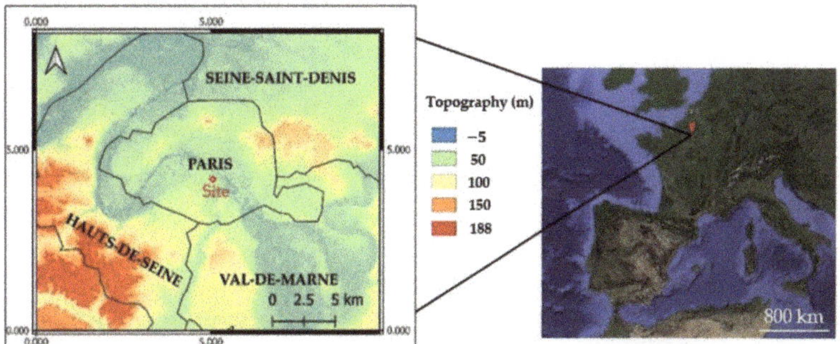

Figure 1. Situation of the study area (a red point) in the city of Paris (France), with the coordinate system of RGF93, Lambert 93. Topographic data source is from the site urs.earthdata.nasa.gov accessed on 24 February 2022.

2.2. Experimental Set Up, Data Acquisition, and Validation

2.2.1. Experimental Set Up

A concrete slab of about 35 m^2 supports eight lysimeters, each one made up of a 1 m^3 pilot rain garden (1 m × 1 m × 1 m) and a cone to increase the impluvium to 4 m^2 (Figure 2). Near the lysimeters, a meteorological station (Figure 2a), which consists of a pyranometer, an anemometer placed at a height of 2 m from the surface, a temperature sensor, and a hygrometer provides climatic data (global radiation, wind speed and direction, air temperature and humidity, and atmospheric pressure).

As the lysimeters are above the soil and therefore not insulated thermally compared to a situation in the ground, a 10 cm of expanded polystyrene insulation was added to all the vertical walls of lysimeters. At the bottom of each lysimeter, a 0.2 m layer of a manufactured alveolar product is installed to store rainwater (Nidaplast® product with a void index of 0.95) and a piezometer is installed to measure the water level in the internal water storage (IWS). The soil, with a thickness of 0.8 m in each lysimeter represents a natural silty-clay soil used in the city's parks and gardens of Paris region; it contains little limestone, 18 to 25%

of clay, with a neutral to basic pH (7.5–8). Weighing cells and a tipping bucket allow the measurement of mass variation and exfiltration at the bottom of each lysimeter respectively (see Figure 3). A pan evaporimeter with a diameter of 1.2 m was also installed to control the quantity of water evaporated.

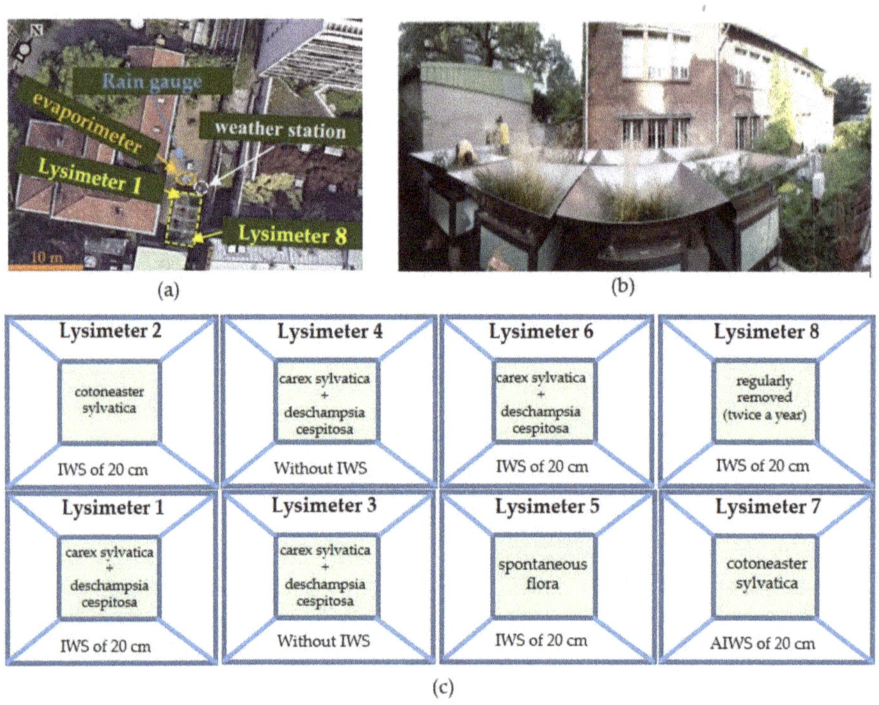

Figure 2. A top and panoramic views of the site in figures (**a**) (Source: google earth) and (**b**), respectively. The figure (**c**) illustrates the positions and the scientific names of vegetation in each lysimeter.

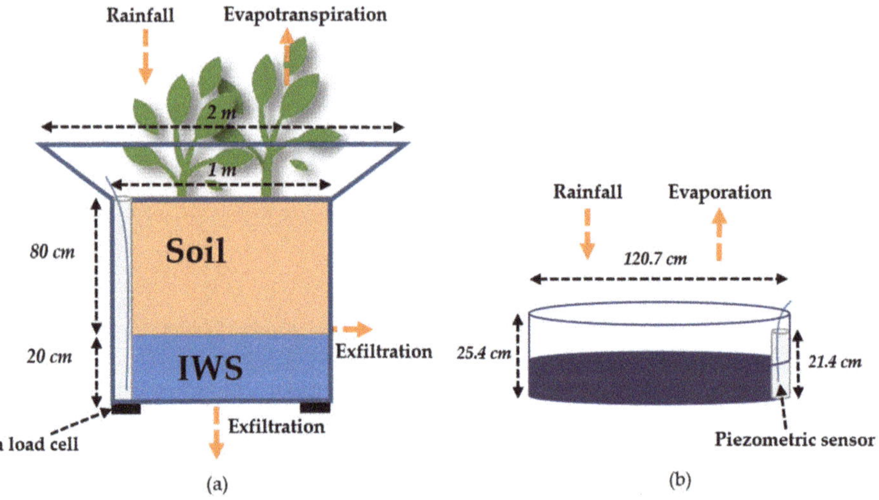

Figure 3. Schematical representation of water fluxes on lysimeters (**a**) and evaporimeter with an overflow of 21.4 cm (**b**). IWS refers to the internal water storage.

The configurations of the lysimeters are numbered from 1 to 8—they differ by the vegetation type and the drainage conditions (presence or not of IWS) (Figure 2c):

- The reference configuration (lysimeters 1 and 6) includes the internal water storage (IWS; i.e., the drainage at the bottom of the lysimeter, which is located just above the alveolar product), with an herbaceous stratum (6 plants of Carex sylvatica and Deschampsia cespitosa, which are native to the Paris region). This configuration is considered as the reference because of the Paris subsoil context (heterogeneous and sensitive areas of gypsum or former mines, etc.), and the importance to anticipate the impact of waterproof systems on climate change;
- Lysimeters 2 and 7 differ from the reference by a modification of the vegetation with a shrub layer (3 Cotoneaster lacteus plants per lysimeter). These plants are from China and are often used in Paris plantations;
- Lysimeters 3 and 4 differ from the reference by the lack of IWS, i.e., the water is evacuated at the bottom of the alveolar product;
- Lysimeter 8 is similar to the reference but without vegetation (spontaneous vegetation is removed twice a year);
- Lysimeter 5 is similar to the reference but with spontaneous vegetation.

2.2.2. Data Acquisition and Validation

The data (Tables 1 and 2) were collected at two-minute time step for a period of about 3 years (24 November 2016 to 26 December 2019). The analysis and the validation of data were carried out at daily steps. For all variables, the maintenance days were removed, whereas maintenance used to be three times a month. The variables involved in the water balance were analysed in the following way. First, for very rainy days, rainfall values were compared with the measurement from a nearby rain gauge of Météo-France (the French meteorological service, situated at 1 km); if our rain gauge data were very different from the reference data of Météo-France, they were considered as non-valid. In addition, the exfiltration data of the lysimeters with reserve were compared with the data of the water level measurement in this reserve. The idea was to have zero exfiltration when the storage is not filled (<20 cm) for lysimeters with IWS.

Table 1. Details of materials used for measurement on each lysimeter (the accuracy is expressed in equivalent mm of water in a lysimeter).

Materials	Variables	Accuracy (mm)
Bucket flow meter (PRÉCIS-MECANIQUE, 3029/2)	Cumulative exfiltration (l)	0.008
Piezometric sensor (PARATRONIC, EN61000-6-2)	Water level (mm) in the IWS	1 mm
Load cells (SKAIM, FT-SK30X-FEG-0603)	Lysimeter's mass (kg)	0.36 mm

After removing the false, the aberrant, and the missing values, over the 1096 days that represented the three years, the percentage of validated data for the precipitation, the exfiltration and the mass variation were, respectively, 82%, 70–83%, and 66–76% (Table A1).

For the pan evaporimeter, in winter days, during rainy periods, the water level measurement (L) frequently reaches its maximum; therefore, an overflow occurs and the level variation is then set to zero. In these periods, the condition is that the water level (L) added to the rainfall should be less than the threshold of the measurement (L_{max}) that has been defined as equal to 170 mm; a maximum value that varied due to the fluctuation of the sensor during maintenance.

Table 2. Details of materials used for measuring meteorological data.

Materials	Variables
Temperature and humidity sensor (LSI-LASTEM, DMA672)	Temperature (°C) and Air humidity (HR en %)
Rain gauge (LSI-LASTEM, DQA131.1)	Rain (mm)
Evaporimeter (Pan, LSI-LASTEM, DYI010)	Water level (mm)
Global radiometer iso cl-2 (LSI-LASTEM, DPA053)	Global incoming solar radiation (Watt/m^2)
Anemometer (LSI-LASTEM, DNA202)	Wind speed (m/s)
Barometer (LSI-LASTEM, DQA24)	Atmospheric pressure (hPa)

2.3. Methods

2.3.1. Water Balance

Daily ET is calculated for each lysimeter based on the following equation:

$$ET = 4*P - Exf - \Delta S \tag{1}$$

with ET the evapotranspiration (mm), P the cumulative rainfall measured with the rain gauge (mm), Exf the cumulative exfiltration (mm), and ΔS the mass variation (mm) of the considered lysimeter.

For the daily evaporation (E, mm) from the pan evaporimeter, it is expressed as the difference between the daily cumulated rainfall (P, mm) and the daily water level variation (ΔL, mm):

$$E = P - \Delta L \tag{2}$$

2.3.2. Evaluation of Measurement Uncertainty

The assessment of the uncertainties associated with the ET estimations is based on the law of the propagation of uncertainties [31]:

$$u(Y)^2 = \sum_{k=1}^{n} u(X_k)^2 \left(\frac{\partial f}{\partial X_k} \right)^2 + 2 \sum_{k=1}^{n-1} \sum_{j=k+1}^{n} u(X_k, X_j) \left(\frac{\partial f}{\partial X_k} \right) \left(\frac{\partial f}{\partial X_j} \right) \tag{3}$$

where f is the function of n measured variables X_k, $u(X_k)$ the standard uncertainty, $u(X_k, X_j) = u(X_k)u(X_j)r(X_k, X_j)$ the estimated covariance of X_k and X_j with $r(X_k, X_j)$ the correlation coefficient.

By applying the Equation (3) to the balance equation (Equation (1)), it gives:

$$u(ET)^2 = \left(u(P)^2 \left(\frac{\partial ET}{\partial P} \right)^2 + u(dM)^2 \left(\frac{\partial ET}{\partial \Delta S} \right)^2 + u(exf)^2 \left(\frac{\partial ET}{\partial exf} \right)^2 \right)$$
$$+ 2 \left(u(P, \Delta S) \left(\frac{\partial ET}{\partial P} \frac{\partial ET}{\partial \Delta S} \right) + u(P, Exf) \left(\frac{\partial ET}{\partial P} \frac{\partial ET}{\partial Exf} \right) + u(\Delta S, Exf) \left(\frac{\partial ET}{\partial \Delta S} \frac{\partial ET}{\partial Exf} \right) \right) \tag{4}$$

To solve the Equation (4), the first hypothesis is that the standard uncertainties associated with the rainfall and exfiltration measurements are at the maximum of a bucket tilt of 0.2 mm and 0.008 mm, respectively. The second assumption was to assume that the uncertainties of rainfall and exfiltration follow uniform laws, which permit their standard uncertainties to be re-estimated by $0.2/\sqrt{3}$ (0.115 mm) and $0.008/\sqrt{3}$ (0.00462 mm), respectively [31]. The standard uncertainty associated with the mass measurement for each lysimeter is 0.36 mm, a value obtained from the manufacturer [32]. The standard uncertainty of the cumulative values is assessed by: $\sqrt{n}*u(ET)$ and the estimated uncertainties are given as a 95% confidence interval.

2.3.3. Comparison Tools

Different statistical tools are used to make comparisons between the different replicates or to compare the observed and modelled data. The non-parametric Wilcoxon rank test for paired samples was performed to compare the significance of differences between replicates and lysimeter configurations. The null hypothesis H_0 of this test suggests the same population for the distributions, while the alternative hypothesis H_1 assumes different distributions. The assumed risk α is taken at 5%. Simple regression models were also used to compare the observed replicas. Cumulations were also made by considering common days with valid data for lysimeters to be compared. Finally, to show the influence of the meteorological variables on ET, the partial least squares (PSL) analysis is performed. The variable important in the projection (VIP, see Appendix A for more details) that resumes the influence of each independent variable in a PSL model was used [33–35]. Indeed, a given variable will have a high importance for VIP > 1, a medium importance for VIP > 0.8, and a low importance for VIP < 0.8 [33,34,36].

2.3.4. Evapotranspiration Formulas

The predictive equations of ET tested here are summarized in Table 3. The two Penman–Monteith models applied on references vegetation (Fao-56 and Météo-France), the Penman and the Priestley–Taylor models will be compared with the estimated ETs from the lysimeters and the evaporation from the evaporimeter.

Table 3. Evapotranspiration (ET)'s formulations used in this study. The FAO and Météo-France formulations are two ways of setting parameters for Penman-Monteith (PM) equation.

Name	Formulas	Hypotheses
Penman [37]	$ET_P = \dfrac{\Delta(Q^*-Q_G)+E_a\gamma}{L_e(\Delta+\gamma)}$	$E_a = 0.35(e_s - e_a)(0.5 + 0.01u)$
PM (FAO-56) [23]	$ET_{PM-FAO-56} = \dfrac{0.408\Delta(Q^*-Q_G)+\frac{900}{T+273}\gamma(e_s-e_a)}{\Delta+\gamma(1+0.34\ u)}$	Well-watered vegetation with a height of 0.12 m, a surface resistance of 70 m/s, a surface emissivity of 1 and an albedo of 0.23.
PM (Météo-France) [38]	$ET_{PM-MF} = \dfrac{0.408\ \Delta(Q^*-Q_G)+\frac{(\gamma)(1297.8+1038.2u)(e_s-e_a)}{T+273}}{\Delta+(\gamma)(1.42+0.336u)}$	Well-watered meadow with a surface resistance of 60 m/s, a surface emissivity of 0.95 and an albedo of 0.2.
Priestley and Taylor [26]	$ET_{PT} = \alpha_{PT}\dfrac{\Delta}{\Delta+\gamma}\dfrac{Q^*}{L_e}$	Defined for saturated soils, the advection coefficient α_{PT} is set to 1.26 [26].

In these equations, terms are defined as follows: Q^* is the net radiation (MJ/d), Q_G the heat flux conducted in the soil (MJ/d), L_e the latent heat of vaporization (KJ/kg), e_s the saturation vapour pressure of air at surface temperature (KPa), e_a the partial vapour pressure of atmosphere (KPa), and u is the wind speed (m.s^{-1}) at a reference level (2 m), Δ the slope of the saturation vapour curve, α_{PT} is the advection coefficient, γ is the psychometric constant, and T refers to the temperature (°K).

3. Results

3.1. Estimated Evapotranspiration

In Figure 4, the meteorological variables measured at the site are presented. All variables are expressed as a daily average, except for exfiltration and rainfall, which are daily cumulated values. Seasonal dynamics specific to the temperate climate are observed for these variables. Global solar radiation (R_G) is higher in summer (up to 265 w/m^2) than in winter (max, 20 w/m^2). The net radiation (Figure 4a) assessed according to Allen et al. [21] is more significant in summer (up to 151 w/m^2) than in winter (max, 49.8 w/m^2).

Temperatures (T) reach the maximum at 34 °C in summer and are sometimes below 0 °C in winter. In contrast to the temperature, the air humidity (H_R, 29–96%) is higher in winter and lower in summer. The air pressure (P_{atm}) shows the same trend as the air humidity but less marked and varying between 980 and 1040 hPa. The wind speed (u) is between 0.1 and 1.6 m/s, higher in winter and lower in summer.

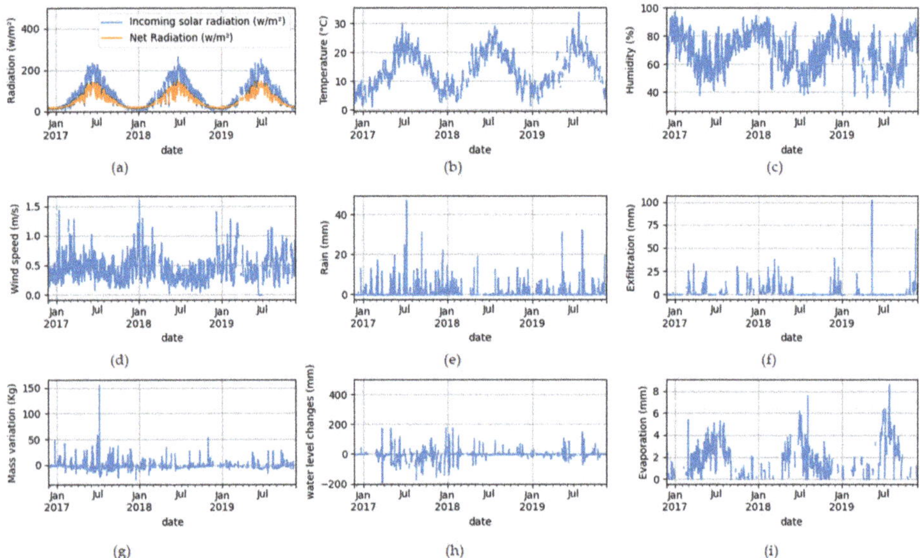

Figure 4. Meteorological variables: (**a**) daily incoming solar radiation and net radiation, (**b**) mean air temperature measured at 2 m, (**c**) relative air humidity, (**d**) wind speed, and (**e**) rainfall, (**f**) exfiltration, lysimeter daily (**g**) mass changes, (**h**) water level variation from the reference lysimeter 1 are added (mm), and (**i**) evaporation estimated from the evaporimeter.

For the variables specific to lysimeters (mass, water level in the IWS, and exfiltration), the reference configuration, i.e., lysimeter 1, is shown in Figure 4f–h. In addition, the Figure 4i gives the estimated evaporation from the evaporimeter with an average of 2.1 mm/d, high values in summer (max, 8.6 mm), and low values in winter.

Validated ET data after processing for the three years (1096 days) vary from 53% to 68% depending on the lysimeter (Table A1). In Figure 5, the validated ET for each lysimeter is presented. The annual dynamics of ET are shown with high daily values that can exceed 10 mm between spring and summer and small values in winter and autumn. These seasonality patterns can be linked to the atmospheric factors described above. The main atmospheric factors affecting ET in these systems are discussed later in Section 3.3.

Daily standard uncertainties and uncertainties at a 95% confidence interval are evaluated for all lysimeters (Table 4). The results uncertainties are in the range ±0.42 to ±0.58 mm for daily ET depending on the lysimeter.

Table 4. Associated daily ET uncertainties for each lysimeter in mm. u(ET) values refer to the standard uncertainty and 1.96 u(ET) the uncertainty for a 95% confidence interval.

	ET 1	ET 2	ET 3	ET 4	ET 5	ET 6	ET 7	ET 8
u(ET)	0.28	0.24	0.24	0.21	0.28	0.29	0.28	0.23
1.96 u(ET)	0.54	0.47	0.47	0.42	0.55	0.58	0.54	0.45

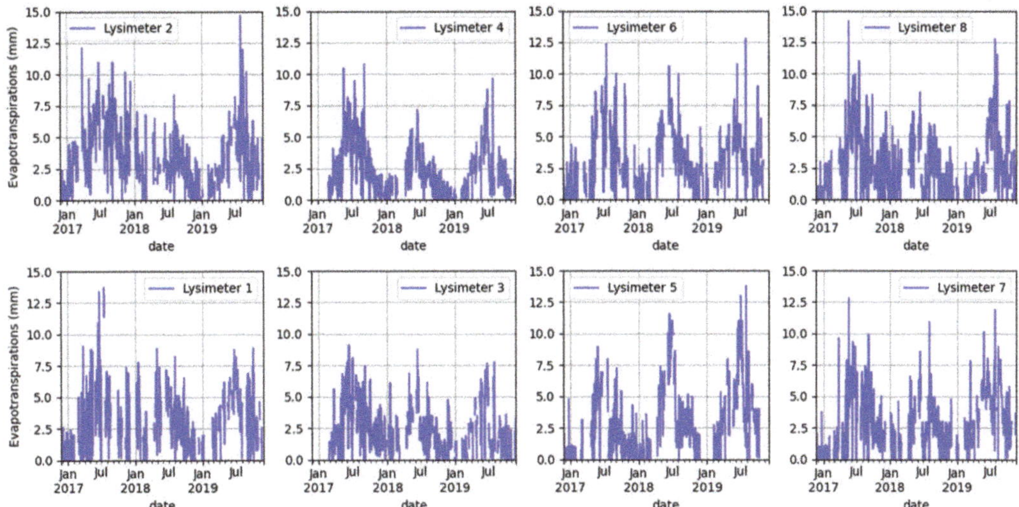

Figure 5. Daily evapotranspiration (ET) validated for all lysimeters.

3.1.1. Comparison of the Replicas

Three pairs of lysimeters (lysimeters 1 and 6, lysimeters 3 and 4, and lysimeters 2 and 7) have the same characteristics: vegetation, presence of storage or not, and the same maintenance planned during the experiment. The aim here is to compare their consistency knowing that they should be similar in term of performance. However, if a major difference is observed, this means that, for identical systems, an external variable, to be identified, is at the origin of this difference.

Based on the regression models and cumulates presented in Figure 6 for each couple of replicas, it is noted that for both lysimeters 2 and 7, if the regression model shows an acceptable fit between the two data sets $r^2 = 0.55$, ET of lysimeter 7 is clearly lower than lysimeter 2 in terms of global trend and cumulative amounts (ET2 = 1967 $\pm$ 11 mm and ET7 = 1662 $\pm$ 13 mm). Lysimeters 1 and 6 have similar trends, and cumulative amounts (ET1 = 1334 $\pm$ 11 mm, ET6 = 1330 $\pm$ 12 mm), even though the determination coefficient is low $r^2 = 0.42$ due to the underestimation and overestimation of lysimeter 6 from 01/2018 to 05/2018 and from 10/2018 to 07/2019, respectively.

For the couple lysimeters 3 and 4, it presents a coefficient of determination $r^2 = 0.57$, similar trends and a slight underestimation of lysimeter 4 in terms of cumulative data (ET3 = 1544 $\pm$ 12 mm, ET4 = 1449 $\pm$ 10 mm).

Another way of comparing these pairs is to perform statistical tests. In Table A2, results of the Wilcoxon rank test are presented. When the test is performed on the whole validated data set (3 years), only lysimeter 1 (the reference) and lysimeter 6 have similar distributions. If the test is performed by season (fall, winter, spring, and summer), different results are obtained. In all seasons, pair 1 and 6 do not show statistically different distributions, lysimeter 3 and 4 show statically different distributions in autumn and winter only while lysimeters 2 and 7 are statistically different except in spring.

It is difficult to conclude that for each replica, both lysimeters evaporated and transpired perfectly in the same way. In addition, the estimated uncertainties on ET for each lysimeter are small compared to the differences between the replicas (Figure 6a–c). However, in view of the above results, it can be said that the couples ET1/ET6 and ET3/ET4 configurations represent an acceptable replica and, the shrub configurations ET2/ET7 cannot be considered as a replica. For lysimeters 2 and 7, the only variable that differs from the two is the exposure to the buildings surrounding the installation.

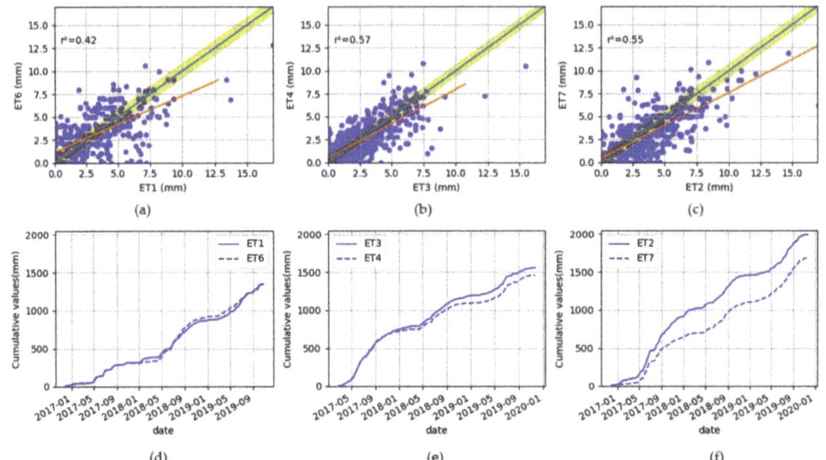

Figure 6. Regressions (**a–c**) and cumulatives (**d–f**) plots comparing the three replicas ET1/ET6 (data = 445), ET3/ET4 (data = 600) and ET2/ET7 (data = 583) for the three years (1096 days). The red line refers to the regression line and the blue line represents the y = x surrounded by the confidence interval (at 95%) that corresponds to the square root of the sums of the squares of the lysimeter uncertainties for each replica (0.79 mm, 0.63 mm, and 0.71 mm, respectively for ET1/ET6, ET3/ET3, and ET2/ET7).

3.1.2. Comparison between Different Configurations

The comparison of all configurations was conducted with the validated common days for the eight lysimeters from November 2016 to November 2019. These days start in mid-spring (83 days), continue throughout the summer (132 days), and end in mid-autumn (65 days). In winter, there are only 25 valid days because of the greater measurement uncertainty during the cold and rainy periods. Table 5 gives the cumulative exfiltration, mass variation, and evapotranspiration based on this common period for lysimeters.

Table 5. Cumulative water balance components (C_{wb}) in mm over the 305 common validated days, for the 8 lysimeters. Cumulated rain (4P) is 679 ± 6 mm.

Lysimeters		1	2	3	4	5	6	7	8
Exf	$C_{wb(Exf)}$	162	86	438	568	204	199	191	250
	% of 4P	24%	13%	65%	84%	30%	29%	28%	37%
ΔS	$C_{wb(\Delta S)}$	−543	−486	−492	−633	−675	−577	−471	−427
	% of 4P	−80%	−72%	−72%	−93%	−99%	−85%	−69%	−63%
ET	$C_{wb(ET)}$	1066 ± 7	1082 ± 6	740 ± 8	750 ± 7	1152 ± 8	1060 ± 7	962 ± 8	864 ± 8
	% of 4P	157%	159%	109%	110%	170%	156%	142%	127%

The exfiltration varies between 13% (lysimeter 2) and 84% (lysimeter 4) of the input rainfall. Free drainage configurations (lysimeter 3 and 4) naturally exfiltrated the most water compared to the others set up with IWS and account for about three times (438 mm and 568 mm) of the standard configurations (lysimeter 1 and 6, 162 and 199 mm respectively). Furthermore, the herbaceous configurations (1 and 6) exfiltrated more than the shrub configurations (2 and 7); although, for lysimeters 6 and 7, this difference is reduced. Finally, the exfiltration capacity of the configuration with regularly removed vegetation (lysimeter 8, 250 mm) is higher compared to the other lysimeters with IWS and that could imply a contribution of the vegetation to the decrease in seepage.

For stock changes (ΔS), the eight settings always have negative values between −633 and −425 mm. Indeed, most of the validated common days are spring and summer days with low rainfall, which are favourable periods for ET. Therefore, for a given day, the mass change is negative meaning that the system (lysimeter) loses water. This explains the negative cumulative ΔS observed here.

3.2. Determinants of ET in Lysimeters

To carry out the analysis in this section, common data of lysimeters were considered in pairs, in order to increase the number of samples and the representativeness of all seasonal periods. These numbers are noted in the text or in the Table A2. It is also important to remind that the experimental set-up was installed to test the impact of three main factors on the water balance in rain gardens. These factors are lysimeter storage (absence or presence of IWS), vegetation type and management, and local meteorological variables.

- Impact of the storage in the lysimeter structure.

Installing an IWS is globally favourable to the ET, and to the reduction in the exfiltration (Table 5). These differences are notable in all seasons. Indeed, the Wilcoxon test between lysimeter 1 and lysimeters 3 and 4 show that the distributions of estimated ET data are different in all seasons and over the whole three years (Table A2). In addition, from Table 5 or Table A7, considering the percentage of ET sum to the collected rainfall (4P), the ET of lysimeter 1 is more compared to the other two lysimeters. Compared to lysimeters 3 and 4, in autumn, winter, spring, and summer, lysimeter 1 evaporates more on average +18%, +37%, +18%, and +87%, respectively. For the three years, it is estimated that more than +31% of ET occurs from a system with IWS compared to those without IWS (3 and 4). These differences are more noticeable in summer, when the water stored in the IWS allows higher soil moisture during dry and hot periods to be maintained.

In Figure 7, the ET in lysimeters 3 or 4 is lower than the references (1 and 6) during a summer period (24 June to 3 July). A same dynamic and quantity can be observed between the water lost from the storage (dH) and the ET in standard lysimeters. In terms of cumulus of ET and water changes (dH) for these 10 days are ET1 = 47 $\pm$ 2 mm, ET6 = 68 $\pm$ 2 mm, ET3 = 27 $\pm$ 1 mm, ET4 = 31.4 $\pm$ 1mm, dH1 = $-41\pm$ 2 mm, and dH6 = $-58 \pm$ 2 mm. In this dry period without rain and exfiltration, for standard lysimeters 1 and 6, the water in the IWS contributes to evapotranspiration by 87 $\pm$ 7% and 85 $\pm$ 5%, respectively. However, in lysimeter 3, ET does not occur at the potential rate and is therefore limited by the water availability.

- The effect of vegetation

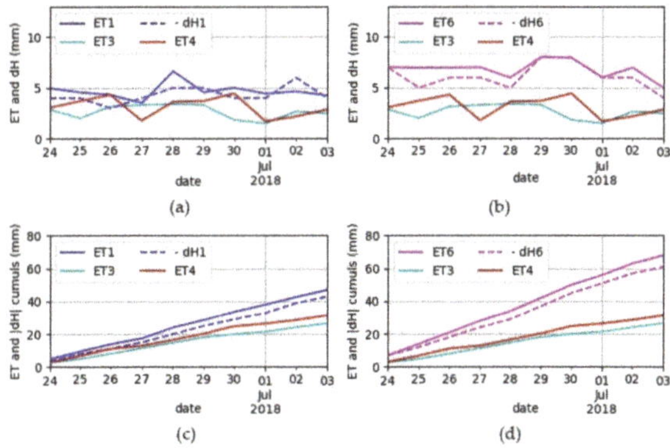

Figure 7. Evapotranspirations (ETs) and water level variations (dH) in the internal water storage for lysimeters 1, 3, and 4 (**a**) and lysimeters 6, 3, and 4 (**b**) during a summer period (24 June to 3 July 2018). (**c**,**d**) show cumulative values for ETs and $|dH|$, respectively, for lysimeters 1, 3, 4, and lysimeters 6, 3, and 4.

Four types of vegetation (herbaceous, shrubs, spontaneous, and removed vegetation) were tested. A comparison of herbaceous plants (1, 6) with shrubs (2, 7) was problematic because while the former can be considered as acceptable replicas, the latter cannot. A priori, it might be possible to compare them according to their closeness. For example, lysimeters 1 vs. 2, and lysimeters 6 vs. 7 are couples that can be used to identify potential differences between herbs and shrubs. Statistically with the Wilcoxon test, on the whole data, there is no difference between lysimeter 1 and 2 distributions except for the fall season (Table A2). For lysimeter 6 and 7, their distributions differ statically for the whole data set. The role of the herbaceous/shrubby vegetation type seems to be difficult to show based on the whole data for the three years.

In terms of cumulative amounts during all period, the spontaneous vegetation (lysimeter 5) produced a lower ET of 4% than the references (difference not significant according to the Wilcoxon test except for fall and winter). However, if we compare quantitatively by season (Tables 5 and A6), the spontaneous vegetation evaporates more than the references in Summer and Spring even if its maximum values are lower than those of lysimeter 1. That is why in the previous comparison in Table 5 (where spring and summer data were dominant) spontaneous vegetation was more important in terms of cumulated ET. Another term to be taken into account in this comparison is the evolution of vegetation. In the first year, for all seasons, lysimeter 1 (reference) shows a higher evaporation while, in the other two drier years (2017–2018 and 2018–2019), the lysimeter 5 (spontaneous vegetation) evaporates more in spring and summer. This could be explained by the fact that spontaneous vegetation adapts more in these periods of water limitations compared to other vegetation. Moreover, the spontaneous vegetation was not well established at the beginning of the experiment and it developed strongly later (Figures A1 and A2). Table A6, which compares the common days between the three years, shows this point. For the summer period 2018 (23 June–4 July), a higher evapotranspiration of spontaneous vegetation is observed confirming the above results (Figure 8a).

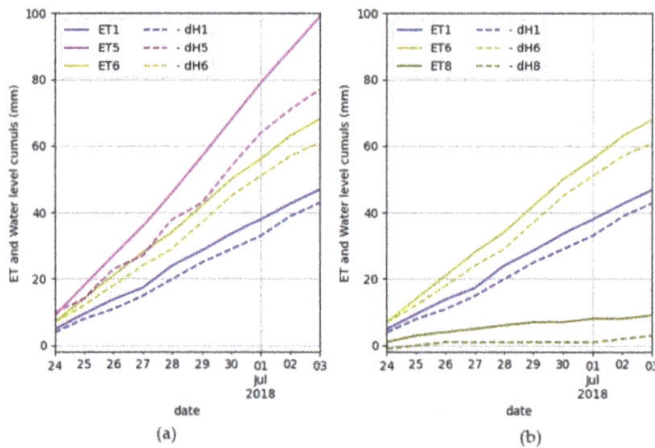

Figure 8. Cumulative curves during a summer period (24 June to 3 July 2018) for evapotranspirations (ET) and water level variations (dH) in the internal water storage (IWS). The herbaceous lysime-ters (1 and 6) are compared with the spontaneous vegetation lysimeter (5) (**a**) and, the regularly removed vegetation one (lysimeter 8) (**b**). Note that for this period, the data for shrubs (2, 7) are not valid.

Finally, the regularly removed vegetation (lysimeter 8) produced a lower evapotranspiration of about −17% than lysimeter 1. Statistically, this difference exists globally and would be more pronounced in autumn, spring, and summer. In the summer of 2018, as the vegetation was removed on 21 June, the difference (ET, dH) is more significant (Figures 8b, A1 and A2) and showing that the plants need to develop sufficiently to be able to properly use the water stored in the IWS.

- Meteorological factors

Apart from the factors related to gardens properties, ET is also subject to meteorological factors. In Table 6, the Pearson coefficients (ratio of covariance to the product of the standard deviations) give an overview of the linear correlation between each estimated daily ET and the measured atmospheric variables.

Table 6. Linear correlation coefficients (Pearson) between estimated evapotranspiration (ET) and measured meteorological variables.

Lysimeters	ET1	ET2	ET3	ET4	ET5	ET6	ET7	ET8
R_G (MJ/d)	0.44	0.30	0.42	0.59	0.68	0.59	0.37	0.42
T (°C)	0.38	0.39	0.29	0.41	0.50	0.48	0.42	0.38
H_R (%)	−0.25	−0.12	−0.21	−0.36	−0.46	−0.36	−0.21	−0.3
u_w (m/s)	0.05	0.08	−0.08	−0.2	−0.06	0.05	0.09	0.02
P_{atm} (hPa)	−0.17	−0.23	−0.08	−0.03	−0.08	−0.13	−0.19	−0.16

A positive correlation between ET and the variables of global solar radiation (0.30 to 0.68) and mean air temperature (0.29 to 0.48) is observed. However, this correlation is of the same order but negative for air humidity (−0.46 to −0.25), and weak for wind speed (−0.06 to 0.08) and atmospheric pressure (−0.23 to 0.08).

A more detailed analysis with PSL models confirms that for all measuring devices (lysimeters and evaporimeter), air temperature and global radiation are the most important variables influencing evapotranspiration with a VIP score greater than one (Figure 9). Moreover, air humidity has a moderate influence on ET in rain gardens (VIP between 0.8 and 1), but for the evaporimeter, it appears as an important determinant for the process (VIP > 1). As previously, wind speed and atmospheric pressure seem to be of low importance. In synthesis, the main atmospheric factors that impact the ET process in these devices are global radiation, air temperature, and air humidity.

- The impact of shading

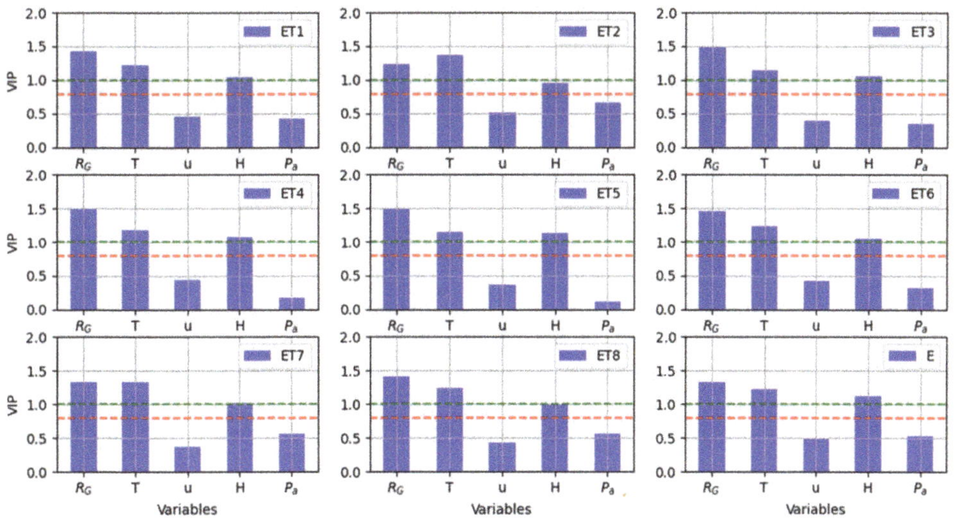

Figure 9. Variable importance in projection (VIP) plots according to partial least squares analysis for lysimeters and evaporimeter. The red and green dashed lines refer respectively to the values of VIP larger than 0.8 and 1. Analysis is conducted based on common data between the explained variables and the evaporation (E) or the lysimeter evapotranspiration (ET).

Another factor that needs to be addressed is the exposure of each lysimeter. The «Rain», which is used to estimate the ET, is susceptible to be impacted by the lysimeter exposition. The closeness and height of the south wall could act as a barrier to rain during a windy period. In such conditions, certain lysimeters could receive less rain than others and so does the rain gauge. Thus, the rainfall measurements from lysimeter 7 and 8 are less consistent with the rain gauge measurements (Figure A3). As a result, the further away from the rain gauge the less important the determination coefficient of the regression model is (Figure A3). Some of other atmospheric conditions can be different between lysimeters, in particular due to the buildings and more precisely the wall in the south, which is close to lysimeters 7 and 8 (less than 1 m). Two main variables can be mentioned:

- The global incident radiation is modified by the evolution of the shading. The shading is variable on the lysimeters, both during the day and seasonally. Shading effect is clearly visible in summer around mid-afternoon on the global radiation measurement with strong decrease in the value (Figure A4);
- The Wind: the linear correlation between wind speed and ET is found to be weak (Table 6). However, the wind could impact the distribution of rainfall on the lysimeters.

These potential modifications would reduce the ET on the southern lysimeters (7 and 8), compared to those to the north. This impact is difficult to assess quantitatively as it would require specific measurements (e.g., 3D site geometry) or extended replicas. Regarding the shrub configuration, it could be suggested that the exposition effect is responsible for the great differences observed between the replicas (lysimeters 2 and 7) that are opposite to each other. Therefore, this difference is estimated on the cumulative ETs (+11%) and also on the scatterplot of the daily ETs (Figure 6c).

3.3. Evapotranspiration Predictive Equations, ETs Estimated from the Evaporimeter, and the Lysimeters

The objective of this section is to compare ET estimated by water budget on lysimeters to two types of reference values: (i) evaporation measured on an open water surface with the pan evaporimeter (Figure 4i) and (ii) ET estimated with potential formulations. In order to increase the amount of available data, the numbers of lysimeters were reduced to the different configurations (1, 2, 3, 5, and 8) and the common validated data over the whole period of study are for 281 days (Table 7).

Table 7. Comparative totals and averages (in mm) of evaporation (E) and evapotranspiration (ET) estimated, respectively, with the evaporimeter and the lysimeters (data = 281).

Seasons (Data)	ET1	ET2	ET3	ET5	ET8	E
Fall (53)	93 ± 4	137 ± 3	62 ± 3	100.0 ± 4	64 ± 3	40.1
Winter (14)	34.3 ± 2	28 ± 2	24 ± 2	14 ± 2	16 ± 2	13
Spring (81)	311 ± 5	294 ± 4	290 ± 4	330 ± 5	27 ± 4	180
Summer (133)	551 ± 6	570 ± 5	370 ± 5	637 ± 6.3	476 ± 5	382.4
Cumulus (281)	988 ± 9	1029 ± 8	746 ± 8	1081 ± 9	836 ± 8	585
Mean	3.5	3.6	2.6	3.8	2.9	2.1

Cumulative ETs indicate that evaporation from the water surface is 585 mm (average 2.1 mm/d), while ET from the lysimeters varies from 746 to 988 mm (average 2.6 and 3.5 mm/d). Compared to the lysimeter 1, this represents a difference of −41%. At the daily step, the E of the evaporimeter is almost systematically lower than the ET of the lysimeters; this is also the case for high values (>8 mm). Compared to the non-IWS configuration (lysimeter 3), the trend is much less marked mainly in summer (when the evaporimeter evaporates more, 382 vs. 370 mm).

Here, the results indicate that the evaporation of the free water surface near the lysimeters is low compared to the ET for rain gardens. This result is not intuitive because the open water surface is always supplied with water and does not offer theoretically a resistance to the ET flux. One hypothesis is that the development of the plants in the lysimeters leads to a larger evapotranspiration surface than what is theoretically perceived, i.e., 1 m^2. Indeed, the surface for evaporation and transpiration for lysimeters is larger than the evaporimeter, so that under certain conditions with no hydric limitations, ET in lysimeters is more important. However, in summer, where the lysimeter 3 does not have IWS, the quantity of evapotranspiration is reduced compared to the evaporimeter. If a short dry period is considered (as mentioned above, 24 June–3 July 2018), the evaporated water from the evaporimeter is slightly higher than the lysimeter 1, which has an IWS (in terms of accumulation 50 ± 2 and 47 mm, respectively, for evaporimeter and lysimeter 1 Figure 10c).

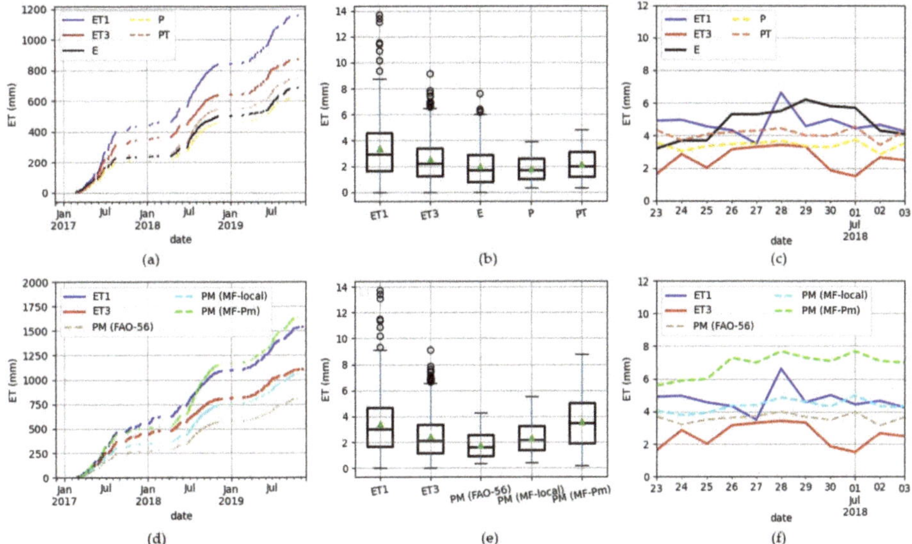

Figure 10. Evapotranspirations (ET) from the lysimeter 1(with internal water storage), the lysimeter 3 (without internal water storage), the evaporimeter, and the potential ET models: (**a–c**) compare the Penman (P) and Priestley–Taylor (PT) potential ETs to ET1 and ET3 (data = 346) and respectively present the cumulative ET values, the boxplots, and the ET dynamics for a dry period (24 June to 3 July 2018); (**d–f**) refer to the ET1, ET3, Potential ET of FAO-56, and Meteo-France models (data = 453) and show, respectively, cumulative ET values, boxplots, and ET dynamics for a dry period (24 June to 3 July 2018). PM and Pm refer respectively to Penman–Monteith and Paris-Montsouris.

The reference (lysimeter 1, with IWS), as well as the lysimeter 3 without IWS are used in comparison with the models because of their closeness to the weather station and the interest in observing the validity of models regarding to the storage presence or not. PT, P, PM (Fao-56), and PM (MF-local.) are potential ETs evaluated with the local meteorological data while PM (MF-Pm) is estimated from the Météo-France equation at the Paris-Montsouris (Pm) station, approximately 2.5 km from the site (Figure 10; Table 8). At Paris-Montsouris, the station is clear and far from the obstacles that can affect the meteorological measurements.

Over the whole study period, ET on the Paris-Montsouris station PM (MF-Pm) is more important in terms of trend and cumulative amount (1639 mm) than the ET estimated locally (1066 mm) with the same Météo-France formulation (Table 8 and Figure 10c–e). This observation illustrates the impacts of the microclimatic variabilities on the assessment of potential ET in urban area. These variations are mainly due to the global incident radiation variable, which over the three years is on average 135 and 80 w/m^2, respectively, at Paris-Montsouris and at the study site. Compared to lysimeters (1, 3) and other models, PM (MF-Pm) is more important particularly in summer and spring.

Table 8. Cumulatives and averages (in mm) evapotranspirations (ET) obtained with the lysimeters (ET1 and ET3) and the potential ET models. P, PT, PM (Fao-56), PM (MF-local.) are Penman, Priestley–Taylor, Fao-56, and Météo-France potential ETs evaluated with the local meteorological data while PM (MF-Pm) is the potential ET evaluated with the Paris-Montsouris station data.

Seasons (Data)	ET1	ET3	P	PT	PM (FAO-56)	PM (MF-local)	PM (MF-Pm)
Fall (109)	240 ± 6	154 ± 5	78	89	93	141	199
Winter (55)	130 ± 4	65 ± 3	32	36	39	58	75
Spring (134)	543 ± 6	475 ± 5	260	312	283	353	536
Summer (155)	646 ± 7	414 ± 6	364	437	396	515	829
Cumulus (453)	1559 ± 11	1107 ± 10	734	874	811	1066	1639
mean	3.4	2.4	1.6	2	1.8	2.4	3.6

For local potential evapotranspirations, i.e., PM (FAO-56), PM (MF-local), P, and PT, a general underestimation (systematically in fall and winter) of lysimeters (1, 3) ETs is observed over the whole simulation period (Figure 10). However, PM (MF-local) seems to be a reasonably good predictor of the ET when IWS is absent (Table 8).

For the June 24 to 3 July 2018 dry period, potential ETs are superior to lysimeter 3 and lower than lysimeter 1, which evaporates and transpires at a considerable rate (Figure 10c,f).

The evaporimeter data (346 observations) were also be compared with potential ET values from Penman and Priestley–Taylor equations. In general, the Priestley–Taylor model (749 mm) overestimates the evaporimeter measurement (686 mm), and the Penman model (627 mm) underestimates it. While the maximum values are significant for the evaporimeter, in terms of average, the three estimates are close (1.9 mm, 1.8 mm, and 2.1 mm, respectively, for E, P and PT). The sensitivity of the Priestley–Taylor equation to the value α_{PT} [39] suggests that modelling ET or E from this equation requires a sensitivity study that would lead to a specific α_{PT} value (and that is not the objective here).

4. Discussion

In this study, the estimated ETs in the pilot rain gardens account for 61–90% of the collected rainfall (Table A3). They are on average 2.4 to 3.78 mm/d, depending on the vegetation maintenance, the presence (3–3.78 mm/d) of an IWS or not (2.4–2.5 mm/d). The uncertainties of ETs are ± 0.42 to ± 0.58 mm. A similar experiment conducted at the Villanova University in Pennsylvania (USA) by Wadzuk et al. [13] showed ET means of 6.1 and 3.1 mm/d, respectively, for IWS or no IWS lysimeters from April to November in 2010 and 2011. The high value of the lysimeter with an IWS compared to what is found here could be explained by the focalised summer period, by the size of the IWS, which is larger in the Wadzuk case (36 cm), by the evaporative demand, and also the inputs (precipitations) to the lysimeter. However, our values are in the same range as those estimated by Hess et al. [12] (4.3 and 2.7–2.9 mm/d, respectively, for IWS or no IWS lysimeters). These two studies showed the importance of ET in the rain gardens and estimate it between 43 and 78% of the collected rainfall.

The impact of the vegetation type (herbaceous and shrubs) was not addressed here because the replicas of shrubs (lysimeters 2 and 7) showed distributions that were statistically significant and there were large differences in terms of cumulative amounts. It has been suggested that this situation is probably due to the shading, which has a great impact on lysimeter 7 by significantly reducing its ET. However, a comparison of the closely related lysimeters 1 and 2 shows a higher evapotranspiration of shrubs configuration (+6%

globally and more significant in autumn based on the Wilcoxon test), which suggests that the vegetation type may be a factor to be considered. Indeed, Nocco et al. [14] investigated the impact of vegetation and vegetation type (grassland, shrubs, turf, and bare soil) in the hydrological performance of free-draining rain gardens in the Midwest of the USA. Their studies were conducted in three months (July, August, and September) and they first found that the effect of vegetation is significant when evaporative demand is also high. In addition, the configuration with grassland showed a higher ET than the others with an average of 9 and 7 mm/d in August and July, respectively. The ET of the bare soil system was lower (4 and 3 mm/d in August and July, respectively), and the ET of the shrubs, in contrast to the other systems that had ETs that decreased as the evaporative demand decreased, had a relatively constant ET around 6 mm/d for the months of July and August. It can be seen that the type of plant not only significantly impacts the ET dynamics, but also the ET cumulative values.

Another factor that was not tested in this study is the effect of soil type. Hess et al. [12] estimated average ETs of 2.9 and 2.7 mm/d, respectively, for free-draining lysimeters with local sandy loam and sandy soil (three-year data). In terms of water balance, the ET with the local soil accounted for 47% and the other 43% of the water balance. These differences show that finer soils are more favourable to ET as they retain much more water than coarse textured soils [15]. However, the issue of soil impact is best addressed in conjunction with the type of vegetation, as vegetation through its development may affect the hydraulic properties of the soil, which are able to affect the ET. Johnston's [40,41] and Le Coustumer et al. [42] illustrate the link between soil and vegetation evolution [14]. The first showed that grassland and shrub rain gardens (without IWS) have significantly lower volumetric soil water content at depths of 0–0.15 and 0.30–0.45 m (3–4 and 10% lower, respectively) compared to turf rain gardens prior to storms, suggesting that vegetation type can impact on the storage capacity of rain gardens. The second one indicated that the type of vegetation through the growth and the morphology of their roots impact the hydraulic conductivity of the soil (with average hydraulic conductivity decreasing by a factor of 3.6 over the 72 weeks of testing) that influences mainly the drainage and the water availability in the garden. For example, Le Coustumer et al. [42] observed that a species with thick roots significantly maintained the permeability of the soil over time. This issue of the link between plants, and soil in these systems, is not limited to the sustainable hydrological services but may well extend to the sustainability of other ecological services (e.g., removal of pollutants, see Glaister et al. [43]) in SUDS.

Until now, it can be argued that if the aim of rain garden design is to maximise ET, it needs to provide an underlying water storage, and select a balanced choice between the vegetation type and the soil type. Further factors to consider are atmospheric factors as shown in the PSL approach, global radiation, air temperature, and humidity impact on the estimated ET of lysimeters. Such factors are responsible for the seasonal variations in ET flux. Similar to the other GIS (e.g., green roofs with Feng et al. [36]), these three variables are known to affect ET and are generally input variables for the models used to simulate ET. Understanding the impacts of atmospheric variables on rain garden ET requires suitable hydrological simulation tools.

Finally, for a better efficiency of hydrological models, the ET process should be well represented. In fact, the ET prediction equations used in these models are based on the concept of potential ET, which account for evaporative demand. These ET models are then coupled with specificities related to vegetation, water availability, and/or local microclimatic conditions (FAO methods [23], WUCOLS [44], and LIMP [45] methods). However, these methods have remained impractical [15,17,27,46–48] as they require measurements of multiple parameters, are derived from the agricultural context, and are less suitable for the urban context. Hess et al. [27] tested the validity of the ASCE-Penman–Monteith [24] and Hargreaves equations in rain garden systems (one system with storage and two systems without storage with free drainage). Without including crop coefficients (estimated ET divided by potential ET) and soil moisture extraction functions, these equations provided an adequate estimate of rain garden ET for all systems on a storm scale. The use of crop coefficients and soil moisture extraction functions in both equations reduced the errors in the ET estimates and increased the predictive power of the equations for all types of weighing lysimeters at the daily scale. In this study, it is found that with potential ETs (FAO and Météo-France) evaluated with local atmospheric variables, lysimeters (one with IWS and three without IWS) were underestimated by the models. Furthermore, if the input data (atmospheric variables) are not local, as shown in Figure 10, then there are issues of urban micrometeorological variability [47] to take into account. However, at a seasonal scale, the PM (MF-local) equation seems to be a good approximation of the ET of the lysimeter without reserve. Monitoring properties that describe the dynamics of the vegetation canopy (stomatal resistance, LAI, roots expansion, etc.) and soil water content would enable more accurate assessment of the impact on plants and the comparison of lysimeters to evaporate and transpire [19].

5. Conclusions

The process of evapotranspiration should be included in the design of green infrastructure systems (GIS) in order to optimise their hydrological functions of stormwater management and their ability to cool the urban area in hot periods. In this study, a comparison of the evapotranspiration capacity between different pilot rain garden configurations, with an impluvium equal to four times the vegetated surface, was carried out, based on data covering a three-year period in Paris (temperate climate, France) that has undergone rigorous validation. The validated periods are less rainy and represent more the summer and spring seasons. It was found that the evapotranspiration flux from rain gardens is significant, with values that can exceed 8 to 12 mm/d in summer period for several days, and is characterized by a marked seasonality with very low values in winter ($\leq$2 mm/d). The installation of an internal water storage at the base is the most favourable determinant to enhance the flux and reduce exfiltration (+28 to 30% if the reference lysimeter with an IWS and those without IWS are compared). The vegetation, here, is a secondary determinant, and less marked (+6% for shrubs compared the reference herbaceous). The spontaneous flora gives more ET than the reference configuration in summer (+8%) and all configurations evaporate and transpire more than the regularly removed vegetation configuration. The positioning of the lysimeters between them (close to or far from buildings) also seems to be a determining factor and, in particular, the shading, which has a reducing effect on ET (the replica that is less exposed to the shade evaporates 15% more than the shaded one).

The experimental set-up used in this work was pertinent, and allowed the observation of water balance components and the assessment of the multi-annual daily ET with admissible uncertainties ($\pm$0.42 to 0.58 mm). Therefore, the seasonal dynamics and the relative significance of each determinant of ET in the rain gardens were highlighted. A possible counterintuitive result in the seasonal analysis was also that the ET values observed on the rain gardens, and particularly for those with an IWS, are higher than the ET from an evaporimeter. Based on the potential ET from a reference station located at 2.4 km from the site, the ET is under-estimated for the setup with an IWS during the winter and fall seasons.

Future studies need to include some aspects in the experimental setup for still better understanding the ET process in rain gardens. First, the location of the experiment should be selected in such a way that local microclimatic factors and especially shading effects are taken into account. Second, monitoring some properties, which describe the dynamics of the vegetation canopy (stomatal resistance, LAI, roots expansion, etc.) [19], and a lysimeter without vegetation could be added to experimentally compare the contribution of plant transpiration and soil evaporation. Finally, these results of ET could be used to investigate the modelling of hydrological processes and more especially on the ET process in urban rain gardens. The use of detailed and physically based hydro-climatic models (as SisPAT [49] et Teb-hydro [50]) should make it possible to better understand and reproduce the process. Nevertheless, the use of this type of models requires a large data set for the parametrization and evaluation steps.

Author Contributions: A.A.O. realized the analysis of the data and wrote the paper. E.B. supervised the study and revised the paper with M.-C.G., B.D. All authors have read and agreed to the published version of the manuscript.

Funding: This work was partially funded by the Paris City Council (grant n° 2019 DPE 57/villede-ParisEJ45026788290), which contributed to the experimental set-up and data acquisition. It was also supported by the OPUR program and the Seine-Normandy Water Agency.

Institutional Review Board Statement: Not applicable.

Informed Consent Statement: Not applicable.

Acknowledgments: Authors are grateful for the support given by M. Ramier David for the data processing, by Md. Tang Jieyu and Bonnefous Hortense for their important contributions to the data analysis and the partners from the Paris City Council(DPE-STEA) and set up the experiment and contributed for data collecting and the OPUR Program.

Conflicts of Interest: The authors declare no conflict of interest.

Appendix A. Assessment of VIP Score from a Partial Least Square (PLS) Model

By resuming the influence of individual X variables on the PLS model, the VIP scores are assessed as the weighted sum of squares of the PLS weights, w, which take into account the amount of explained y variance in each extracted latent variable [33,34]. The VIP score for a given variable j^{th} is given according to Farrès et al. [33]:

$$VIP_j = \sqrt{\frac{\sum_{f-1}^{F} w_{jf}^2 SSY_f J}{SSY_{total} \cdot F}} \tag{A1}$$

where w_{jf} is the weight value for j variable and f component, SSY_f is the sum of squares of explained variance for the f_{th} component and J number of X variables, SSY_{total} is the total sum of squares explained of the dependent variable, and F is the total number of components. The w_{jf}^2 gives the importance of the jth variable in each f_{th} component, and VIP_j is a measure of the global contribution of j variable in the complete PLS model.

$$SSY_f = b_f^2 t'_f t_f SSY_{total} = b^2 T'T. \tag{A2}$$

where T is the X scores matrix and b is the PLS inner relation vector of coefficients.

Appendix B. Figures

Figure A1. The vegetation in the eight lysimeters on 21 June 2018 (Source: DPE-STEA, Paris council). It can be observed that the spontaneous vegetation (lysimeter 5) is more developed compared to the other settings and that today the vegetation in lysimeter 8 has been removed. Moreover, the shrub configurations (2,7) are not well developed compared to the other configurations.

Figure A2. The vegetation in the eight lysimeters on 20 September 2018 (Source: DPE-STEA, Paris council). It can be observed particularly a development of the shrubs (lysimeter 2 and 7) and a revegetation of lysimeter 8 three months after all the plants have been removed.

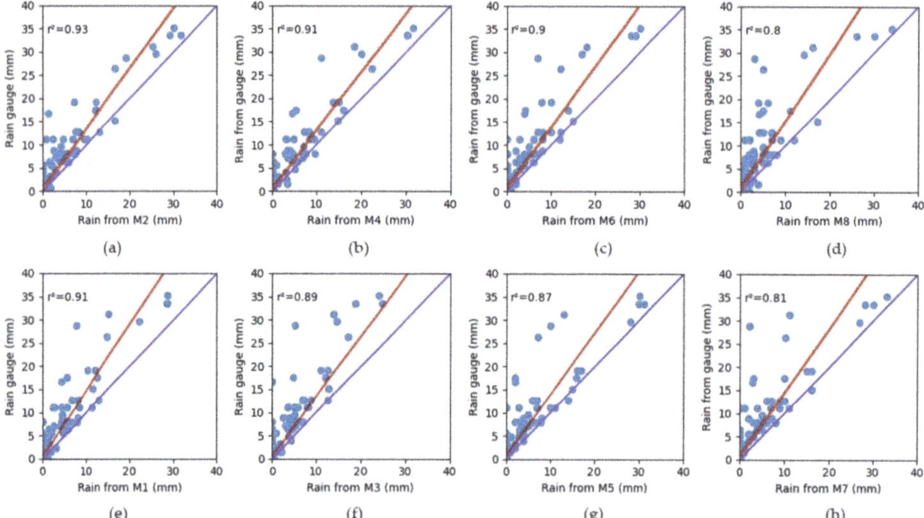

Figure A3. Comparison of the rain measurements resulting from the rain gauge data and by the mass variation in the system (rainfall are estimated as any increase in the total mass (lysimeter mass + exfiltration) of the lysimeter). The rainfall estimated from the lysimeter masses is lower compared to the rain gauge. Figure (**a**–**h**) represent, respectively, lysimeters 2, 4, 6, 8, 1, 3, 5, and 7.

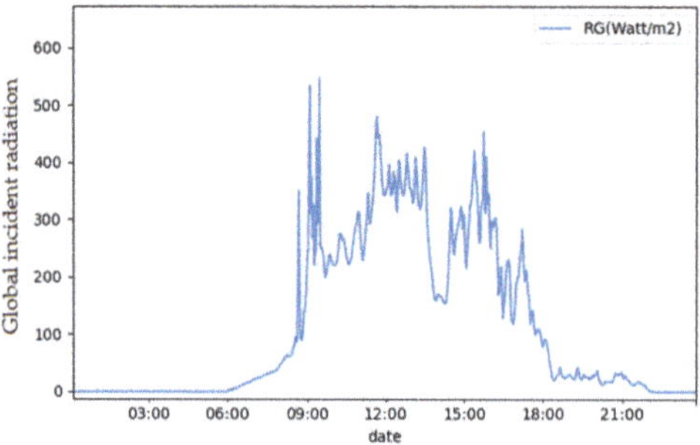

Figure A4. Global incident radiation (w/m^2) of 14 June 2018 measured at a time step of 2 min.

Appendix C. Tables

Table A1. Validated data after processing for the three years (2016–2019). The numbers and the symbol "E" refer respectively to lysimeters and evaporimeter. For season, it was considered fall (22 September to 21 December), winter (21 December to 20 March), spring (20 March to 21 June), and summer (21 June to 22 September). The total data are for 1096 days.

Parameter	Seasons	1	2	3	4	5	6	7	8	E	Rain
ET	Fall	151	197	181	186	166	161	163	207	126	239
	Spring	145	145	169	169	151	131	154	173	143	198
	Summer	161	198	203	198	192	192	199	214	185	246
	Winter	124	152	101	111	138	146	150	151	74	224
	Total	581 (53%)	692 (63%)	654 (59%)	664 (60.6%)	647 (59%)	630 (57.5%)	666 (60.8%)	745 (68%)	528 (48.2%)	907 (82.7%)
Exf	Fall	197	231	215	215	183	200	218	231	-	
	Spring	213	207	229	221	200	161	192	217	-	
	Summer	195	232	246	249	230	235	251	251	-	
	Winter	174	196	149	161	163	202	196	209	-	
	Total	779 (71%)	866 (79%)	839 (76.6%)	846 (77.2%)	776 (71%)	798 (73%)	857 (78.2%)	908 (83%)	-	
ds	Fall	196	208	205	192	213	202	182	217	-	
	Spring	186	195	193	193	195	197	200	197	-	
	Summer	214	215	213	203	217	211	204	221	-	
	Winter	190	202	190	143	209	180	189	181	-	
	Total	786 (71.8%)	820 (75%)	801 (73%)	731 (66.7%)	834 (76%)	790 (72%)	775 (70%)	816 (74%)	-	
dL	Fall	197	193	-	-	215	197	196	212	-	
	Spring	198	188	-	-	164	191	188	196	-	
	Summer	200	204	-	-	204	174	214	205	-	
	Winter	175	182	-	-	213	209	189	202	-	
	Total	770 (70.2%)	767 (70%)	-	-	796 (72.6%)	771 (70.3%)	787 (71.8%)	815 (74.4%)	-	

Table A2. A Wilcoxon test (wt) results comparing lysimeters replicas and different configurations to the reference one the lysimeter 1. The test is performed based on the whole validated data and seasons. For the seasons, it is considered fall (22 September to 21 December), winter (21 December to 20 March), spring (20 March to 21 June), and summer (21 June to 22 September). Note that the coloured boxes represent *p*-values, which are superior to 5%.

	Lysimeters		Validated Data	Seasonal Comparison			
				Fall (273 Days)	Winter (272 Days)	Spring (276 Days)	Summer (276 Days)
			p_v	p_v	p_v	p_v	p_v
Comparison of the Replicas	1, 6	wt	0.87	0.35	0.333	0.42	0.511
		n	(445)	(112/273)	(82/272)	(106/276)	(145/276)
	3, 4	wt	0.007	0.007	0.004	0.05	0.26
		n	(600)	(160/273)	(87/272)	(166/276)	(187/276)
	2, 7	wt	2×10^{-16}	8.45×10^{-7}	1.72×10^{-6}	0.08	1.2×10^{-06}
		n	(583)	(154/273)	(116/272)	(127/276)	(186/276)
Different settings compared to the reference (1 or 6)	1, 3	wt	1.16×10^{-20}	4.5×10^{-7}	0.0001	0.01	5×10^{-12}
		n	(464)	(113/273)	(58/272)	(135/276)	(186/276)
	1, 4	wt	1.15×10^{-62}	4.14×10^{-20}	1.04×10^{-15}	3.98×10^{-12}	1.01×10^{-19}
		n	(460)	(117/273)	(57/272)	(133/272)	(153/272)
	1, 5 (data)	wt	0.1	0.0061	1.78×10^{-6}	0.15	0.17
		n	(475)	(121/273)	(86/272)	(118/276)	(150/276)
	1, 2	wt	0.098	0.046	0.3844	0.81	0.36
		n	(503)	(136/273)	(89/272)	(121/276)	(157/276)
	1, 8	wt	2×10^{-8}	8×10^{-5}	0.134	2×10^{-4}	0.01
		n	(515)	(136/273)	(86/272)	(134/276)	(159/276)
	6, 7	wt	0.001	0.59	0.288	5×10^{-6}	0.97
		n	(538)	(125/273)	(111/272)	(123/276)	(179/276)
	6, 5	wt	0.08	0.06	0.0037	0.89	0.34
		n	(506)	(122/273)	(94/272)	(122/276)	(168/276)

Table A3. Proportions of evapotranspiration (ET) to the rainfall (4P) received in each lysimeter. The number of data (n) refers to the common validated data between the rainfall and the considered ET of the lysimeter (on a total of 1096 days).

Lysimeter (n)	1 (557)	2 (674)	3 (636)	4 (630)	5 (624)	6 (601)	7 (656)	8 (718)
ET (mm)	1705.7	2295.9	1605.3	1491.5	1846.3	1846	1919.9	1924.6
P (mm)	502	637.6	643.8	613.8	650.6	614.8	556.6	752.4
4P (mm)	2008	2550.4	2575.2	2455.2	2602.4	2459.2	2226.4	3009.6
%ET	85%	90%	62%	61%	71%	75%	86%	64%

Table A4. Cumulative ET in mm for the lysimeters in common validated days (305/1098).

Year	Seasons	Days	ET1	ET2	ET3	ET4	ET5	ET6	ET7	ET8
2016 to 2017	Fall	12	28.9 ± 1.8	45.2 ± 1.6	21.8 ± 1.6	27.01 ± 1.4	12.7 ± 1.9	21.6 ± 2	25.9 ± 1.9	20.6 ± 1.5
	Winter	0	0	0	0	0	0	0	0	0
	Spring	27	93.1 ± 2.8	118.8 ± 2.4	82 ± 2.4	87 ± 2.2	79.7 ± 2.8	104 ± 3	83.1 ± 2.8	91.7 ± 2.3
	Summer	15	76.1 ± 2	97.7 ± 1.8	60.2 ± 1.8	61.8 ± 1.6	47.6 ± 2.1	70.5 ± 2.2	79.5 ± 2.1	58.6 ± 1.7
2017 to 2018	Fall	39	82.8 ± 3.4	80.9 ± 2.9	41.1 ± 2.9	36.8 ± 2.6	89.2 ± 3.4	70 ± 3.6	55 ± 3.4	44.3 ± 2.8
	Winter	15	45.8 ± 2	39.7 ± 1.8	23.4 ± 1.8	15.2 ± 1.6	10.1 ± 2.1	14.5 ± 2.2	22.8 ± 2.1	22.8 ± 1.7
	Spring	25	108.2 ± 2.7	88.6 ± 2.3	88 ± 2.3	90.1 ± 2.1	137 ± 2.7	135.9 ± 2.9	83.2 ± 2.7	110 ± 2.2
	Summer	70	234.8 ± 4.5	229.9 ± 3.9	169 ± 3.9	160 ± 3.5	321.3 ± 4.6	300 ± 4.8	205.7 ± 4.6	184.9 ± 3.7
2017 to 2018	Fall	14	44.7 ± 2	48.6 ± 1.7	22.7 ± 1.7	25.9 ± 1.5	35 ± 2	53.9 ± 2.1	45.7 ± 2	35.26 ± 1.7
	Winter	10	22.07	14.3 ± 1.7	10.9 ± 1.5	9.03 ± 1.5	19.09 ± 1.7	24.8 ± 1.8	20.63 ± 1.7	9.5 ± 1.4
	Spring	31	122.14 ± 3	99 ± 2.6	103.9 ± 2.6	89.44 ± 2.3	127.7 ± 3	114.5 ± 3.2	117.5 ± 3	53.5 ± 2.5
	Summer	47	207.3 ± 3.7	218.8 ± 3.2	116 ± 3.2	147.7 ± 2.8	271.4 ± 3.7	150.3 ± 3.9	222.3 ± 3.7	232.2 ± 3.1

Table A5. Cumulative ET and P in mm for the lysimeters in common validated days for reference (herbaceous) and shrubs configurations. The couples' lysimeter 1 vs. lysimeter 2 and lysimeter 1 vs. lysimeter 7 are presented.

Year	Seasons	Days (1 vs. 2)	Rain (P)	ET1	ET2	Days (1 vs. 7)	Rain (P)	ET1	ET7
2016–2017	Fall	40	25.8	71.17	103.98	42	23.4	67.61	65.57
	Winter	38	28.8	37.36	74.68	49	55	57.31	34.00
	Spring	57	59.8	212.24	274.06	51	87	209.13	189.99
	Summer	36	37.8	193.27	216.7	34	39.2	155.91	158.99
2017–2018	Fall	61	58.4	110.44	125	50	26.6	97.09	70.02
	Winter	33	41.6	87.6	80	29	39.6	82.01	38.02
	Spring	26	20.4	112.76	90	31	23.6	129.87	100.30
	Summer	72	17.6	240	232	72	17.6	239.94	208.93
2018–2019	Fall	33	54	80.15	78.89	28	52.4	68.71	96.95
	Winter	18	5.4	33.53	22.21	24	10.2	37.74	36.18
	Spring	38	19.8	157.63	129.17	41	15.4	179.63	175.68
	Summer	49	17.2	217.12	230.57	49	17.2	217.12	230.03
2016–2018	Fall	136	138.8	264.47	315.93	121	103	233.96	234.83
	Winter	89	75.8	158.5	176.89	102	104.8	177.06	108.21
	Spring	121	100	482.64	493.24	123	126	518.63	465.97
	Summer	157	72.6	650.33	679.27	155	74	612.97	597.95
Total		503	387.2	1555.95	1665.34	501	407.8	1542.63	1406.97

Table A6. Cumulative ET and P in mm for the lysimeters in common validated days for reference (lysimeter 1), spontaneous (lysimeter 5), and regularly removed (lysimeter 8) vegetation configurations.

Year	Seasons	Days (1 vs. 5)	P	ET1	ET5	Days (1 vs. 8)	P	ET1	ET8
2016–2017	Fall	45	42	81.0	35.8	39	25.8	69.9	51.7
	Winter	38	51.6	43.7	18.1	33	23.6	28.2	43.8
	Spring	45	66.4	187.4	162.1	60	66.8	227.3	224.2
	Summer	30	22.2	163.3	88.4	38	39.2	205.7	156.0
2017–2018	Fall	61	60.2	112.4	113.6	61	50.8	114.5	72.5
	Winter	34	44.8	89.1	22.8	30	36.8	84.8	55.0
	Spring	32	23.6	134.4	165.8	32	23.6	134.4	139.3
	Summer	72	17.6	239.9	323.8	72	17.6	239.9	189.4
2018–2019	Fall	15	35	48.4	35.9	35	56.4	79.3	70.8
	Winter	14	9.2	29.1	24.7	23	7.2	36.8	18.8
	Spring	41	20.6	179.4	197.8	42	20.6	184.5	82.3
	Summer	48	13	212.7	275.3	49	17.2	217.1	234.9
2016–2018	Fall	121	137.2	241.9	185.3	136	133.6	264.2	195.0 (14.5%)
	Winter	86	105.6	161.9	65.6	86	67.6	149.8	117.6
	Spring	118	110.6	501.3	525.7	134	111	546.1	445.7
	Summer	150	52.8	616.0	687.5	159	746	662.7	580.3
Total		475	406.2	1521.0	1464.2	515	386.2	1622.9	1338.6

Table A7. Cumulative ET and P in mm for the lysimeters in common validated days for reference (lysimeter 1) and non-internal water storage configurations (lysimeters 3 and 4).

Year	Seasons	Days (1 vs. 3)	P	ET1	ET3	Days (1 vs. 4)	P	ET1	ET4
2016–2017	Fall	26	47.6	64.9	59.0	26	37	59.1	54
	Winter	13	17.6	22.4	11.7	14	17.6	25.9	13
	Spring	64	87	242.0	215.5	62	73.4	236.8	207.6
	Summer	38	39.2	205.7	143.6	36	26.4	205.5	161.3
2017–2018	Fall	53	47.6	105.3	56.4	59	48.4	113.6	46
	Winter	24	30.6	73.4	34.9	21	21	59.5	21.3
	Spring	31	22.6	133.7	110.6	31	22.6	133.7	104.3
	Summer	72	17.6	239.9	169.5	70	9.2	234.8	160.1
2018–2019	Fall	32	50	75.1	44.1	30	49	72.2	35.7
	Winter	21	5	36.5	19.3	22	5.2	36.5	12.8
	Spring	40	20.6	169.5	151.8	40	20.6	169.5	124.2
	Summer	48	13	212.7	117.7	47	13	207.3	147.7
2016–2018	Fall	113	145.8	248.0	162.4	117	135	247.5	135.9
	Winter	58	53.2	132.3	65.9	57	43.8	122	47.2
	Spring	135	130.2	545.2	477.9	113	116.6	540	436.2
	Summer	158	69.8	658.3	430.8	153	48.6	647.6	469.2
Total		464	399	1583.9	1137.0	460	344	1557.1	1088.4

References

1. Katsifarakis, K.L.; Vafeiadis, M.; Theodossiou, N. Sustainable Drainage and Urban Landscape Upgrading Using Rain Gardens. Site Selection in Thessaloniki, Greece. *Agric. Agric. Sci. Procedia* **2015**, *4*, 338–347. [CrossRef]
2. Andrés-Doménech, I.; Anta, J.; Perales-Momparler, S.; Rodriguez-Hernandez, J. Sustainable Urban Drainage Systems in Spain: A Diagnosis. *Sustainability* **2021**, *13*, 2791. [CrossRef]
3. Ferrans, P.; Torres, M.N.; Temprano, J.; Rodríguez Sánchez, J.P. Sustainable Urban Drainage System (SUDS) Modeling Supporting Decision-Making: A Systematic Quantitative Review. *Sci. Total Environ.* **2022**, *806*, 150447. [CrossRef] [PubMed]
4. Jato-Espino, D.; Toro-Huertas, E.I.; Güereca, L.P. Lifecycle Sustainability Assessment for the Comparison of Traditional and Sustainable Drainage Systems. *Sci. Total Environ.* **2022**, *817*, 152959. [CrossRef]
5. Ishimatsu, K.; Ito, K.; Mitani, Y.; Tanaka, Y.; Sugahara, T.; Naka, Y. Use of Rain Gardens for Stormwater Management in Urban Design and Planning. *Landsc. Ecol. Eng.* **2017**, *13*, 205–212. [CrossRef]
6. Zhang, L.; Ye, Z.; Shibata, S. Assessment of Rain Garden Effects for the Management of Urban Storm Runoff in Japan. *Sustainability* **2020**, *12*, 9982. [CrossRef]
7. Dussaillant, A.R.; Wu, C.H.; Potter, K.W. Richards Equation Model of a Rain Garden. *J. Hydrol. Eng.* **2004**, *9*, 219–225. [CrossRef]
8. Bortolini, L.; Zanin, G. Reprint of: Hydrological Behaviour of Rain Gardens and Plant Suitability: A Study in the Veneto Plain (North-Eastern Italy) Conditions. *Urban For. Urban Green.* **2019**, *37*, 74–86. [CrossRef]
9. Muthanna, T.M.; Viklander, M.; Thorolfsson, S.T. Seasonal Climatic Effects on the Hydrology of a Rain Garden. *Hydrol. Process.* **2008**, *22*, 1640–1649. [CrossRef]
10. Jennings, A.A.; Berger, M.A.; Hale, J.D. Hydraulic and Hydrologic Performance of Residential Rain Gardens. *J. Environ. Eng.* **2015**, *141*, 04015033. [CrossRef]
11. Denich, C.; Bradford, A. Estimation of Evapotranspiration from Bioretention Areas Using Weighing Lysimeters. *J. Hydrol. Eng.* **2010**, *15*, 522–530. [CrossRef]
12. Hess, A.; Wadzuk, B.; Welker, A. Evapotranspiration in Rain Gardens Using Weighing Lysimeters. *J. Irrig. Drain. Eng.* **2017**, *143*, 04017004. [CrossRef]
13. Wadzuk, B.M.; Hickman, J.M.; Traver, R.G. Understanding the Role of Evapotranspiration in Bioretention: Mesocosm Study. *J. Sustain. Water Built Environ.* **2015**, *1*, 04014002. [CrossRef]
14. Nocco, M.A.; Rouse, S.E.; Balster, N.J. Vegetation Type Alters Water and Nitrogen Budgets in a Controlled, Replicated Experiment on Residential-Sized Rain Gardens Planted with Prairie, Shrub, and Turfgrass. *Urban Ecosyst.* **2016**, *19*, 1665–1691. [CrossRef]
15. Ebrahimian, A.; Wadzuk, B.; Traver, R. Evapotranspiration in Green Stormwater Infrastructure Systems. *Sci. Total Environ.* **2019**, *688*, 797–810. [CrossRef]
16. Cascone, S.; Coma, J.; Gagliano, A.; Pérez, G. The Evapotranspiration Process in Green Roofs: A Review. *Build. Environ.* **2019**, *147*, 337–355. [CrossRef]
17. Hess, A.; Wadzuk, B.; Welker, A. Evapotranspiration Estimation in Rain Gardens Using Soil Moisture Sensors. *Vadose Zone J.* **2021**, *20*, e20100. [CrossRef]
18. Saher, R.; Stephen, H.; Ahmad, S. Urban Evapotranspiration of Green Spaces in Arid Regions through Two Established Approaches: A Review of Key Drivers, Advancements, Limitations, and Potential Opportunities. *Urban Water J.* **2021**, *18*, 115–127. [CrossRef]
19. Shahtahmassebi, A.R.; Li, C.; Fan, Y.; Wu, Y.; Lin, Y.; Gan, M.; Wang, K.; Malik, A.; Blackburn, G.A. Remote Sensing of Urban Green Spaces: A Review. *Urban For. Urban Green.* **2021**, *57*, 126946. [CrossRef]
20. Li, X.; Chen, W.Y.; Sanesi, G.; Lafortezza, R. Remote Sensing in Urban Forestry: Recent Applications and Future Directions. *Remote Sens.* **2019**, *11*, 1144. [CrossRef]
21. Lubczynski, M.W.; Gurwin, J. Integration of Various Data Sources for Transient Groundwater Modeling with Spatio-Temporally Variable Fluxes—Sardon Study Case, Spain. *J. Hydrol.* **2005**, *306*, 71–96. [CrossRef]
22. Monteith, J.L. Evaporation and Environment. *Symp. Soc. Exp. Biol.* **1965**, *19*, 205–234. [PubMed]
23. Allen, R.G.; Pereira, L.S.; Raes, D.; Smith, M. Crop Evapotranspiration-Guidelines for Computing Crop Water Requirements-FAO Irrigation and Drainage Paper 56. *Fao Rome* **1998**, *300*, D05109.
24. Walter, I.A.; Allen, R.G.; Elliott, R.; Jensen, M.E.; Itenfisu, D.; Mecham, B.; Howell, T.A.; Snyder, R.; Brown, P.; Echings, S.; et al. ASCE's standardized reference evapotranspiration equation. Watershed Management and Operations Management. 2000, pp. 1–11. Available online: https://ascelibrary.org/doi/book/10.1061/9780784404997 (accessed on 30 January 2022).
25. Hargreaves, G.H.; Allen, R.G. History and Evaluation of Hargreaves Evapotranspiration Equation. *J. Irrig. Drain. Eng.* **2003**, *129*, 53–63. [CrossRef]
26. Priestley, C.H.B.; Taylor, R.J. On the Assessment of Surface Heat Flux and Evaporation Using Large-Scale Parameters. *Mon. Weather Rev.* **1972**, *100*, 81–92. [CrossRef]
27. Hess, A.; Wadzuk, B.; Welker, A. Predictive Evapotranspiration Equations in Rain Gardens. *J. Irrig. Drain. Eng.* **2019**, *145*, 04019010. [CrossRef]
28. Nezeys, A.; Reboul, S.; Saison, O.; Baillet, M. Le Plan Pluie à Paris: La Nécessaire Dimension Environnementale. In *Stratégie/Strategy-Démarche Intégrée & Développement Durable/Integrated Approach & Sustainable Development*; GRAIE: Lyon, France, 2016.

29. Masson, V.; Lion, Y.; Peter, A.; Pigeon, G.; Buyck, J.; Brun, E. "Grand Paris": Regional Landscape Change to Adapt City to Climate Warming. *Clim. Chang.* **2013**, *117*, 769–782. [CrossRef]
30. Agence Parisienne du Climat (APC) et Météo-France. Brochure « L'îlot de Chaleur Urbain à Paris, un Microclimat au Cœur de la Ville ». Agence Parisienne du Climat. 2018. Available online: https://www.apc-paris.com/publication/brochure-lilot-chaleur-urbain-a-paris-microclimat-coeur-ville (accessed on 30 January 2022).
31. Bertrand-Krajewski, J.-L. (Ed.) *Mesures en Hydrologie Urbaine et Assainissement*; Éditions Tec & Doc: Paris, France, 2000; ISBN 978-2-7430-0380-7.
32. Capteurs de Pesage et Pesons—Scaime. Available online: https://fr.scaime.com/capteurs-de-pesage (accessed on 16 November 2021).
33. Farrés, M.; Platikanov, S.; Tsakovski, S.; Tauler, R. Comparison of the Variable Importance in Projection (VIP) and of the Selectivity Ratio (SR) Methods for Variable Selection and Interpretation. *J. Chemom.* **2015**, *29*, 528–536. [CrossRef]
34. Gosselin, R.; Rodrigue, D.; Duchesne, C. A Bootstrap-VIP Approach for Selecting Wavelength Intervals in Spectral Imaging Applications. *Chemom. Intell. Lab. Syst.* **2010**, *100*, 12–21. [CrossRef]
35. Mehmood, T.; Liland, K.H.; Snipen, L.; Sæbø, S. A Review of Variable Selection Methods in Partial Least Squares Regression. *Chemom. Intell. Lab. Syst.* **2012**, *118*, 62–69. [CrossRef]
36. Feng, Y.; Burian, S.; Pardyjak, E. Observation and Estimation of Evapotranspiration from an Irrigated Green Roof in a Rain-Scarce Environment. *Water* **2018**, *10*, 262. [CrossRef]
37. Penman, H.L.; Keen, B.A. Natural Evaporation from Open Water, Bare Soil and Grass. *Proc. R. Soc. Lond. Ser. Math. Phys. Sci.* **1948**, *193*, 120–145. [CrossRef]
38. Vannier, O.; Braud, I. Calcul d'une Évapotranspiration de Référence Spatialisée Pour La Modélisation Hydrologique à Partir Des Don-nées de La Réanalyse SAFRAN de Météo-France. Research Report, IRSTEA, France. 2012, p. 22. Available online: https://hal.inrae.fr/hal-02593413 (accessed on 30 January 2022).
39. Marasco, D.E.; Culligan, P.J.; McGillis, W.R. Evaluation of Common Evapotranspiration Models Based on Measurements from Two Extensive Green Roofs in New York City. *Ecol. Eng.* **2015**, *84*, 451–462. [CrossRef]
40. Johnston, M.R. Vegetation Type Alters Rain Garden Hydrology through Changes to Soil Porosity and Evapotranspiration. Ph.D. Thesis, University of Wisconsin, Madison, WI, USA, 2011; pp. 130–139.
41. Johnston, M.R.; Balster, N.J.; Thompson, A.M. Vegetation Alters Soil Water Drainage and Retention of Replicate Rain Gardens. *Water* **2020**, *12*, 3151. [CrossRef]
42. Le Coustumer, S.; Fletcher, T.D.; Deletic, A.; Barraud, S.; Poelsma, P. The Influence of Design Parameters on Clogging of Stormwater Biofilters: A Large-Scale Column Study. *Water Res.* **2012**, *46*, 6743–6752. [CrossRef]
43. Glaister, B.J.; Fletcher, T.D.; Cook, P.L.M.; Hatt, B.E. Interactions between Design, Plant Growth and the Treatment Performance of Stormwater Biofilters. *Ecol. Eng.* **2017**, *105*, 21–31. [CrossRef]
44. Costello, L.R.; Jones, K.S. *WUCOLS, Water Use Classification of Landscape Species: A Guide to the Water Needs of Landscape Plants*; University of California Press: Sacramento, CA, USA, 2014; Volume 96.
45. Romero, C.C.; Dukes, M.D. *Residential Benchmarks for Minimal Landscape Water Use*; Agricultural and Biological Engineering, Department University of Florida UF Water Institute: Gainesville, FL, USA, 2010; pp. 1–49.
46. DiGiovanni, K.; Montalto, F.; Gaffin, S.; Rosenzweig, C. Applicability of Classical Predictive Equations for the Estimation of Evapotranspiration from Urban Green Spaces: Green Roof Results. *J. Hydrol. Eng.* **2013**, *18*, 99–107. [CrossRef]
47. DiGiovanni-White, K.; Montalto, F.; Gaffin, S. A Comparative Analysis of Micrometeorological Determinants of Evapotranspiration Rates within a Heterogeneous Urban Environment. *J. Hydrol.* **2018**, *562*, 223–243. [CrossRef]
48. Jahanfar, A.; Drake, J.; Sleep, B.; Gharabaghi, B. A Modified FAO Evapotranspiration Model for Refined Water Budget Analysis for Green Roof Systems. *Ecol. Eng.* **2018**, *119*, 45–53. [CrossRef]
49. Braud, I.; Dantas-Antonino, A.C.; Vauclin, M.; Thony, J.L.; Ruelle, P. A Simple Soil-Plant-Atmosphere Transfer Model (SiSPAT) Development and Field Verification. *J. Hydrol.* **1995**, *166*, 213–250. [CrossRef]
50. Stavropulos-Laffaille, X. Pour Une Analyse Des Impacts Du Changement Climatique Sur l'hydrologie Urbaine: Modélisation Hydro-Microclimatique de Deux Bassins Versants Expérimentaux de l'agglomération Nantaise. Ph.D. Thesis, Ecole Centrale de Nantes, Nantes, France, 2019.

hydrology

Essay

RASPOTION—A New Global PET Dataset by Means of Remote Monthly Temperature Data and Parametric Modelling

Aristoteles Tegos [1,*], Nikolaos Malamos [2] and Demetris Koutsoyiannis [1]

[1] Department of Water Resources and Environmental Engineering, National Technical University of Athens, Heroon Polytechneiou 5, GR-157 80 Athens, Greece; dk@itia.ntua.gr

[2] Department of Agriculture, University of Patras, GR-272 00 Amaliada, Greece; nmalamos@upatras.gr

* Correspondence: tegosaris@yahoo.gr

Abstract: Regional estimations of Potential Evapotranspiration (PET) are of key interest for a number of geosciences, particularly those that are water-related (hydrology, agrometeorology). Therefore, several models have been developed for the consistent quantification of different time scales (hourly, daily, monthly, annual). During the last few decades, remote sensing techniques have continued to grow rapidly with the simultaneous development of new local and regional evapotranspiration datasets. Here, we develop a novel set T maps over the globe, namely RASPOTION, for the period 2003 to 2016, by integrating: (a) mean climatic data at 4088 stations, extracted by the FAO-CLIMWAT database; (b) mean monthly PET estimates by the Penman–Monteith method, at the aforementioned locations; (c) mean monthly PET estimates by a recently proposed parametric model, calibrated against local Penman–Monteith data; (d) spatially interpolated parameters of the Parametric PET model over the globe, using the Inverse Distance Weighting technique; and (e) remote sensing mean monthly air temperature data. The RASPOTION dataset was validated with in situ samples (USA, Germany, Spain, Ireland, Greece, Australia, China) and by using a spatial Penman–Monteith estimates in England. The results in both cases are satisfactory. The main objective is to demonstrate the practical usefulness of these PET map products across different research disciplines and spatiotemporal scales, towards assisting decision making for both short- and long-term hydro-climatic policy actions.

Keywords: RASPOTION; potential evapotranspiration; parametric model; remote sensing; hydrological calibration

Citation: Tegos, A.; Malamos, N.; Koutsoyiannis, D. RASPOTION—A New Global PET Dataset by Means of Remote Monthly Temperature Data and Parametric Modelling. *Hydrology* **2022**, *9*, 32. https://doi.org/10.3390/hydrology9020032

Academic Editor: Luca Brocca

Received: 24 December 2021
Accepted: 7 February 2022
Published: 10 February 2022

Publisher's Note: MDPI stays neutral with regard to jurisdictional claims in published maps and institutional affiliations.

1. Introduction

Evapotranspiration (ET) is a crucial element of the hydrological cycle relevant in a wide range of geosciences, since it represents the combined water losses from soil surface and vegetation. It is influenced by several meteorological variables such as air temperature, solar radiation, wind speed, and relative humidity. The literature proposes several approaches to quantify the process in terms of actual evapotranspiration, potential evapotranspiration (PET) or reference evapotranspiration. By definition, PET refers to "the rate at which evapotranspiration would occur from a large area completely and uniformly covered with growing vegetation, which has access to an unlimited supply of soil water, and without advection or heating effects" [1]. PET is different from the actual evapotranspiration, which also depends on the actual soil water supply, mainly driven by the precipitation regime. In recent decades, advanced methods have been introduced for ET and PET estimation, the most recent being the remote sensing technique, incorporating aerial and satellite imagery [2–5]. Generally, the classification of remote sensing for ET assessment includes four groups referred to as empirical, direct, residual, inference and deterministic models [6]. The most well-known approach for the actual evapotranspiration estimation for daily and monthly time step is the modified surface energy balance algorithm for land (SEBAL) model [7]. A limited number of studies have focused on the global PET assessment utilizing remote sensing tools. Specifically, the global distribution of potential evaporation has been calculated from the Penman–Monteith equation using satellite and assimilated data for a 24-month period, i.e., January 1987 to December 1988 [8].

The Parametric model is a temperature-based model that requires only temperature data and utilizes a parsimonious expression for the potential evapotranspiration (PET) estimation [9]. It replaces some of the variables and constants that are used in the standard Penman–Monteith model by regionally varying parameters, which are estimated through calibration [10–12]. The large-scale Parametric model application was satisfactory, and it outperformed the efficiency of several simplified models such as Hargreaves, Thornthwaite, Oudin, and Jensen–Haise.

In this study, a new global PET monthly dataset is introduced by applying the Parametric model using the remote sensing data (LANDSAT) of mean air temperature, provided by a recent remote mean temperature dataset from 2003 to 2016. As most global applications refer to the actual evapotranspiration assessment [2–5], this dataset may contribute to hydrological balance modelling and agrometeorological applications.

2. Materials and Methods

The Parametric model employs physically consistent parameters distributed over the globe, overcoming the main weakness of the Penman–Monteith model, which is the necessity of simultaneous observations of four meteorological variables [10–14].

The modified Parametric model implements two instead of three parameters, namely parameter a' in the numerator and parameter c' in the denominator of the formula:

$$\mathrm{PET} = \frac{a' R_a}{1 - c' T} \tag{1}$$

where PET is the potential evapotranspiration (mm), R_a (kJ m^{-2}) is the extra-terrestrial radiation, a' (kg kJ^{-1}), and c' (°C^{-1}) are the calibrated parameters and T (°C) is the monthly mean air temperature. As already stated in Tegos et al. [12], the parameters have some physical correspondence to the Penman–Monteith equation, since the product $a' R_a$ represents the overall energy term (i.e., incoming minus outgoing solar radiation), while the quantity $1 - c' T$ approximates the denominator term of the Penman–Monteith formula. More information about the parameters a' and c' and their spatial patterns across the globe can be found in [12].

The model was applied globally using the values of parameters a' and c' at the locations of 4088 stations of the FAO-CLIMWAT database, which presented positive efficiency according to the Nash–Sutcliffe criterion during calibration (Figure 1). These values were interpolated over the globe using the inverse distance weighting (IDW) technique into a geographical information system (GIS) [12].

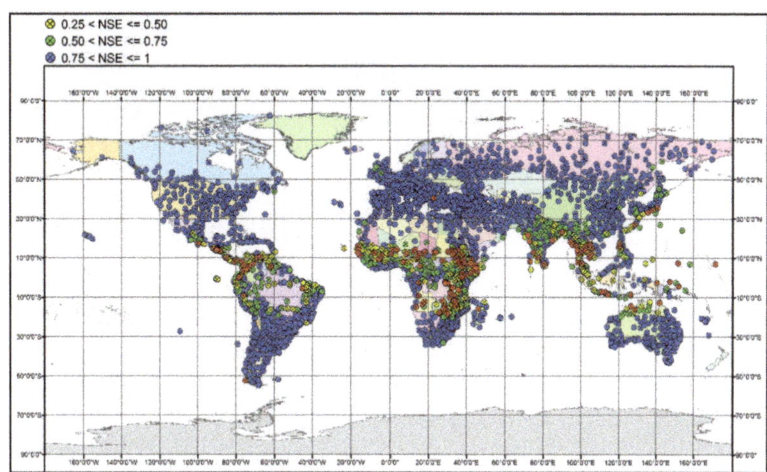

Figure 1. CLIMWAT meteorological stations network and distribution of efficiency (from [12]).

The extra-terrestrial radiation (R_a) monthly raster datasets were derived from the respective daily values using an analytical mathematical expression [12], from $-90°$ to $+90°$ of latitude with a step of $0.05°$ for normal and leap years (Figure 2), taking into consideration the polar daylight and polar night periods [15].

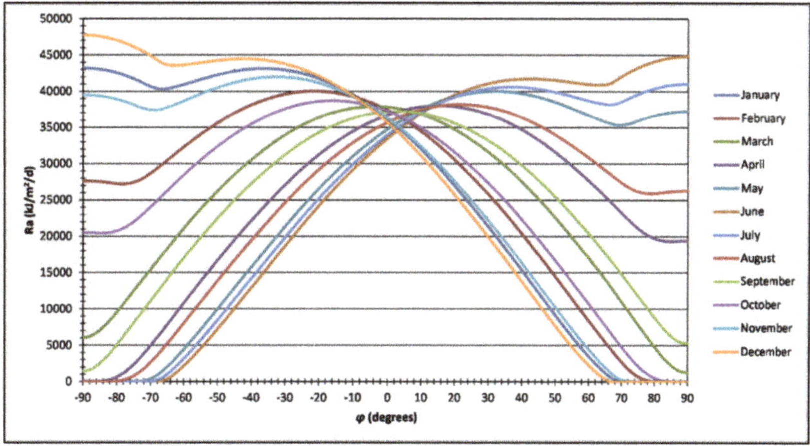

Figure 2. Mean monthly extra-terrestrial radiation (R_a) for latitudes $-90°$ to $+90°$.

The mean air temperature values, covering a period from 2003 to 2016, were acquired as raster datasets from the recent analysis presented in [16]. Figure 3 shows the spatial temperature variation across the globe in June 2011.

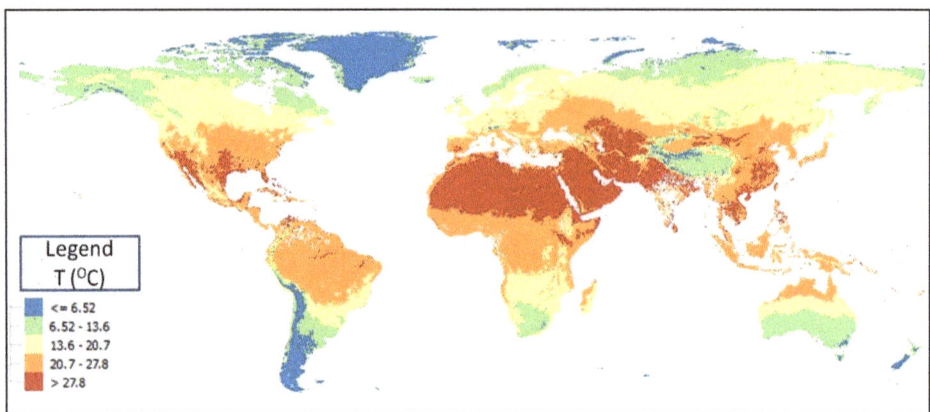

Figure 3. Monthly air temperature in June 2011.

All three layers of information were embedded in GIS and constituted a framework that permitted quality control screening by application of reasonable thresholds to exclude the outliers in the PET values. The maps obtained using the parametric method were produced in a GIS environment by applying Equation (1) with the required raster datasets, i.e., parameters a' and c', extra-terrestrial radiation R_a, and monthly mean air temperature T.

3. Results

3.1. PET Global Mapping

Following the above presented methodology, a monthly PET global dataset was acquired covering the period 2003–2016. Figure 4, visualizes the PET distribution for a representative month (August 2011) globally.

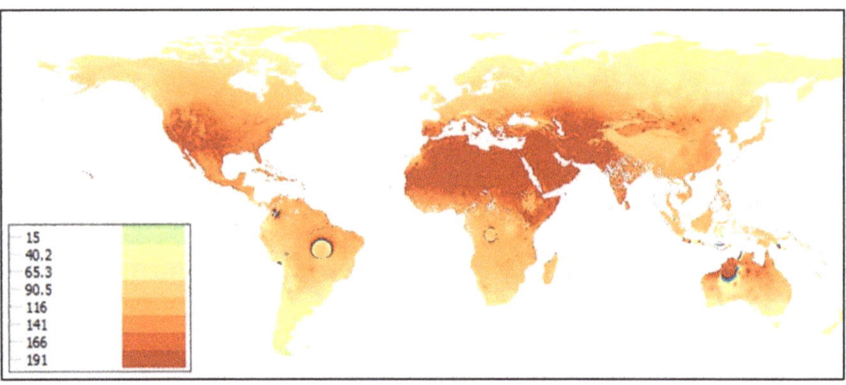

Figure 4. Global PET map (mm/month) for August 2011.

In Eurasia, where PET monthly values range from 19 to 239 mm, PET increases from north to south. The latter is well explained from the similar variation of temperature and extra-terrestrial radiation. The highest values were observed in the Middle East, where extremely arid climatic conditions occur. A pattern similar to Eurasia was observed in North America, with the highest values at regions near the equator (e.g., Mexico) and lowest in Canada, Alaska and Greenland. In South America, PET decreases from north to south. Some inconsistencies in the area of Amazon and some peculiarly low values in the area of equator can be explained from the limitations of the Parametric model to represent the combined effect on PET estimation of relative humidity and wind speed, which are key drivers of the evapotranspiration processes across these areas, influencing the net incoming solar radiation and the evaporation demand, as detailed in Tegos et al. [12].

High monthly PET values were acquired in the equatorial zone for Africa, mainly in the lower Congo, where the hydro-meteorological observations were limited in the Parametric model calibration. The decreasing trend from north to south in Oceania follows the pattern of radiation and air temperature variation.

3.2. Validation

The RASPOTION remote sensing dataset has been compared against the Penman–Monteith timeseries, estimated at 26 meteorological stations across the globe, listed in Table 1. The validation was carried out for long-term monthly Penman–Monteith samples by examining the coefficient of efficiency (CE) due to Nash and Sutcliffe [17] in different countries, namely, the USA, Germany, Spain, Ireland, Greece, Australia, and China, and with different climatic regimes. Penman–Monteith timeseries were retrieved by different databases such as CIMIS network (https://cimis.water.ca.gov/), European Climate Assessment data set (http://eca.knmi.nl/), Australian Bureau of Meteorology (http://www.bom.gov.au/watl/eto/, accessed on 1 February 2022), the Irish Meteorological Service—Met Éireann (https://www.met.ie/)—and a previously published paper [18]. The new RASPOTION dataset shows an excellent performance across different climatic regimes. Only two stations (Shanxi, Sydney Airport) demonstrate a moderate performance; however, the coefficient of efficiency was above a threshold that can safely allow its further operational use.

Table 1. Validation dataset.

Station	Country	Period	CE
Kostakioi	Greece	04/2008–07/2013	89
Mace Head	Ireland	10/2010–11/2016	90.9
Zaragoza	Spain	01/2003–11/2009	92.8
Alicante	Spain	01/2003–10/2009	92.4
Munchen	Germany	01/2003–06/2013	90.0
Karshue	Germany	01/2003–08/2009	89.2
Hamburg	Germany	01/2003–06/2013	93.7
Frankfurt	Germany	01/2003–06/2013	96.7
Dusseldorf	Germany	01/2003–06/2013	94.7
Dresden	Germany	01/2003–06/2013	92.9
Bremen	Germany	01/2003–06/2013	94.9
Angermunde	Germany	01/2003–06/2013	95.2
Aachen	Germany	01/2003–11/2005	91.5
Tulelake	USA	01/2003–11/2005	79.2
Meloland	USA	01/2003–06/2013	89.0
Manteca	USA	01/2003–06/2013	93.1
Temecula	USA	01/2003–06/2013	84.5
Buntigville	USA	01/2003–06/2013	89.6
Mc Arthur	USA	01/2003–06/2013	89.5
Davis	USA	01/2003–06/2013	93.1
Tunnack Firestation	Australia	01/2009–12/2014	90.5
Adelaide airport	Australia	01/2009–12/2014	83.2
Sydney Airport	Australia	01/2009–12/2014	43.2
Alice Springs	Australia	01/2009–12/2014	84.1
Albany airport	Australia	01/2009–12/2014	90.6
Shanxi	China	01/2003–12/2014	22.3

Following RASPOTION's evaluation performance at a point basis across the globe, an advanced comparison is presented here that demonstrates its performance against spatial PET monthly maps provided by the Environmental Agency throughout England. The latter dataset has been developed using the daily FAO-56 PET method in an extensive gauge meteorological network across England, and spatial mapping with the use IDW method. A free data access is provided by DEFRA Data Services Platform. Figure 5 shows the PET monthly variation across England. PET ranges from 25 mm/month to 135 mm/month and increases from North to Southeast.

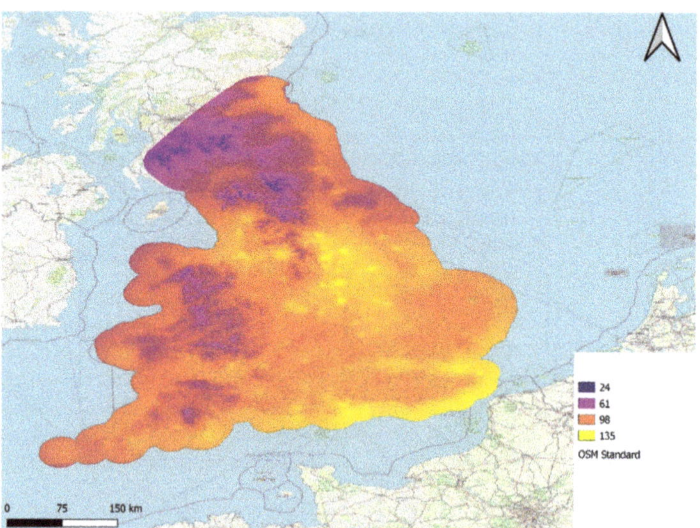

Figure 5. PET monthly map (DEFRA- June 2011 PET in mm/month).

By comparing RASPOTION's June 2011 raster map with the DEFRA PET map a very good performance is achieved, as the actual difference for the majority of England (around 80%) is up to ±9 mm/month (Figure 6), corresponding to a range of up to 12% of overestimation and 9% underestimation of actual PET, respectively (Figure 7).

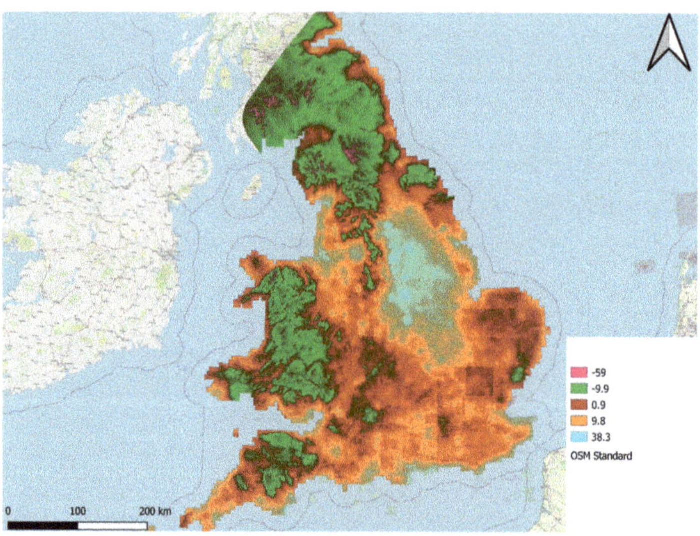

Figure 6. Spatial difference RASPOTION against DEFRA PET (mm/month).

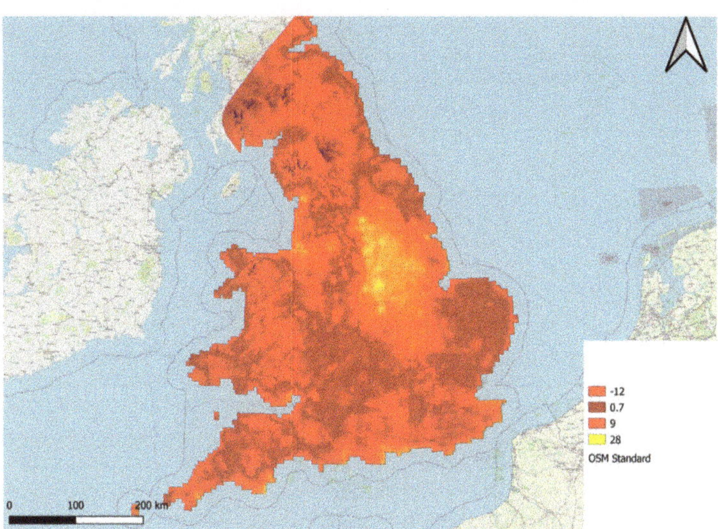

Figure 7. Ratio difference RASPOTION against DEFRA PET.

4. Discussion

In previous studies, extended discussion has shown that the Parametric PET framework provides satisfactory results with some limitations [12], also outlined here. In some areas with high values of relative humidity and wind speed, the existing PET parametric approach fails to efficiently reproduce the PET characteristics and further improvements to the parametric approach are recommended locally. Those areas are located in sub-Saharan Africa (Kongo, Zimbabwe, Rwanda and Uganda) and South America, adjacent to the Amazon River catchment (Brazil). It should be noted that the FAO-CLIMWAT data were scarce and poor in Sub-Saharan and South American territories. In these territories the calibrated parameters a' and c' are indicative and we recommend further calibration should take place. The missing hydrometeorological variables and the lack of calibration was also shown in a recent study by dos Santos et al. [19], where the parametric model displayed a moderate performance in subtropical areas. Long-term measurements of relative humidity, wind speed, temperature and radiation in conjunction with a new calibration of a revised parametric formula would lead to substantial improvements in the existing parametric estimates. The work should be accompanied and supported by a thorough statistical investigation between the climatic factors (i.e., radiation, wind speed, relative humidity) to identify the dependence of PET with any other variable except temperature.

Nonetheless, hydrologists, agronomists and other scientists with potential interest in this dataset could make efficient use of it, in about 80% of the earth's territory based on our previous studies [12]. Taking into account the fact that other global PET datasets are not available, the potential benefits of our new dataset may include the following:

(1) In the applications of physically based hydrological models which use PET as an input in catchment modelling. This acquires higher usefulness as we are moving forward toward global scale hydrological models. It is already highlighted by several researchers that the use of accurate PET estimates is of great importance for the reproduction of physically based hydrological responses [20] and its use in calibrating complex physically based hydrological models [21,22].

(2) In the crop-water demand assessment. The integration of the monthly PET and the cropping pattern quantifies the monthly water needs according to vegetation. Accurate PET estimates with fewer demands of meteorological data, combined with

modern techniques such as the use of a drone for micro-farming data gathering [23] greatly support the food–water nexus.

(3)   In climatological studies and drought assessment. The use of reliable PET models greatly influences several well-known indexes such as Palmer Drought Severity Index [24], aridity index [9] and Surface Wetness Index [25].

(4)   In building and evaluating environmental resilience indexes for developing and defining new multi-dimensional approaches. Such approaches would ensure an accurate decision-making basis for sustainable ecosystem management [26].

5. Conclusions

As part of the PET Parametric model, a new global PET monthly dataset based on remote-sensing temperature data is introduced, covering the period 2003–2016. This global dataset was produced using the Parametric formula which uses as input variables extra-terrestrial radiation and mean air temperature. The remote temperature data have been taken from a freely available dataset provided by Hooker et al. [16]. Previous analyses with this approach showed satisfying performance through validation under several climatic regimes and different validation procedures. In regions where the available hydro-meteorological information was scarce or insufficient, the modelling results were weak in terms of PET's physical interpretation. In these areas the RASPOTION dataset should be used with caution. The dataset is open and freely accessible from http://www.itia.ntua.gr/2167/ (accessed on 1 February 2022), where a total number of 168 monthly global raster files (GeoTIFF) are stored. Overall, for the majority of the Earth's surface, a reliable monthly PET dataset is compiled and made available to scientists across different research disciplines in order to assist scientific studies into the global hydrological cycle and decisions for both short- and long-term hydro-climatic policy actions. Future research, consisting of exploration of model parameters and their clustering across the globe, will provide area specific parameters, thereby excluding the need for local calibration and further enabling the use of the Parametric framework.

Author Contributions: A.T.; methodology, data mining, draft reporting N.M.; draft reviewing R_a modelling, D.K.; founding the parametric method, draft reviewing/editing, supervision. All authors have read and agreed to the published version of the manuscript.

Funding: This research received no external funding.

Data Availability Statement: The dataset is open and freely accessible from http://www.itia.ntua.gr/2167/ (accessed on 1 February 2022), where a total number of 168 monthly global raster files (GeoTIFF) are stored.

Acknowledgments: The manuscript is an invited paper as part of the Special Issue "Advances in Evaporation and Evaporative Demand" organized by Hydrology journal. We are grateful to Scientific Editor for handling it and Assistant Editor for inviting us to submit our work. We are also thankful to two anonymous reviewers for the constructive comments which helped us to improve our manuscript substantially. We, finally, thank Brendan Larkin for his final English proofreading and suggested corrections.

Conflicts of Interest: The authors declare no conflict of interest.

References

1.   Dingman, S.L. *Physical Hydrology*; MacMillan Publishing Company: New York, NY, USA, 1994.
2.   Mu, Q.; Zhao, M.; Running, S.W. Improvements to a MODIS global terrestrial evapotranspiration algorithm. *Remote Sens. Environ.* **2011**, *115*, 1781–1800. [CrossRef]
3.   Ghilain, N.; Gellens-Meulenberghs, F. Assessing the impact of land cover map resolution and geolocation accuracy on evapotranspiration simulations by a land surface model. *Remote Sens. Lett.* **2014**, *5*, 491–499. [CrossRef]
4.   Vinukollu, R.K.; Wood, E.F.; Ferguson, C.R.; Fisher, J.B. Global estimates of evapotranspiration for climate studies using multi-sensor remote sensing data: Evaluation of three process-based approaches. *Remote Sens. Environ.* **2011**, *115*, 801–823. [CrossRef]

5.   Yuan, W.; Liu, S.; Yu, G.; Bonnefond, J.-M.; Chen, J.; Davis, K.J.; Desai, A.R.; Goldstein, A.H.; Gianelle, D.; Rossi, F.; et al. Global estimates of evapotranspiration and gross primary production based on MODIS and global meteorology data. *Remote Sens. Environ.* **2010**, *114*, 1416–1431. [CrossRef]

6.   Nouri, H.; Beecham, S.; Kazemi, F.; Hassanli, A.M.; Anderson, S. Remote sensing techniques for predicting evapo-transpiration from mixed vegetated surfaces. *Hydrol. Earth Syst. Sci. Discuss.* **2013**, *10*, 3897–3925.

7.   Bhattarai, N.; Dougherty, M.; Marzen, L.J.; Kalin, L. Validation of evaporation estimates from a modified surface energy balance algorithm for land (SEBAL) model in the south-eastern United States. *Remote Sens. Lett.* **2012**, *3*, 511–519. [CrossRef]

8.   Choudhury, B.J. Global pattern of potential evaporation calculated from the Penman-Monteith equation using satel-lite and assimilated data. *Remote Sens. Environ.* **1997**, *61*, 64–81. [CrossRef]

9.   Stefanidis, S.; Alexandridis, V. Precipitation and Potential Evapotranspiration Temporal Variability and Their Relationship in Two Forest Ecosystems in Greece. *Hydrology* **2021**, *8*, 160. [CrossRef]

10.   Tegos, A.; Efstratiadis, A.; Koutsoyiannis, D. A Parametric Model for Potential Evapotranspiration Estimation Based on a Simplified Formulation of the Penman-Monteith Equation. *Evapotranspiration Overv.* **2013**, 143–165. [CrossRef]

11.   Tegos, A.; Malamos, N.; Koutsoyiannis, D. A parsimonious regional parametric evapotranspiration model based on a simplification of the Penman–Monteith formula. *J. Hydrol.* **2015**, *524*, 708–717. [CrossRef]

12.   Tegos, A.; Efstratiadis, A.; Malamos, N.; Mamassis, N.; Koutsoyiannis, D. Evaluation of a Parametric Approach for Estimating Potential Evapotranspiration Across Different Climates. *Agric. Agric. Sci. Procedia* **2015**, *4*, 2–9. [CrossRef]

13.   Tegos, A.; Malamos, N.; Efstratiadis, A.; Tsoukalas, I.; Karanasios, A.; Koutsoyiannis, D. Parametric Modelling of Potential Evapotranspiration: A Global Survey. *Water* **2017**, *9*, 795. [CrossRef]

14.   Tegos, A.; Mamassis, N.; Koutsoyiannis, D. Estimation of potential evapotranspiration with minimal data dependence. In *EGU General Assembly Conference Abstracts*; European Geosciences Union: Vienna, Austria, 2009; Volume 11. [CrossRef]

15.   Whiteman, C.D.; Allwine, K.J. Extraterrestrial solar radiation on inclined surfaces. *Environ. Softw.* **1986**, *1*, 164–169. [CrossRef]

16.   Hooker, J.; Duveiller, G.; Cescatti, A. A global dataset of air temperature derived from satellite remote sensing and weather stations. *Sci. Data* **2018**, *5*, 180246. [CrossRef] [PubMed]

17.   Nash, J.E.; Sutcliffe, J.V. River flow forecasting through conceptual models part I—A discussion of principles. *J. Hydrol.* **1970**, *10*, 282–290. [CrossRef]

18.   Yuanyuan, W.; Jun, Z.; Yanqiang, W. Spatiotemporal variation characteristics of surface evapotranspiration in Shanxi Province based on MOD16. *Prog. Geogr.* **2020**, *39*, 255–264.

19.   dos Santos, A.A.; Moretti de Souza, J.L.; Rosa, S.L.K. Evapotranspiration with the Moretti-Jerszurki-Silva model for the Brazilian subtropical climate. *Hydrol. Sci. J.* **2021**, *66*, 2267–2279. [CrossRef]

20.   Seiller, G.; Anctil, F. How do potential evapotranspiration formulas influence hydrological projections? *Hydrol. Sci. J.* **2016**, *61*, 2249–2266. [CrossRef]

21.   Immerzeel, W.A.; Droogers, P. Calibration of a distributed hydrological model based on satellite evapotranspiration. *J. Hydrol.* **2008**, *349*, 411–424. [CrossRef]

22.   López López, P.; Sutanudjaja, E.H.; Schellekens, J.; Sterk, G.; Bierkens, M.F. Calibration of a large-scale hydrological model using satellite-based soil moisture and evapotranspiration products. *Hydrol. Earth Syst. Sci.* **2017**, *21*, 3125–3144. [CrossRef]

23.   Alexandris, S.; Psomiadis, E.; Proutsos, N.; Philippopoulos, P.; Charalampopoulos, I.; Kakaletris, G.; Papoutsi, E.-M.; Vassilakis, S.; Paraskevopoulos, A. Integrating Drone Technology into an Innovative Agrometeorological Methodology for the Precise and Real-Time Estimation of Crop Water Requirements. *Hydrology* **2021**, *8*, 131. [CrossRef]

24.   van der Schrier, G.; Jones, P.D.; Briffa, K.R. The sensitivity of the PDSI to the Thornthwaite and Penman-Monteith parameterizations for potential evapotranspiration. *J. Geophys. Res. Earth Surf.* **2011**, *116*. [CrossRef]

25.   Yang, Q.; Ma, Z.; Zheng, Z.; Duan, Y. Sensitivity of potential evapotranspiration estimation to the Thornthwaite and Penman–Monteith methods in the study of global drylands. *Adv. Atmos. Sci.* **2017**, *34*, 1381–1394. [CrossRef]

26.   Fan, X.; Hao, X.; Hao, H.; Zhang, J.; Li, Y. Comprehensive Assessment Indicator of Ecosystem Resilience in Central Asia. *Water* **2021**, *13*, 124. [CrossRef]

 *hydrology*

Article

Stochastic Analysis of Hourly to Monthly Potential Evapotranspiration with a Focus on the Long-Range Dependence and Application with Reanalysis and Ground-Station Data

Panayiotis Dimitriadis 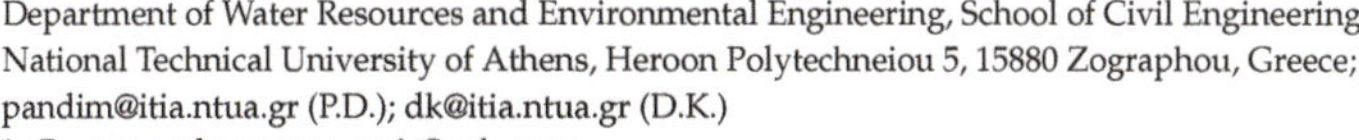, Aristoteles Tegos * and Demetris Koutsoyiannis

Department of Water Resources and Environmental Engineering, School of Civil Engineering, National Technical University of Athens, Heroon Polytechneiou 5, 15880 Zographou, Greece; pandim@itia.ntua.gr (P.D.); dk@itia.ntua.gr (D.K.)
* Correspondence: tegosaris@yahoo.gr

Abstract: The stochastic structures of potential evaporation and evapotranspiration (PEV and PET or ETo) are analyzed using the ERA5 hourly reanalysis data and the Penman–Monteith model applied to the well-known CIMIS network. The latter includes high-quality ground meteorological samples with long lengths and simultaneous measurements of monthly incoming shortwave radiation, temperature, relative humidity, and wind speed. It is found that both the PEV and PET processes exhibit a moderate long-range dependence structure with a Hurst parameter of 0.64 and 0.69, respectively. Additionally, it is noted that their marginal structures are found to be light-tailed when estimated through the Pareto–Burr–Feller distribution function. Both results are consistent with the global-scale hydrological-cycle path, determined by all the above variables and rainfall, in terms of the marginal and dependence structures. Finally, it is discussed how the existence of, even moderate, long-range dependence can increase the variability and uncertainty of both processes and, thus, limit their predictability.

Keywords: potential evapotranspiration; stochastic simulation; marginal structure; long-range dependence; Hurst–Kolmogorov dynamics

Citation: Dimitriadis, P.; Tegos, A.; Koutsoyiannis, D. Stochastic Analysis of Hourly to Monthly Potential Evapotranspiration with a Focus on the Long-Range Dependence and Application with Reanalysis and Ground-Station Data. *Hydrology* **2021**, *8*, 177. https://doi.org/10.3390/hydrology8040177

Academic Editor: Andrea Petroselli

Received: 1 November 2021
Accepted: 29 November 2021
Published: 1 December 2021

Publisher's Note: MDPI stays neutral with regard to jurisdictional claims in published maps and institutional affiliations.

1. Introduction

Evapotranspiration is a paramount element in hydrology, with relevance in many aspects of the geosciences. From hydrological and agronomic perspectives, the potential evapotranspiration (PET) and (potential) evaporation (PEV) are key for water balance estimation, the assessment of crop water demand, and integrated rainfall-runoff modelling. PET [1] is defined as "the amount of water transpired in a given time by a short green crop, completely shading the ground, of uniform height and with adequate water status in the soil profile". A particular (reference) case thereof is the reference evapotranspiration (ETo), which refers to "the rate of evapotranspiration from a hypothetical reference crop with an assumed crop height of 0.12 m, a fixed surface resistance of 70 s/m, and an albedo of 0.23, closely resembling the evapotranspiration from an extensive surface of green (cool season) grass of uniform height, actively growing, well-watered, and completely shading the ground" [2]. Evaporation is the physical process by which liquid water enters the atmosphere as water vapor. In what follows, when we refer to all of the above processes, we use the acronym PE. We also note that PE may be different from the actual evapo(transpi)ration (in cases where there is not adequate water availability).

For the PE assessment, historically, many models have been developed highlighting the Penman–Monteith model as the most suitable [3]. One of the main shortcomings of estimating PE with the Penman–Monteith model is the requirement of a significant number of meteorological inputs such as, without distinction, temperature, radiation, relative humidity, and wind speed [4–6]. In the case that we require a synthetic PE timeseries for risk management (e.g., in a Monte Carlo simulation framework), when the above meteorological inputs are not available for the requested period, then one may use a stochastic model that preserves the important statistical attributes of PE. Additionally, due to the physical complexity of assessing the PE, stochastic modelling provides a solid scientific ground for further consideration in several fields of PE assessment, and the stochastic analysis can contribute to the PE physical interpretation along with other hydrometeorological processes because stochastics is proven as a collection of mathematical tools able to give physical explanations [7]. As highlighted in the aforementioned work [7], the role of stochastics is crucial: (a) to infer dynamics (laws) from past data; (b) to formulate the complex natural system equations; (c) to estimate the involved parameters; and (d) to test any hypothesis regarding the dynamics. There are only limited works providing a thorough stochastic analysis in PE timeseries, even though the necessity of stochastic modelling is of paramount importance. Based on the published literature, a seasonal ARIMA model and Winters' exponential smoothing model [8] have been investigated for their applicability for forecasting weekly reference crop ETo [9]. Both models demonstrated satisfactory results compared to a simple PE model. Pandey et al. [10] provided a stochastic analysis in assessing black gram evapotranspiration regimes using a long-term pan-evaporation dataset of 23 years in Udaipur, India. Black gram is an important crop of the Udaipur region, and the lack of long-term crop demand assessment led to the need for stochastic analysis using pan-evaporation gauges to predict daily black gram evapotranspiration. As noted by the authors, the new stochastic model for black gram evapotranspiration was found to predict daily black gram evapotranspiration with high accuracy ($R^2 = 0.94$). Dynamic stochastic modelling, with a focus on the marginal probability distribution function (known as cumulative distribution function), has been also used for quantifying the PE uncertainty associated with irrigation scheduling [11–13]. Recently, an application of vine copulas with a focus on the short-term structure of the daily evaporation process has been presented [14]. Rainfall-runoff approaches have been presented using stochastic inputs of precipitation and PE to overcome the lack of Penman–Monteith estimates and long-term gauge inputs [15,16].

A substantial amount of previous works have focused on the trend PE assessment [17,18] in conjunction with the well-known term, evaporation paradox [19,20]. The later has been defined as the assumption that, under warming climate and higher temperatures, increased PE rates are expected; however, gauge data show the opposite because observations across the U.S. and the globe show a decreasing trend in pan evaporation. Recent studies recommend the revision of common trend tests through re-evaluation of the statistical significance of an observed trend in a timeseries by assuming a model exhibiting the scaling hypothesis [21], which is shown to be apparent in most key hydrological-cycle processes [22] and provides a more accurate modelling framework than a trend-based approach [23].

The stochastic structure of the PE process, ranging from hourly to climatic scales, is studied here in terms of Hurst–Kolmogorov (HK) dynamics, which describes all processes exhibiting the Hurst phenomenon (i.e., with a power-law autocorrelation function at large scales). Additionally, we focus on the marginal structure of the PE process as fitted through the Pareto–Burr–Feller (PBF) distribution function [24], which includes a large variety of tail-behaviors [25]. Both marginal and second-order dependence structures of the HK dynamics are estimated and compared to the ones identified from global-scale analyses in other key hydrometeorological processes that form the hydrological-cycle path driven by atmospheric turbulence [26], such as temperature, wind, solar radiation, and relative humidity [22,27–29].

Because observations for the PE process are usually found on monthly or daily resolutions, here we use two datasets. The first dataset comprises PET timeseries with monthly resolution extracted from the California Irrigation Management Information System (CIMIS) network in California, comprising 41 ground stations. For the second dataset, we extracted gridded reanalysis PEV data of hourly resolution. In particular, we retrieved the reanalysis data for the grid points in the same area of the network of the ground stations, so that we could compare its stochastic structures to the PET records, with a focus on the long-range dependence (LRD) behavior.

In Section 2, we introduce the methodology on the estimation of the marginal and second-order dependence structures, while in Section 3, we present the statistical characteristics of the selected stations as well as the results obtained from the analysis, with a focus on the marginal and the dependence structures. Finally, in Sections 4 and 5, we summarize our findings, and we discuss how the results may be consistent with the ones obtained from the hydrological-cycle path under HK dynamics, expanding from Gaussian to Pareto-type tail behavior, and from fractal and intermittent behavior at small scales to LRD behavior at large scales.

2. Metrics of Marginal and Dependence Structures

The estimators and models applied for both the marginal and the second-order dependence structures are part of the stochastic framework of the HK dynamics, with a focus on the LRD behavior [30–35], and they have been applied to turbulent and key hydrological-cycle processes of global networks with resolutions spanning from small scales (relevant to the fractal behavior) to climatic scales (for a review, see [26]).

It has been shown that a flexible probability distribution function, which seems to fit well a great variety of key hydrological-cycle processes [25,26], with tail-behaviors ranging from Gaussian to Pareto, is the PBF distribution function [24,36–38], i.e.:

$$F(x) = P\{\underline{x} \le x\} = 1 - \left(1 + \zeta\xi\left(\frac{x-d}{\lambda}\right)^\zeta\right)^{-\frac{1}{\zeta\xi}} \tag{1}$$

where $x > d$, d is a location parameter (in units of x), ζ and ξ are dimensionless shape parameters, and λ is a scale parameter (in units of x). It is noted that here the Dutch convention is adopted, where underlined symbols denote random variables and stochastic processes.

The estimation of the parameters of the PBF distribution function for the identification of the marginal structure of the PE process is based on the first four statistical moments, and particularly on the central moments and coefficients (i.e., mean, variance, skewness, and kurtosis). It is stressed that, although the estimation from the classical moments of high order are unknowable, especially in the presence of heavy tails and LRD [25], the hourly PEV and the monthly PET processes are expected to be close to a light-tail behavior and, therefore, the estimation of skewness and kurtosis coefficients could be, in approximation, reliably estimated from data.

For the dependence structure of the PE processes, we select the climacogram metric, which is defined as the variance of the averaged process at the scale domain [7]. i.e.:

$$\gamma(k) := \mathrm{Var}\left[\int_0^k \underline{x}(y)\mathrm{d}y\right]/k^2 \tag{2}$$

where k is the scale (in units of $\underline{x}$). (See discussion on the origins of the name, mathematical definitions, etc., in [26,39])

It has been shown that the climacogram estimator at the scale domain is a more powerful estimator than the autocovariance function at the lag domain or the power-spectrum at the frequency domain [34], while its classical estimator adjusted for bias is defined as [40]:

$$\underline{\hat{\gamma}}(\kappa\Delta) = \frac{1}{\lfloor n/\kappa\rfloor}\sum_{i=1}^{\lfloor n/\kappa\rfloor}\left(\underline{x}_i^{(\kappa)} - \underline{\hat{\mu}}\right)^2 + \gamma(\lfloor n/\kappa\rfloor\kappa\Delta) \tag{3}$$

where $\kappa = k/\Delta$ is the dimensionless scale, Δ is the time resolution of the process, $\hat{\mu}$ is the mean of the process, $\lfloor n/\kappa \rfloor$ is the integer part of n/κ, and $\underline{x}_i^{(\kappa)}$ is the i-th element of the averaged sample of the process at scale κ, i.e.:

$$\underline{x}_i^{(\kappa)} = \frac{1}{\kappa} \sum_{j=(i-1)\kappa+1}^{i\kappa} \underline{x}_j \tag{4}$$

For the climacogram model, contained in the above estimator, we select a generalization of the HK model (for details and more sophisticated models, see [25,26]), which has been shown to well simulate processes from sub-hourly to over-annual resolutions, and from short- to long-term scales associated with fractal and LRD behaviors that exclude the drop of variance at the intermediate scales:

$$\gamma(k) = \frac{a}{\left(1 + (k/q)^{2M}\right)^{(1-H)/M}} \tag{5}$$

where a is the variance of the process, q is a scale parameter (in units of the scale k), M is the fractal parameter, and H is the Hurst parameter indicative of the LRD of the process, i.e., for $0.5 < H < 1$ the process exhibits LRD behavior, while for $0 < H < 0.5$ it exhibits an anti-persistent behavior, and for $H = 0.5$ a white-noise behavior. Here, the standardized climacogram is used, i.e., $\hat{\gamma}(k)/\hat{\gamma}(1)$, because the effect of the sample variance is already accounted for through the marginal fitting. We also note that a Gaussian process with $q \to 0$ and $M = 0.5$ coincides with the well-known fractional Gaussian noise model (e.g., [41]).

3. Data Extraction and Processing

For the analysis of the hourly PEV process, we use the reanalysis ensemble data extracted (access date at 29/10/2021; with coordinates S32-N42 and W115-E125) from the ERA5 [42] of the Centre for Medium-Range Weather Forecasts (ECMWF; https://cds.climate.copernicus.eu/ accessed on 1 October 2021) across California (Figure 1) and for the period 1979–today (Table 1).

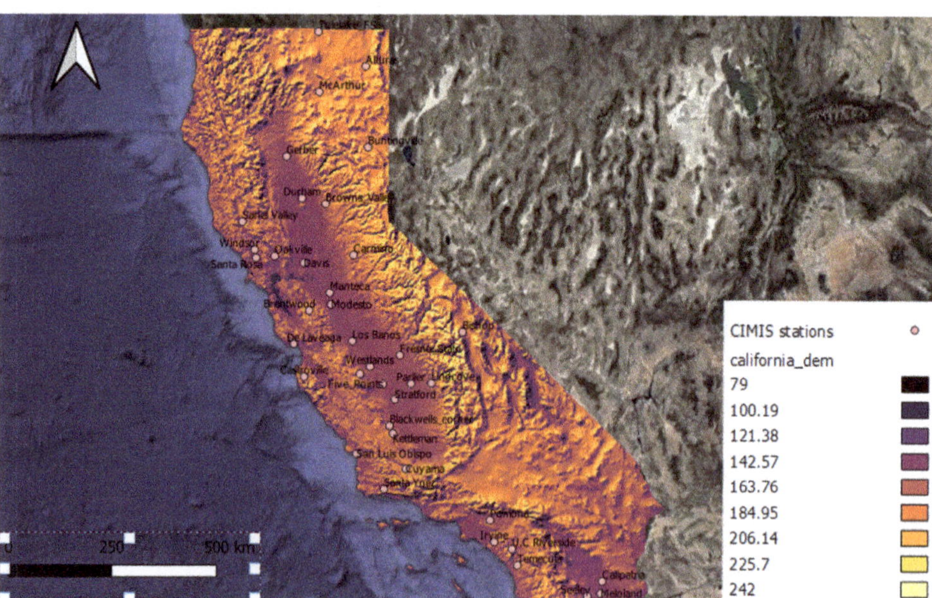

Figure 1. Map of Locations of the selected stations.

Table 1. Information on the selected stations and the reanalysis data.

Sequence Number	Name	Process	Temporal Resolution	Time Period	Number of Data Values	Mean (mm)	Standard Deviation (mm)	Skewness Coefficient
1	Five Points	PET	monthly	1982–2013	363	131.5	73.7	0.0
2	Davis	PET	monthly	1982–2013	372	120.6	68.7	0.0
3	Firebaugh Teles	PET	monthly	1982–2013	370	118.1	68.9	0.1
4	Gerber	PET	monthly	1982–2013	370	117.3	67.9	0.1
5	Durham	PET	monthly	1982–2013	369	107.8	61.7	0.1
6	Carmino	PET	monthly	1982–2013	369	116.8	68.8	0.3
7	Stratford	PET	monthly	1982–2013	369	128.2	75.4	0.0
8	Castorville	PET	monthly	1982–2013	368	79.9	32.0	0.1
9	Kettleman	PET	monthly	1982–2013	368	130.4	73.9	0.0
10	Bishop	PET	monthly	1983–2013	363	125.5	60.9	0.0
11	Parlier	PET	monthly	1983–2013	362	112.5	66.0	0.1
12	Calipatria	PET	monthly	1983–2013	360	151.2	65.2	−0.1
13	Mc_Arthur	PET	monthly	1983–2013	357	101.2	66.2	0.2
14	UC_Riverside	PET	monthly	1985–2013	337	121.9	47.0	0.1
15	Brentwood	PET	monthly	1985–2013	327	115.8	68.1	0.1
16	San_Luis_Obispo	PET	monthly	1986–2013	327	107.5	39.5	−0.1
17	Blackwells_corner	PET	monthly	1987–2013	321	128.9	73.1	0.2
18	Los Banos	PET	monthly	1988–2013	301	119.8	70.4	0.1
19	Buntigville	PET	monthly	1986–2013	325	112.9	67.8	0.1
20	Temecula	PET	monthly	1986–2013	320	113.5	39.9	0.0
21	Santa_Ynez	PET	monthly	1986–2013	320	105.1	46.3	0.0
22	Seeley	PET	monthly	1987–2013	314	159.7	69.1	−0.1
23	Manteca	PET	monthly	1987–2013	308	109.7	64.7	0.1
24	Modesto	PET	monthly	1987–2013	312	110.7	64.9	0.1
25	Irvine	PET	monthly	1987–2013	309	105.0	39.4	0.1
26	Oakville	PET	monthly	1989–2013	292	103.8	55.5	0.0
27	Pomona	PET	monthly	1989–2013	291	103.4	44.7	0.1
28	Frenso_State	PET	monthly	1988–2013	297	117.7	71.2	0.1
29	Santa_Rosa	PET	monthly	1990–2013	282	93.9	50.9	0.0
30	Browns_Valley	PET	monthly	1989–2013	291	112.2	65.4	0.1
31	Lindcove	PET	monthly	1989–2013	290	110.4	65.9	0.1
32	Meloland	PET	monthly	1989–2013	283	153.3	66.5	−0.1
33	Alturas	PET	monthly	1989–2013	291	97.0	60.7	0.3
34	Cuyama	PET	monthly	1989–2013	289	128.4	61.4	0.1
35	Tulelake	PET	monthly	1990–2013	291	96.4	60.6	0.2
36	Goleta_foothills *	PET	monthly	1990–2013	197	99.1	34.8	0.0
37	Windsor	PET	monthly	1990–2013	266	96.4	53.6	0.1
38	De_Laveaga	PET	monthly	1990–2013	274	88.6	39.4	−0.1
39	Westlands	PET	monthly	1992–2013	255	131.2	76.0	0.0
40	Sanel_Valley	PET	monthly	1990–2013	269	107.2	62.8	0.1
41	Santa_Monica	PET	monthly	1993–2013	246	99.1	34.9	0.0
42	CIMIS (overall)	PET	monthly	1983–2013	12985	114.4	63.5	0.2
44	ERA5	PEV	hourly	1979–2021	0.93×10^6	0.08	0.11	1.5

* There is a large gap in timeseries from 1995–2001.

Observations for the PE processes are usually available in monthly or daily resolutions and usually only for short periods, while a global gridded dataset based on the ERA5 data has been recently released [43]. Here, we use two datasets and compare the marginal and dependence structures of the reanalysis PEV timeseries with the PET timeseries of coarser monthly resolution, extracted from a network of 41 ground stations (see details in Table 1 and Figures 2 and 3). Particularly, the monthly Penman–Monteith dataset of the CIMIS network is used, in which reference evapotranspiration and potential evapotranspiration coincide due to local surface and vegetation conditions. The samples of 41 meteorological stations (https://cimis.water.ca.gov/, accessed on 1 October 2021) are well-distributed across California (Figure 1) for the period 1983–2013 (Table 1), which corresponds to a maximum of 372 monthly values. The meteorological network has been developed in co-operation with Davis University, and the local environment of the meteorological stations allow accurate estimation of the PET.

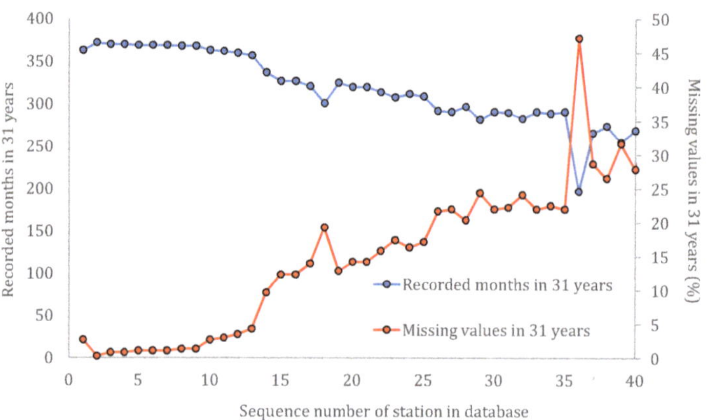

Figure 2. Total recorded and missing values of the PET timeseries for each station.

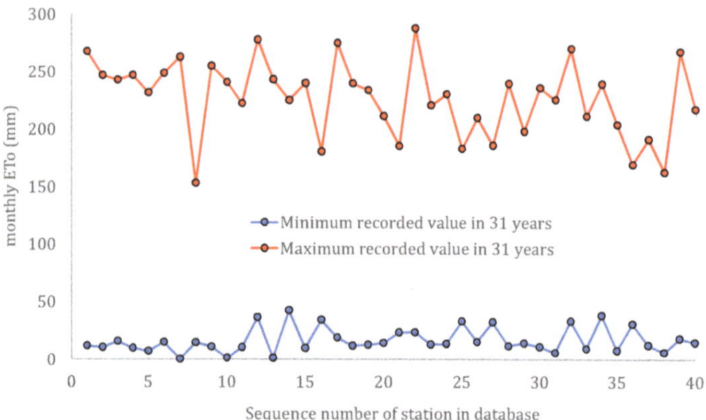

Figure 3. Minimum and maximum values of the PET timeseries for each station.

To account for the impact of the double periodicity (diurnal and seasonal) of the PE processes on the dependence structure, we simulate the transformed process by applying a double standardization on the original timeseries. Particularly, we subtract the hourly and monthly means (Figures 4 and 5) and then we divide with the hourly and monthly standard deviations (Figures 6 and 7). Other transformation methods could be applied that take into consideration higher moments (e.g., [26]) such as skewness (Figure 8) and kurtosis

(Figure 9) coefficients, or even more sophisticated ones [44]; however, as can be derived from Table 1 and Figures 6 and 7, the PE processes (especially the aggregated PET process) is close to a light-tail distribution, and therefore we do not expect any significant differences by applying those methods. After the double standardization, we de-standardize each timeseries based on the total mean and standard deviation of the original timeseries (Table 1 and Figure 10). Finally, we fit the marginal and dependence models described in the previous section to each transformed timeseries, and the results are depicted and described in the next section.

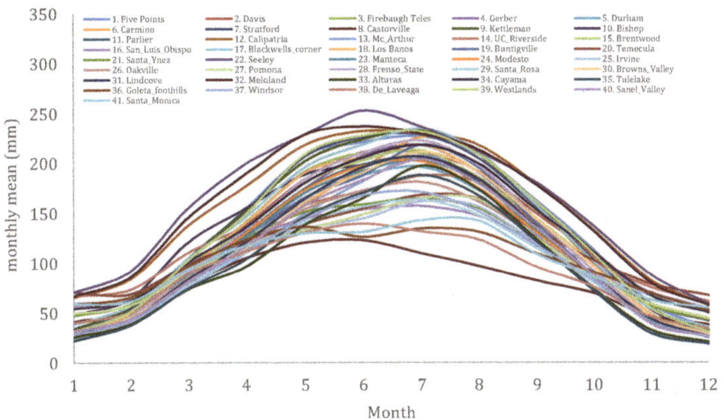

Figure 4. Monthly means of the PET timeseries for each station.

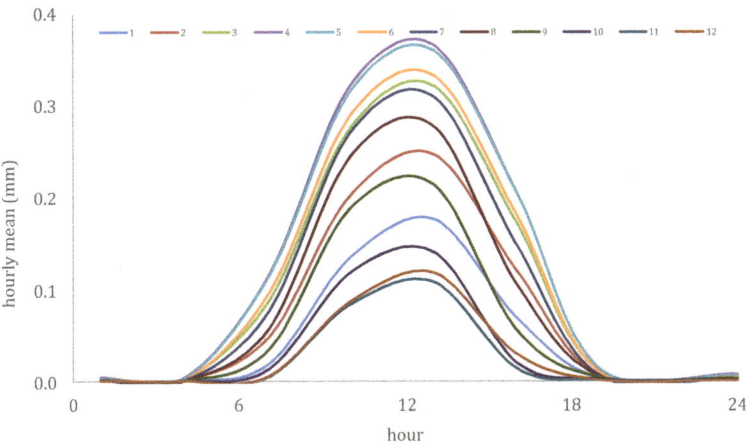

Figure 5. Hourly means of the PEV timeseries for each month.

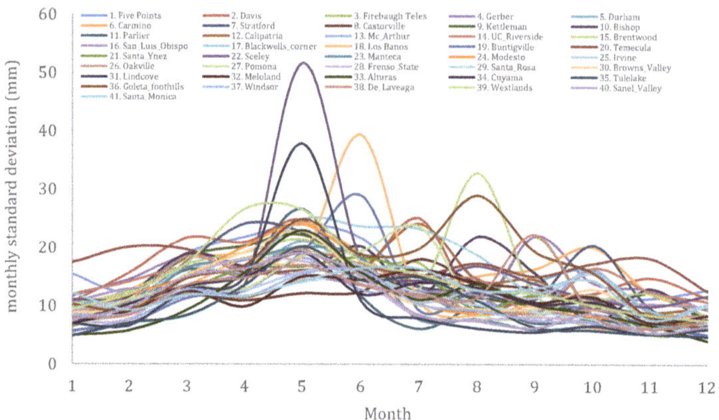

Figure 6. Monthly standard deviations of the PET timeseries for each station.

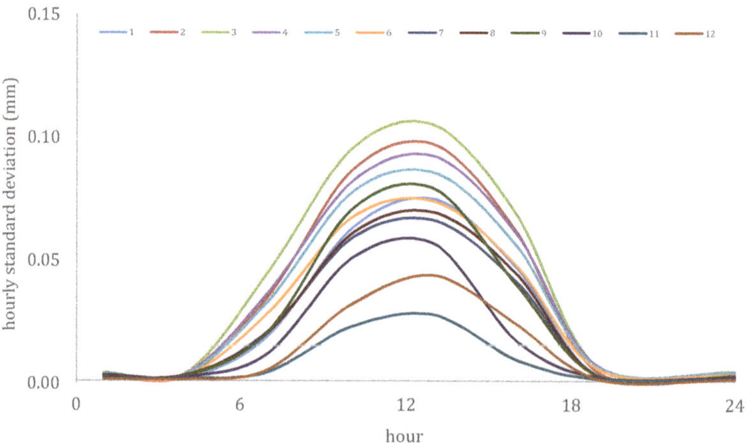

Figure 7. Hourly standard deviations of the PEV timeseries for each month.

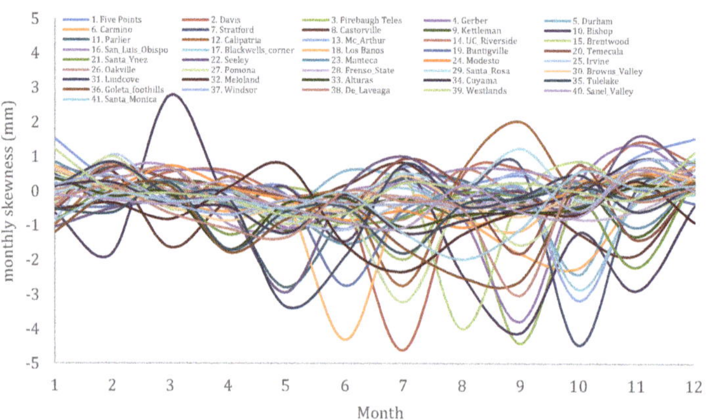

Figure 8. Monthly skewness coefficients of the PET timeseries for each station.

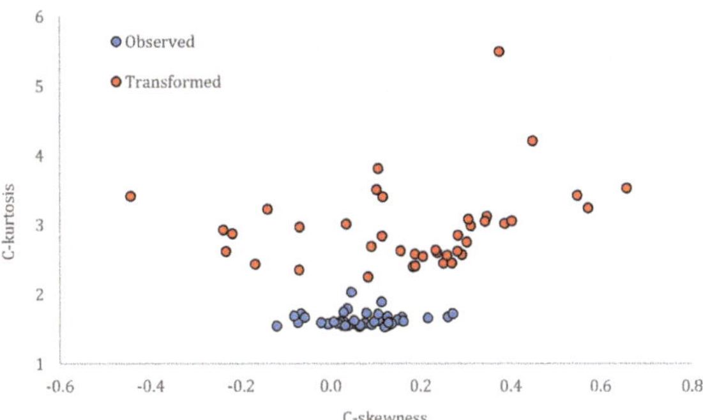

Figure 9. Skewness and kurtosis coefficients of the PET timeseries for all stations.

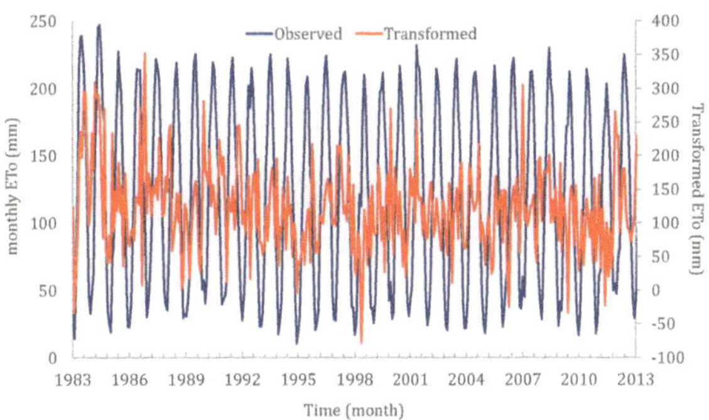

Figure 10. Observed and transformed PET timeseries at the Davis station.

4. Results

The PBF marginal distribution function is fitted to each transformed timeseries (e.g., Figure 11), and the parameters for all transformed timeseries can be seen in Table 2. Note that the fit of the PBF to all timeseries is exceptionally good. From Table 2, it can be observed that the transformed PEV and PET processes exhibit a light-tail behavior. The average values of the shape parameters are estimated as $\xi \approx 0.04$ and $\zeta \approx 5.7$ for the CIMIS dataset, and $\xi = 0.08$ and $\zeta = 7.6$ for the ERA5 transformed timeseries. It has been shown [25] that the tail index, ξ, does not depend on the averaging scale. Therefore, the slight differences in the estimated values are either due to statistical uncertainty or to differences in the nature of the data.

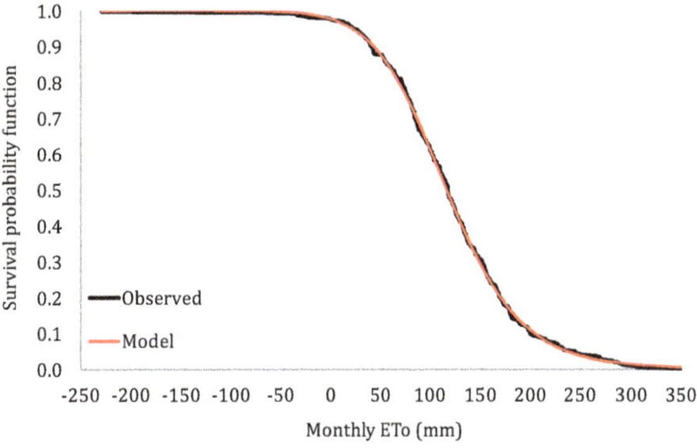

Figure 11. Observed and theoretical results of the PBF marginal distribution function of the PET transformed timeseries at the Davis station.

Table 2. Parameters of the marginal probability distribution function for all transformed timeseries of each station (note that the squared correlation coefficient is $R^2 > 0.99$ for all models). The symbols ξ, ζ, λ, and d correspond to Equation (1).

Sequence Number	Name	ξ	ζ	λ (mm)	d (mm)
1	Five Points	0.100	4.5	240.0	−105.0
2	Davis	0.094	9.2	355.1	−236.0
3	Firebaugh Teles	0.071	8.0	353.2	−227.5
4	Gerber	0.078	8.2	349.6	−227.9
5	Durham	0.063	6.3	285.2	−167.6
6	Carmino	0.034	5.8	326.1	−191.1
7	Stratford	0.054	8.0	427.1	−286.5
8	Castorville	0.049	5.4	132.7	−45.7
9	Kettleman	0.018	4.7	308.4	−154.0
10	Bishop	0.067	10.4	304.9	−171.3
11	Parlier	0.042	6.4	326.8	−199.4
12	Calipatria	0.093	7.7	285.2	−131.2
13	Mc_Arthur	0.038	6.8	340.7	−223.3
14	UC_Riverside	0.071	4.2	145.7	−14.8
15	Brentwood	0.072	7.4	344.0	−221.1
16	San_Luis_Obispo	0.077	4.6	138.0	−25.1
17	Blackwells_corner	0.001	5.1	352.9	−196.8
18	Los Banos	0.056	7.2	367.4	−235.1
19	Buntigville	0.025	6.0	327.6	−193.9
20	Temecula	0.074	5.1	145.7	−26.0
21	Santa_Ynez	0.012	5.1	199.9	−78.5
22	Seeley	0.085	7.7	285.3	−116.6
23	Manteca	0.046	4.3	233.0	−107.6
24	Modesto	0.013	3.7	231.3	−100.0
25	Irvine	0.031	4.1	132.3	−15.5
26	Oakville	0.002	3.5	188.2	−65.3
27	Pomona	0.025	6.0	208.3	−91.1
28	Frenso_State	0.031	3.3	210.4	−72.4
29	Santa_Rosa	0.026	3.6	169.3	−61.1
30	Browns_Valley	0.004	4.4	279.5	−143.7
31	Lindcove	0.045	6.2	315.7	−190.4
32	Meloland	0.029	5.2	268.0	−94.7
33	Alturas	0.019	5.0	259.4	−142.3
34	Cuyama	0.030	6.8	343.4	−197.5
35	Tulelake	0.022	5.2	262.2	−146.8
36	Goleta_foothills	0.017	4.7	129.3	−18.5
37	Windsor	0.001	3.1	166.1	−51.4
38	De_Laveaga	0.001	5.5	195.3	−90.9
39	Westlands	0.018	3.2	230.2	−76.6
40	Sanel_Valley	0.001	6.6	385.2	−252.7
41	Santa_Monica	0.016	4.7	144.3	−33.4
42	CIMIS (meanl)	0.040	5.7	260.8	−132.4
43	ERA5-PEV	0.076	7.6	0.63	−0.54

Additionally, the combined climacogram from all the empirical ones for the CIMIS transformed timeseries is depicted in Figure 12 and compared to the one from the ERA5 transformed timeseries depicted in Figure 13. It can be observed that a Hurst–Kolmogorov behavior is detected in both data sources, with a Hurst parameter of approximately 0.65. Specifically, the estimated parameters for the CIMIS dataset are $H = 0.64$ and $q = 1.17$ months (M is assumed to be 0.5 because the empirical climacogram is very close to an fGn process), and for the ERA5 timeseries they are $H = 0.69$, $q = 19.7$ h, and $M = 0.8$.

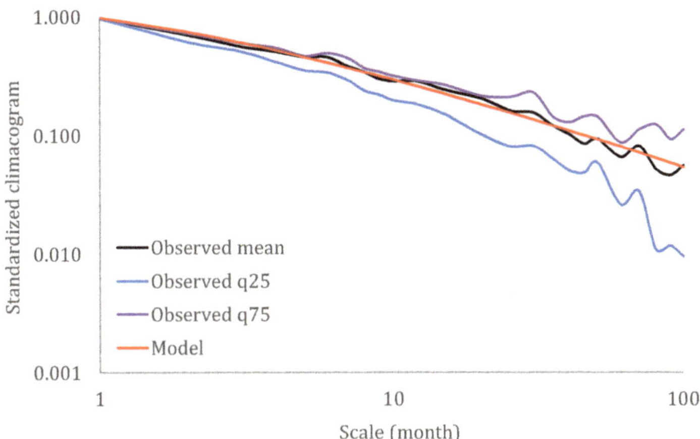

Figure 12. Observed and theoretical climacograms through the HK model for all the available PET transformed timeseries adjusted for bias, with the 25% and 75% quantiles (note that the coefficient of determination for the model is $R^2 = 0.993$).

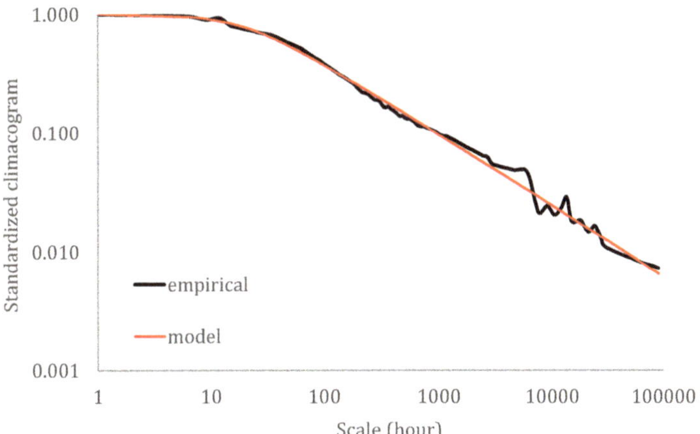

Figure 13. Observed and theoretical climacograms through the HK model for the ERA5 transformed timeseries adjusted for bias (note that the coefficient of determination for the model is $R^2 = 0.997$).

5. Discussion

Here we discuss how the above results can contribute to the existing literature relating to the potential evaporation and evapotranspiration from the point of view of stochastics and, in particular, of the HK dynamics.

The stochastic analysis of the potential evaporation (PEV) and potential evapotranspiration (PET) presented is useful (a) to highlight the stochastic similarities between them,

(b) to quantify the variability and uncertainty of these processes, and (c) to develop a stochastic model capable of simulating important stochastic characteristics, for purposes such as forecasting and risk management. The PEV timeseries is extracted in hourly resolution as a reanalysis ensemble over California and through the ERA5 network, while for the PET, the high-quality CIMIS dataset with 41 stations is used over the same area for comparison.

The analysis of the above three tasks is performed based on the stochastic metrics and Hurst–Kolmgorov (HK) dynamics. Moreover, the marginal structures and second-order dependence structures are compared to the structures of each other and of other key hydrological-cycle processes such as temperature, relative humidity, wind speed, streamflow, and precipitation, as analyzed from a global network of stations in [25].

In particular, and similar to the global analysis, it is illustrated how the Pareto–Burr–Feller (PBF) probability distribution function may well describe the marginal structure of both the hourly PEV and monthly PET. Additionally, both processes are shown to exhibit a light-tail behavior. However, it is noted that the shape parameters of the PBF (i.e., ξ and ζ), which characterize the type of the tail, are slightly smaller in the CIMIS data (i.e., overall mean from stations 0.04 and 5.7, respectively) as compared to the reanalysis data (i.e., 0.08 and 7.6, respectively), indicating a heavier tail for the latter.

Additionally, it is found that, similarly to the other key hydrological-cycle processes mentioned above, both PEV and PET processes exhibit long-range dependence, with a Hurst parameter of medium strength. In particularly, H is estimated as 0.65 and 0.68 for the PET and PEV processes, respectively, which is weaker than the ones for temperature, relative humidity, solar radiation, and wind speed (0.80–0.85 [25]) and stronger than the one for precipitation (i.e., 0.61 [25]) for the examined range of scales spanning from the hourly resolution to the climatic scales. This can be interpreted as an indication that the PET and PEV processes have a wider predictability time window than precipitation's, and narrower than the rest (i.e., entailing a higher degree of long-term unpredictability).

As a final remark, the need to apply a suitable stochastic model to reproduce important characteristics, such as LRD behavior, is stressed. The work shows the robust use of a stochastic framework to simulate the variability and uncertainty of a hydrometeorological process in emerging new practices and challenges:

- Stochastic modelling of evapotranspiration at a fine time scale (e.g., hourly) is considered to be useful for numerous agronomist applications because it is strongly connected to the forecast of the plant water demands. In recent years of micro-farm techniques, the stochastic modelling of evapotranspiration, with sound physical-interpretation, has tracked the attention of the scientific community in order to simulate more accurately the water-food-energy nexus.
- A proper stochastic model for the simulation of the evapotranspiration should be based at a wide range of spatio-temporal scales and meteorological conditions; thus, a global-scale analysis is important in order to identify stochastic similarities so as to improve the simulation techniques.
- Stochastic simulation of the error analysis between the modelled and the measured Penman–Monteith assessment could highly contribute to improving potential evapotranspiration estimates.
- Stochastic PET modeling could offer a solid probabilistic frame for identifying the long-term trend of hydrometeorological components in horizons greater than the available records and thus is of potential interest for climatological studies.

6. Conclusions

A stochastic model is presented for hourly potential evaporation (PEV) and monthly potential evapotranspiration (PET) based on the ERA5 hourly reanalysis data and the Penman–Monteith model applied to the well-known CIMIS network.

It was found that both the marginal probability distributions of PEV and PET are light-tailed when estimated through the Pareto–Burr–Feller distribution function. Additionally, the long-range dependence of both the PEV and PET is found to be of moderate strength, quantified through a Hurst parameter of 0.64 and 0.69, respectively.

The above results reveal the stochastic similarities between the ground and reanalysis data series. Additionally, the results are shown to be consistent to the hydrological-path of the marginal and dependence structures of Hurst–Kolmogorov dynamics. In particular, both PET and PEV can be placed between the stochastic structures of temperature, relative humidity, solar radiation, and wind speed (i.e., strong LRD and light- to medium-tail) and the precipitation's structures (i.e., weak LRD and heavy tail). Finally, it is discussed how the existence of, even moderate, long-range dependence and tail distribution increase the variability and uncertainty of both processes, and thus limit their predictability.

Author Contributions: Conceptualization, P.D.; methodology, P.D. and D.K.; formal analysis, P.D.; investigation, P.D. and A.T.; data curation, A.T.; writing—review and editing, D.K. All authors have read and agreed to the published version of the manuscript.

Funding: This research received no external funding.

Informed Consent Statement: Informed consent was obtained from all subjects involved in the study.

Data Availability Statement: Hersbach, H. et al., (2018) was downloaded from the Copernicus Climate Change Service (C3S) Climate Data Store. The results contain modified Copernicus Climate Change Service information 2020. Neither the European Commission nor ECMWF is responsible for any use that may be made of the Copernicus information or data it contains.

Acknowledgments: We thank both the anonymous reviewers for their suggestions and comments, and the editors for handling the manuscript.

Conflicts of Interest: The authors declare no conflict of interest.

References

1. Alexandris, S.; Proutsos, N. How significant is the effect of the surface characteristics on the Reference Evapotranspiration estimates? *Agric. Water Manag.* **2020**, *237*, 106181. [CrossRef]
2. Doorenboos, J.; Pruitt, W.O. *Guidelines for Predicting Crop Water Requirements. FAO Irrigation and Drainage Paper No.24*; Food and Agricultural Organization of the United Nations: Rome, Italy, 1977.
3. McMahon, T.A.; Finlayson, B.L.; Peel, M.C. Historical developments of models for estimating evaporation using standard meteorological data. *Wiley Interdiscip. Rev. Water* **2016**, *3*, 788–818. [CrossRef]
4. Tegos, A.; Efstratiadis, A.; Koutsoyiannis, D. *A Parametric Model for Potential Evapotranspiration Estimation Based on a Simplified Formulation of the Penman-Monteith Equation*; InTech: Rijeka, Croatia, 2013; pp. 143–165.
5. Tegos, A.; Malamos, N.; Koutsoyiannis, D. A parsimonious regional parametric evapotranspiration model based on a simplification of the Penman–Monteith formula. *J. Hydrol.* **2015**, *524*, 708–717. [CrossRef]
6. Tegos, A.; Malamos, N.; Efstratiadis, A.; Tsoukalas, I.; Karanasios, A.; Koutsoyiannis, D. Parametric Modelling of Potential Evapotranspiration: A Global Survey. *Water* **2017**, *9*, 795. [CrossRef]
7. Koutsoyiannis, D. HESS opinions. A random walk on water. *Hydrol. Earth Syst. Sci.* **2010**, *14*, 585–601. [CrossRef]
8. Winters, P.R. Forecasting Sales by Exponentially Weighted Moving Averages. *Manag. Sci.* **1960**, *6*, 324–342. [CrossRef]
9. Mohan, S.; Arumugam, N. Forecasting weekly reference crop evapotranspiration series. *Hydrol. Sci. J.* **1995**, *40*, 689–702. [CrossRef]
10. Pandey, P.K.; Pandey, V.; Singh, R.; Bhakar, S.R. Stochastic Modelling of Actual Black Gram Evapotranspiration. *J. Water Resour. Prot.* **2009**, *1*, 448–455. [CrossRef]
11. Rhenals, A.E.; Bras, R.L. The irrigation scheduling problem and evapotranspiration uncertainty. *Water Resour. Res.* **1981**, *17*, 1328–1338. [CrossRef]
12. Uliana, E.M.; Da Silva, D.D.; Fraga, M.D.S.; Lisboa, L. Estimate of reference evapotranspiration through continuous probability modelling. *Eng. Agríc.* **2017**, *37*, 257–267. [CrossRef]
13. Khanmohammadi, N.; Rezaie, H.; Montaseri, M.; Behmanesh, J. Regional probability distribution of the annual reference evapotranspiration and its effective parameters in Iran. *Theor. Appl. Climatol.* **2018**, *134*, 411–422. [CrossRef]
14. Pouliasis, G.; Torres-Alves, G.A.; Morales-Napoles, O. Stochastic Modeling of Hydroclimatic Processes Using Vine Copulas. *Water* **2021**, *13*, 2156. [CrossRef]
15. Pham, M.T.; Vernieuwe, H.; De Baets, B.; Willems, P.; Verhoest, N.E.C. Stochastic simulation of precipitation-consistent daily reference evapotranspiration using vine copulas. *Stoch. Environ. Res. Risk Assess.* **2016**, *30*, 2197–2214. [CrossRef]
16. Pham, M.T.; Vernieuwe, H.; De Baets, B.; Verhoest, N.E.C. A coupled stochastic rainfall–evapotranspiration model for hydrological impact analysis. *Hydrol. Earth Syst. Sci.* **2018**, *22*, 1263–1283. [CrossRef]

17. Zhang, Y.; Liu, C.; Tang, Y.; Yang, Y. Trends in pan evaporation and reference and actual evapotranspiration across the Tibetan Plateau. *J. Geophys. Res. Atmos.* **2007**, *112*. [CrossRef]
18. Stefanidis, S.; Alexandridis, V. Precipitation and Potential Evapotranspiration Temporal Variability and Their Relationship in Two Forest Ecosystems in Greece. *Hydrology* **2021**, *8*, 160. [CrossRef]
19. Hobbins, M.; Ramírez, J.A.; Brown, T.C. Trends in pan evaporation and actual evapotranspiration across the conterminous U.S.: Paradoxical or complementary? *Geophys. Res. Lett.* **2004**, *31*. [CrossRef]
20. Cong, Z.T.; Yang, D.W.; Ni, G.H. Does evaporation paradox exist in China? *Hydrol. Earth Syst. Sci.* **2009**, *13*, 357–366. [CrossRef]
21. Tegos, A.H.; Tyralis, D.; Koutsoyiannis, D.; Hamed, K.H. An R function for the estimation of trend signifcance under the scaling hypothesis—Application in PET parametric annual time series. *Open Water J.* **2017**, *4*, 66–71.
22. Koutsoyiannis, D. Revisiting the global hydrological cycle: Is it intensifying? *Hydrol. Earth Syst. Sci.* **2020**, *24*, 3899–3932. [CrossRef]
23. Iliopoulou, T.; Koutsoyiannis, D. Projecting the future of rainfall extremes: Better classic than trendy. *J. Hydrol.* **2020**, *588*, 125005. [CrossRef]
24. Koutsoyiannis, D.; Dimitriadis, P.; Lombardo, F.; Stevens, S.; Tsonis, A.A. From Fractals to Stochastics: Seeking Theoretical Consistency in Analysis of Geophysical Data. In *Advances in Nonlinear Geosciences*; Springer International Publishing: Berlin/Heidelberg, Germany, 2017; pp. 237–278.
25. Koutsoyiannis, D. *Stochastics of Hydroclimatic Extremes—A Cool Look at Risk*; National Technical University of Athens: Athens, Greece, 2021; 330p.
26. Dimitriadis, P.; Koutsoyiannis, D.; Iliopoulou, T.; Papanicolaou, P. A Global-Scale Investiga-tion of Stochastic Similarities in Marginal Distribution and Dependence Structure of Key Hydrological-Cycle Processes. *Hydrology* **2021**, *8*, 59. [CrossRef]
27. Glynis, K.T.; Iliopoulou, P.; Dimitriadis, D. Koutsoyiannis, Stochastic investigation of daily air temperature extremes from a global ground station network. *Stoch. Environ. Res. Risk Assess.* **2021**, *35*, 1585–1603. [CrossRef]
28. Katikas, L.; Dimitriadis, P.; Koutsoyiannis, D.; Kontos, T.; Kyriakidis, P. A stochastic simulation scheme for the long-term persistence, heavy-tailed and double periodic behavior of observational and reanalysis wind time-series. *Appl. Energy* **2021**, *295*, 116873. [CrossRef]
29. Koudouris, G.P.; Dimitriadis, T.; Iliopoulou, N.; Mamassis, N.; Koutsoyiannis, D. A stochastic model for the hourly solar radiation process for application in renewable resources management. *Adv. Geosci.* **2018**, *45*, 139–145. [CrossRef]
30. Hurst, H.E. Long-Term Storage Capacity of Reservoirs. *Trans. Am. Soc. Civ. Eng.* **1951**, *116*, 770–799. [CrossRef]
31. Kolmogorov, A.N. *Wiener Spirals and Some Other Interesting Curves in a Hilbert Space. Selected Works of Kolmogorov, A.N.*; Tikhomirov, V.M., Ed.; Wolters Kluwer: Dordrecht, The Netherlands, 1991; pp. 303–307.
32. Koutsoyiannis, D. Hurst–Kolmogorov dynamics as a result of extremal entropy production. *Phys. A Stat. Mech. Appl.* **2011**, *390*, 1424–1432. [CrossRef]
33. Koutsoyiannis, D. Hurst-Kolmogorov dynamics and uncertainty. *J. Am. Water Resour. Assoc.* **2011**, *47*, 481–495. [CrossRef]
34. Dimitriadis, P.; Koutsoyiannis, D. Climacogram versus autocovariance and power spectrum in stochastic modelling for Markovian and Hurst–Kolmogorov processes. *Stoch. Environ. Res. Risk Assess.* **2015**, *29*, 1649–1669. [CrossRef]
35. Dimitriadis, P.; Iliopoulou, T.; Sargentis, G.-F.; Koutsoyiannis, D. Spatial Hurst–Kolmogorov Clustering. *Encyclopedia* **2021**, *1*, 77. [CrossRef]
36. Singh, S.K.; Maddala, G.S. A Function for Size Distribution of Incomes: Reply. *Economic* **1978**, *46*, 461. [CrossRef]
37. Burr, I.W. Cumulative Frequency Functions. *Ann. Math. Stat.* **1942**, *13*, 215–232. [CrossRef]
38. Feller, W. Law of large numbers for identically distributed variables. *Introd. Probab. Theory Appl.* **1971**, *2*, 231–234.
39. Beran, J. Statistical Methods for Data with Long-Range Dependence. *Stat. Sci.* **1992**, *7*, 404–416.
40. Dimitriadis, P.; Koutsoyiannis, D. Stochastic synthesis approximating any process dependence and distribution. *Stoch. Environ. Res. Risk Assess.* **2018**, *32*, 1493–1515. [CrossRef]
41. Mandelbrot, B.B.; Van Ness, J.W. Fractional Brownian motions, fractional noises and applications. *J. Soc. Ind. Appl. Math.* **1968**, *10*, 422–437. [CrossRef]
42. Hersbach, H.; Bell, B.; Berrisford, P.; Biavati, G.; Horányi, A.; Muñoz-Sabater, J.; Nicolas, J.; Peubey, C.; Radu, R.; Rozum, I.; et al. *ERA5 Hourly Data on Single Levels from 1979 to Present*; Copernicus Climate Change Service (C3S) and Climate Data Store (CDS): Bonn, Germany, 2018. [CrossRef]
43. Singer, M.B.; Asfaw, D.T.; Rosolem, R.; Cuthbert, M.O.; Miralles, D.G.; MacLeod, D.; Quichimbo, E.A.; Michaelides, K. Hourly potential evapotranspiration at 0.1° resolution for the global land surface from 1981-present. *Sci. Data* **2021**, *8*, 224. [CrossRef]
44. Vavoulogiannis, S.; Iliopoulou, T.; Dimitriadis, P.; Koutsoyiannis, D. Multiscale temporal irreversibility of streamflow and its stochastic modelling. *Hydrology* **2021**, *8*, 63. [CrossRef]

 *hydrology*

Review

Evapotranspiration Trends and Interactions in Light of the Anthropogenic Footprint and the Climate Crisis: A Review

Stavroula Dimitriadou and Konstantinos G. Nikolakopoulos *

Department of Geology, University of Patras, 26504 Rion, Greece; sdhm@upatras.gr
* Correspondence: knikolakop@upatras.gr; Tel.: +30-261-099-759-2

Abstract: Evapotranspiration (ET) is a parameter of major importance participating in both hydrological cycle and surface energy balance. Trends of ET are discussed along with the dependence of evaporation to key environmental variables. The evaporation paradox can be approached via natural phenomena aggravated by anthropogenic impact. ET appears as one of the most affected parameters by human activities. Complex hydrological processes are governed by local environmental conditions thus generalizations are difficult. However, in some settings, common hydrological interactions could be detected. Mediterranean climate regions (MCRs) appear vulnerability to the foreseen increase in ET, aggravated by precipitation shifting and air temperature warming, whereas in tropical forests its role is rather beneficial. ET determines groundwater level and quality. Groundwater level appeared to be a robust predictor of annual ET for peatlands in Southeast Asia. In semi-arid to arid areas, increases in ET have implications on water availability and soil salinization. ET-changes after a wildfire can be substantial for groundwater recharge if a canopy-loss threshold is surpassed. Those consequences are site-specific. Post-fire ET rebound seems climate and fire-severity-dependent. Overall, this qualitative structured review sets the foundations for interdisciplinary researchers and water managers to deploy ET as a means to address challenging environmental issues such as water availability.

Keywords: actual evapotranspiration; potential evapotranspiration; reference evapotranspiration; evaporation; evaporation paradox; global dimming; wind stilling; forest fires; groundwater

Citation: Dimitriadou, S.; Nikolakopoulos, K.G. Evapotranspiration Trends and Interactions in Light of the Anthropogenic Footprint and the Climate Crisis: A Review. *Hydrology* **2021**, *8*, 163. https://doi.org/10.3390/hydrology8040163

Academic Editors: Aristoteles Tegos and Nikolaos Malamos

Received: 16 September 2021
Accepted: 27 October 2021
Published: 1 November 2021

Publisher's Note: MDPI stays neutral with regard to jurisdictional claims in published maps and institutional affiliations.

1. Introduction

The importance of evapotranspiration (ET) is demonstrated by its participation in the hydrological cycle (as a hydrological process) and in the surface energy balance (as a flux) [1]. Taking into account that a high percentage of the precipitated water is evaporated and transpired (e.g., 65% Ireland [2]; 62% Greece [3]) it is obvious that water budgets are dictated by the fluctuations of ET and subsequently by the dependency of ET on several environmental parameters [4–6]. ET according to researchers is a component that is not perfectly understood yet. Thus, it should be thoroughly studied as a major key parameter involving numerous mechanisms, mediating fluctuations of other variables, and controlling processes or causing considerable problems after intense disturbances by human activity or climate change.

1.1. Types of ET

Actual evapotranspiration (AET), which constitutes the actual water amount evaporated and transpired under the existing environmental conditions of a specific area, is challenging to measure. Thus, many studies attempt to obtain potential evapotranspiration (PET), pan evaporation (PE), or reference evapotranspiration (RET) values depending on their specific methodological approaches and research objectives. PET determines the evaporative demand of the atmosphere [7]. It can be defined as the amount of water (in mm of water depth) that can be evaporated by the soil of a land surface and transpired by the plants of the specific area, under the occurring conditions, providing that water supply is not a limitation. Usually, PET constitutes the upper limit of AET. As Lv et al. (2019) [8] underline, PET is higher than precipitation in arid areas, thus AET is close to the latter, whereas precipitation is higher than PET in humid areas, therefore, PET is close to AET. Pan evaporation (PE), meaning the depth of the water evaporated from the wet surface of an evaporation pan, can serve according to Sun et al. (2018) [9] as a proxy of PET since, besides their differences, they both quantify the evaporative demand. The usual types of pans are the following: class-A evaporation pan (d = 120.1 cm), Colorado sunken pan (area equal to 0.846 m^2), Φ20 evaporation pan (d = 20 cm), large-pans (area equal to 20 m^2), and floating evaporation pans [9–11]. The latter two are preferentially used to estimate the evaporation from free waterbodies such as lakes and dam reservoirs [12]. PE measurements have been used in several studies as reference or "truth data" [13,14]. Reference evapotranspiration (RET) is defined as the evaporation rate from a reference surface of grass with a height of 0.12 m, surface resistance of 70 m/s, and an albedo value of 0.23 [10]. In addition, alfalfa is another reference surface used by the Food and Agriculture Organization (FAO) http://www.fao.org/3/x0490e/x0490e0b.htm#alfalfa%20based%20 crop%20coefficients (accessed on 6 October 2021), [15]. According to Jiang et al. (2019) [16], variations in RET is the resultant of the integrated effect of climatic variables, thus RET in several cases reflects the impact of climate change on meteorological and hydrological cycles (e.g., increase in RET of water fed crops in arid or semi-arid areas indicates a high risk of drought). AET values for crops are determined from RET values using an appropriate crop coefficient [10]. RET and AET depend, among other variables, on air temperature which is in turn dictated to a large degree by the incoming solar radiation, thus RET and AET are prone to be affected by the ongoing rise in air temperature compared to the pre-industrial era, a phenomenon known as global warming. The former is supported by the findings of several studies which indicate that RET has been increased during the last 50 years in numerous regions of the globe [17].

1.2. Parameters Affecting ET

There are several parameters affecting ET such as climatological-meteorological, hydrogeological, topographical, and physiological. The parameters mainly affecting ET as a climate variable (i.e., PET, RET types) are solar radiation, air temperature, humidity, and wind speed, whereas AET mainly depends on water availability [10].

Sensitivity analysis regarding ET is the procedure that investigates the change in ET values caused by the change of a specific variable in the employed models. In other words, it identifies the parameters which dictate ET variability and the order (i.e., the ascending degree) the former affects ET [18]. Research on the sensitivity of PET to climatological-meteorological factors is of major importance since it aims to explain the hydrological cycle at different regions [7]. RET is considered as an integrated measure of four key climatological-meteorological variables: radiation, wind speed, air temperature, and atmospheric humidity [19]. Differentiations are detected among studies concerning the variables which employ parameters such as air temperature (T), radiation (net radiation, sunshine hours, or sunlight duration), and atmospheric humidity (relative humidity, vapor pressure, vapor pressure deficit) [7,19]. The vapor pressure deficit (VPD) is defined as the difference between the saturation and actual vapor pressure for a specific period of time [10]. Relative humidity represents the degree of saturation of the air as a ratio of the actual to the saturated vapor pressure at the same temperature [10].

1.3. Developments in ET Measurement and Estimation

As the relevant research continued over decades more sophisticated interpretations were presented, incorporating the mediating factors of PET sensitivity, such as topographic parameters (e.g., shading) and characteristics (e.g., complex terrain), climatic conditions (e.g., different climatic zones), and the timing of certain weather episodes. Furthermore, partitioning ET into components in complex land covers such as forests refined the assessment of the accuracy of several methods which have been developed to estimate forest ET types (AET, PET). The latter is of great importance since forest contribution to global climate responses to disturbances is critical [20]. Tie et al. (2018) [21] asserted that forest ET can be divided into three components: understory ET (soil evaporation and transpiration from understory vegetation), transpiration, and "interception loss from the overstory canopy" (i.e., the evaporation of the water intercepted by the overstory canopy).

The spatial scale of forest ET estimation and the observation height can differentiate the estimated contributions by each component depending on the geological parameters of the study area. Amongst the most frequently used methods, the catchment water balance method (annual or longer temporal scale) might overestimate forest ET by underestimating subsurface runoff if fractured bedrock occurs [21], suggesting the role of lithology, tectonics, and also of erosion (related to climate change) to ET estimation. Upscaled sap flow methods estimate transpiration and demonstrate diurnal lag for actual ecosystem transpiration compared to the eddy covariance method. Soil water budget methods give point-scale estimations of understory ET and overstory transpiration and reflect the general trend and dynamics of forest ET [21].

The inclusion of canopy evaporation or interception loss in the late years refined hydrological modeling. Interception expresses the difference between gross rainfall and rain passing through the crowns [22]. According to the literature, canopy evaporation alters the microclimate of the field by reducing VPD which in turn reduces the evaporative demand. Transpiration is suppressed during sprinkler irrigation and this is a reason why a number of researchers assert that intersection loss does not constitute a loss, since, in the case of a dry canopy, transpiration would occur instead [5]. Canopy storage is the amount of water held on the canopy. Bart and Tague (2017) [23] suggested that reduced postfire ET values in catchments across California were due to the reduced canopy interception, as a result of canopy removal. Bulcock and Jewitt (2012) [4] after investigating interception in a humid forest in the Seven Oaks area in South Africa consisting of pinus, acacia, and euca-lyptus species, found that the former parameter, often neglected in estimations, accounted for 40% of the gross precipitation loss. Interception includes the water evaporated both during and after a rainfall or an irrigation event, over a certain period [4,5]. Interception is divided into canopy interception and litter interception which have been reported to reach 26.6% and 13.4% of the total evapotranspiration, respectively [4]. The latter is primarily dependent on the storage capacity which in turn varies with rainfall intensity, constituting a parameter that should be taken into consideration in modeling for improving the accuracy of the results, since intensity variation is a common denominator of rainfall events in the frame of climate change. Canopy interception depends on PET, rainfall intensity, duration, and storage capacity. Moreover, it has been documented that broad leaves are associated with high litter interception [4]. On the other hand, in some cases, the plant species with the highest leaf area index (LAI) had the lowest canopy interception because of the angle and the smooth surface of their leaves, which both reduced water retention. LAI is defined as the cumulative one-sided area of leaves per unit area [4]. Rainfall is also intercepted in urban areas by building walls and roofs and urban trees [24].

Isotope technics are useful tools to detect hydrological processes and useful alterna-tives to upscaling methods in the partitioning of evaporation and transpiration in arid and semi-arid areas [25]. Recent implementations over the arid Upper Yellow River and Qilian Mountains in China showed that the stable isotopes ^{18}O and ^{2}H can reflect the characteris-tics of water sources and evaporation [26,27]. Given that the light-isotope evaporation rate is high compared to heavy-isotope evaporation, evaporation in lakes and surface water bodies can be easily detected since the condensed water is enriched in heavy isotopes whereas precipitation water is depleted (of heavy isotopes) [21]. In the same direction, the depth of the soil where evaporation takes place between rainfall or irrigation events could be identified by the stable-isotope vertical profile of soil [26].

ET is a part of complex mechanisms, thus both its measurement and estimation are challenging. Measurement techniques include lysimeters and micrometeorological methods for fluxes such as Bowen Ratio and Eddy Covariance often used as reference estimates for remote sensing-based algorithms validation [5,28]. Empirical equations have been developed, adapted, and widely used over the decades. However, climate change has fueled the need for empirical formulae to be tested in terms of accuracy and recalibrate at each area of application due to ongoing alterations in climatic variables and shifts in climatic patterns. Valipour et al. (2017) [29] tested 15 empirical models widely used in the literature for RET estimation and deducted that the radiation—based formulae were adapted to climate change better than the temperature-based ones in Iran. Specifically, radiation-based formulae appeared to be more accurate in arid, semi-arid, and Mediterranean areas, whereas temperature-based formulae outperformed the rest in very humid areas [29]. Besides direct measurement and the employment of well-established empirical formulae and empirical models (e.g., Stephens–Stewart's model, Griffith's model) based on meteorological data from stations, ET has been obtained via remote sensing data (MODIS, Landsat, etc.) either as remote sensing derived-products (e.g., MODIS products) [30], or can be estimated by surface energy balance models, employing satellite and ground-based data, such as ALEXI, METRIC, SEBAL, SEBS, STSEB models [31–36] or via empirical and physical-based methods [37–42]. ET time-series are used to calibrate hydrological models such as Sacramento, SWAT, and CropPwat [10,43,44]. Models employing complex algorithms such as general circulation models (GCMs) have also been used for long-term projections of evaporation, although questions for their reliability for future projections of ET had been raised [45]. Zhao et al. (2019) [46] developed a method for post-processing seasonal GCM outputs to predict monthly and seasonal RET. Several models on heuristic and fuzzy-logic science for estimations of PE and RET and machine learning algorithms such as combined neural networks, genetic algorithm model, linear genetic programming, fuzzy genetic, adaptive neuro-fuzzy inference system, artificial neural networks, multi-layer perceptron neural network, co-active neuro-fuzzy inference system, radial basis neural network and self-organizing map neural network showed high accuracy in different climate zones [15,47–51].

1.4. Objectives of the Review

What are the latest trends in ET globally? In what ways do climate change and anthropogenic footprint (e.g., air pollution, land use/ land cover (LU/LC) changes) affect ET? What are the interactions reported between ET and the main hydrological components (e.g., groundwater, streamflow)? How do wildfires affect ET and how do ET pre-fire values lead to forest fire risk identification? The objective of the present study is the attempt to respond to the aforementioned scientific questions by combining reported findings by eligible studies in a holistic way and underline any potential conflicts. In other words, the aim of the present study is twofold: First to qualitatively review the footprint emerging from ET trends over the latest decades in areas with different environmental conditions in the context of the ongoing climate change. Second, to focus on critical components such as the anthropogenic impact on ET, the mechanisms in which ET participates regarding forest land-cover and wildfires, croplands (irrigation and cultivation practices), groundwater (quantity and quality), and ambient air. Studies on climate change and water cycle usually address ET as a secondary component whereas studies concerning ET are focused on very specific objectives (i.e., measurement or estimation or sensitivity analysis of one form of ET under specific spatiotemporal conditions, development of a specific model, or testing an algorithm) thus, they do not combine different aspects and roles of ET. Acknowledging the contribution of the former types of research on ET, this is an attempt to compare, link, and synthesize findings around the world and extract useful conclusions on the role of ET. To the authors' knowledge, such an integrated and holistic synthesis of ET mechanisms, complex interactions, services, and impacts based on the latest research findings and conclusions does not yet exist. This qualitative review aims to constitute a useful background for interdisciplinary scientists and a reference point for water managers.

2. Materials and Methods

The methodology which has been followed was based on the criteria of a systematic review as developed by Boaz et al. (2002) [52] usually followed in environmental sciences [53], adapted to the qualitatively synthesizing character of the present review. Those criteria are elaborated as follows:

i   Procedure based on protocols: The collecting of literature was based on two combined criteria: the most recent bibliography would be analyzed, from reliable repositories (e.g., Scopus (https://www.scopus.com/ (accessed on 10 May 2021); PubMed https://www.ncbi.nlm.nih.gov/pmc/ (accessed on 10 May 2021); and Science Direct https://www.sciencedirect.com/ (accessed on 10 May 2021)) and scientific reports (https://www.ipcc.ch/ (accessed on 12 May 2021); https://iahs.info/Publications-News.do; http://www.fao.org/ (accessed on 13 May 2021))). Studies were scrutinized, similarities and differences among them were marked, and elaborating literature was sought to verify every piece of information before an association to be made or a conclusion to be reached. All the references of the articles were checked to validate the background of every study before employing them. Some of the cited articles were selected in a scheme of snowball collection of studies and added to the references after following the same procedure. In the process, several studies were neglected if the aforementioned criteria were not met.

ii   Since this review has a holistic approach, a number of research questions were posed in order to serve as axons of the review. The question that constituted the common denominator of all stages of the review was "Is there quantifiable evidence that a relationship occurs between ET and a specific meteorological factor or process"?

iii   Identification of relevant research: 141 research articles of trustworthy peer-reviewed scientific journals obtained from literature repositories were employed in an iterating way already described.

iv   Validation of the quality of the used research: Cross-referencing of every study was carried out and multiple studies with similar findings were sought to aim to strengthen the validity of the conclusions.

v   Synthesis of the findings of the employed studies: findings were synthesized in a deductive way, where reported cases with similar climate conditions, vegetation, and type of disturbance were examined to find out if the same relationship between ET and one other party (meteorological factor or process) occurs (e.g., relationship between the number of the years for ET to reach pre-fire levels and fire severity for eucalyptus forests in Mediterranean climate regions (MCRs)).

vi   Objectivity was reached by comparing corresponding methodologies and results and seeking verification from multiple sources (e.g., PE trends for the same region for overlapping time periods).

vii   Updated information: the conclusions of the review can be easily updated as ET trends are presented in tables and the relationships and interactions are clearly thematically presented in paragraphs (e.g., ET and wildfires, ET affects groundwater recharge, etc.).

The used methodology is schematically presented in Figure 1.

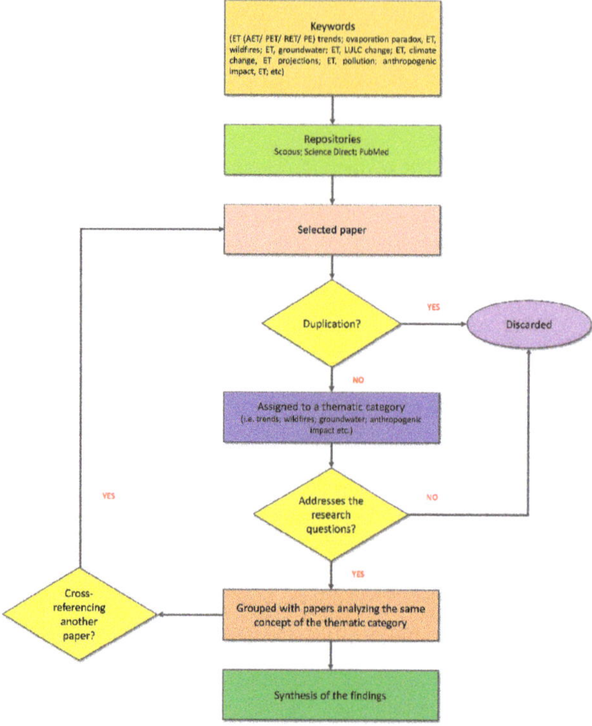

Figure 1. Flowchart of the review's methodology.

3. The Conflict of Increasing and Decreasing Trends of ET Types

The Evaporation paradox is the reported decreasing trends in PE (or RET) until the mid-1980s (or mid-1990s for regions in USA and Australia; Table 1) which contradicted the anticipated long-term increasing trends in the atmospheric evaporative demand, in a concurrently warming atmosphere [54,55]. The latter phenomenon is important since PE is considered "a clue to the direction of the change in AET" [56]. There is a considerable number of studies on ET trends over the last decades reporting decreasing or increasing trends or ET rebound after a critical period of time [9,16,57–87], (Table 1). These temporal breakpoints have been associated with anthropogenic impacts on regional climate (e.g., air pollution due to industrialization) and global phenomena (e.g., wind stilling, global dimming, and brightening) (see Figure S1). In the direction to investigate the reasons for which some trends seem conflicting at first sight, sensitivity analysis of ET (PE, RET) on key meteorological factors has been applied by researchers to determine the governing factors in each case (Table 1).

Table 1. Trends in ET (i.e., RET, PE, and LE) categorized as decreasing, increasing, of insignificant variability (no trend), and of high variability (both increasing and decreasing trends—no pattern), and the dominant climate variable affecting ET for each area, as obtained by the latest studies.

ET Type	Period of Analysis	Study Area	Dominant Climate Variable Affecting the Trend	Reference
		Decreasing trends		
LE	1950–2000	Southern Canadian Prairies	Wind speed	Burn and Hesch, 2007 [57]
RET	1982–2013	NW China (Gobi Desert)	Wind speed	Wang et al., 2017 [19]
		SE China	Sunlight duration	Wang et al., 2017 [19]
PE	1960–1991	China	Wind speed + solar radiation	Liu et al., 2011 [61]
RET	1961–1996	SW Chin (Western-Sichuan Plateau, Sichuan Basin, Yunnan-Guizhou Plateau, and Guangxi Basin	Sunshine hours + wind speed	Jiang et al., 2019 [16]
RET	Until early 1980s	Greece	Global dimming	Papaioannou et al., 2011 [64]
PE	Until 1979	Nigeria (4 climate zones)	n.d.	Ogolo, 2011 [66]
RET	1979–2000	India (NW, whole)	Net radiation + wind speed	Jhajharia et al., 2009 [62]
RET	1965–2005	C. Iran	Wind speed	Dinpashoh et al., 2011 [76]
PE	1961–2010	Mexico	Wind speed + solar radiation	Breña-Naranjo et al., 2017 [72]
PE	after 1970	Thailand	Wind speed	Limjirakan & Limsakul, 2012 [78]
PE	1975–1999	Australia (whole)	Wind speed	Johnson and Sharma, 2010 [46]
PE	1975–1994	Southern and Western Australia	Wind speed	Stephens et al., 2018 [87]
PE	1990–2016	Central, Northern Australia	Wind speed	Stephens et al., 2018 [87]
		Increasing trends		
LE	1950–2000	Northern Canadian Prairies	VPD	Burn and Hesch, 2007 [57]
RET	1975–2006	Turkey	Air temperature + relative humidity	Dadaser-Celik et al., 2016 [68]
RET	1961–2016	Slovenia (2 mountainous sites)	Solar radiation	Maček et al., 2018 [63]
RET	After late 1980s	Greece	Global warming + brightening	Papaioannou et al., 2011 [64]
PE	After 1979	Nigeria (4 climate zones)	n.d.	Ogolo, 2011 [66]
RET	1986–2007	NW Iran	n.d.	Azizzadeh and Javan, 2015 [75]
RET	1965–2005	Iran (NW, NE)	Wind speed	Dinpashoh et al., 2011 [76]
E, ET	1992–2009	S. Florida USA	Air humidity	Abtew et al., 2011 [79]
RET	1961–1982	NW China (Gobi Desert)	Wind speed	Wang et al. (2017) [19]
PE	1992–2007	China	Air temperature	Liu et al. (2011) [61]
RET	1997–2016	SW China (Western Sichuan Plateau, Sichuan Basin, Yunnan-Guizhou Plateau, and Guangxi Basin)	Air temperature + relative humidity	Jiang et al., 2019 [16]
RET	1951–2020	China, Upper Yangtze River Basin	Relative humidity	Wang et al. (2021) [59]
PE	2008–2014	China (Lower Yellow River)	Heat waves and droughts	Sun et al., 2018 [9]
PET [1]	2020–2080	Ireland (Shannon River Basin)	n.d.	Gharbia et al., 2018 [2]
PET [1]	2071–2100	Italy (High Plain Veneto and Friuli)	n.d.	Baruffi et al., 2015 [88]
PE	1975–2002	Australia (whole)	Solar radiation	Roderick & Farquhar, 2004 [85]
PE	1975–2004	Australia (whole)	Wind speed	Rayner, 2007 [86]
PE	1975–1990	Central, Northern Australia	Wind speed	Stephens et al., 2018 [87]
PE	1994–2016	Southern and Western Australia	Air temperature	Stephens et al., 2018 [87]
		Insignificant variability		
PE	1964–1998	Israel	Global dimming	Cohen et al., 2002 [67]
PE	1975–2000	W. Turkey (Buyuk Menderes Basin)	n.d.	Yeşilırmak, 2013 [69]
PE	1973–2014	Uruguay	n.d.	Vicente-Serrano et al., 2018 [71]
		High variability [2]		
PE	1950–2002	Conterminous U.S.	Radiation + advection	Hobbins et al., 2004 [54]
PE	1980–2009	Conterminous U.S.	1 of 4 variables [3] depending on season	Hobbins, 2012 [82]
PE	2030, 2050, 2070 [1]	Australia (whole)	Radiation + advection	Johnson and Sharma, 2010 [46]

Note: [1] Projected PET values (GCMs). [2] Both increasing and decreasing trends. [3] Air Temperature, Specific Humidity, Downwelling Shortwave Radiation and Wind Speed. "n.d." stands for "not defined".

4. ET Affects Groundwater Recharge

Climatic change is defined, as any change in climatic conditions, as a result of natural or anthropogenic causes [89]. By altering ET and groundwater-recharge rates climate change has the potential to affect both the quantity and the quality of groundwater [89]. Precipitation, snow thaw, interactions with surface water bodies (e.g., rivers, lakes, and wetlands) are the main sources of groundwater recharge [89]. Thus, alterations in precipitation patterns, ET, and air temperature affect groundwater recharge. An increase in precipitation frequency and intensity would contribute to runoff and global warming would rise ET rates [89]. Although there are projections that the overall water recharge could potentially increase as a result of climate change (e.g., due to snow-packs thaw), it is rather that changes in water supplies, storage, and baseflow depend on regional conditions [89]. There are arid or semi-arid regions where recharge rates are low and water demand is high [90]. Groundwater recharge would potentially increase over regions where snow thaw occurs (depending on infiltration, lateral recharge, etc) [90]. Furthermore, as soon as snowpack thaws soil temperature rises, both photosynthesis and water use by vegetation increase [91]. Soil humidity serves as a mediating factor. However, even if the amount of water increases, water availability will still be limited due to the expected increase in evaporation [89]. In addition, precipitation in arid regions is anticipated to be even more scarce in the future [92].

Future implications in groundwater recharge are critical not only for arid or semi-arid regions. As Gharbia et al. (2018) [2] indicated, 65% of gross precipitation over Shannon River Basin, Ireland, is annually evaporated or transpired. The projected PET values reflect an increasing trend of 0.9–1.3% by 2020 and up to 13.5% by 2080 with serious implications on water availability [2]. Baruffi et al. (2015) [88] projected evaporation of High Plain Veneto and Friuli in Italy (300–600 m altitude) for 2071–2100 and found a 25% increase in PET during winter, 15% during summer, and more than 20% during fall. Although projected gross precipitation is 20% higher compared to the reference period (1971–2000), summertime rainfall is expected to be lower by 15%. As a result, runoff is expected to increase by 60% in winter and decrease up to 45% in summer [88]. Groundwater storage is projected to be reduced by 70% in the former area [88]. According to Lionello and Scarascia (2018) [70], winter precipitation in South Mediterranean areas is predicted to decrease, thus, aquifer recharge during the hydrological year, along with the increasing evaporative demand, is expected to aggravate.

Forests can enhance both ET and infiltration rates, thereby reducing surface runoff and enhancing groundwater recharge [8]. It has been documented that in tropical forests soil moisture in the top 1–1.5 m layer is lower than 34 mm, hence deeper soil moisture and groundwater contribute to the transpiration demand of vegetation during the dry season [93]. Moreover, in tropical forests the rapid soil saturation during the rainy season which follows vegetation removals greater than 45% due to disturbances such as wildfires causes post-fire floods, deteriorating the high water deficit during the dry season [94]. ET in tropical forests is a beneficial process since it reduces excess humidity, (indirectly) enhances infiltration during rainfalls, and moderates flood peaks during the rainy season [93,94].

5. ET and Wildfires

5.1. AET Rates after Wildfires

Wildfires differentiate the hydrological cycle and the surface energy fluxes by altering the microclimate of the subject area along with the soil structure and soil properties [95]. Latent heat flux (λE) represents the energy flux that is directly converted into AET in the atmosphere. λE is the form in which evaporation participates in the surface energy balance. This component varies between burnt and control (unburnt) sites [95]. As a rule, after a fire event, AET decreases due to the removal of tree crowns and understory (thus the rise of albedo value). Häusler et al. (2018) [95] studied the difference of ET between fire-subject sites and control sites of eucalyptus forest cover in NC Portugal. They reported that the post-fire disturbances in the water cycle constituted by limited water vapor and higher water demand. Not only AET but also the λE ratio from the canopy and soil dramatically changed post-fire. Specifically, 80% from canopy and 20% from soil (pre-fire) became 30% and 70%, respectively, shortly after the fire [95]. Fire severity has been associated with AET differences. Pre-fire ET levels of eucalyptus forest of MCRs were reached 2 years after the fire for low-to-moderate fire severity [95], after 5–8 years for moderate severity, and after 8–12 years for high severity [96]. The observed paradox for certain plant species such as eucalyptus is that 2–3 years after the wildfire AET was 29% higher than in unburnt sites in South Australia [96], and 2% higher after 4 years in Portugal. This paradox has been attributed to the epicormic regrowth and the partial regrowth of foliage (temporarily higher LAI) which increased water demand. Unlike eucalyptus forest, the AET of burnt sites of pine forest cover in SE Spain remained lower than the corresponding AET of the control sites for 11 years after the fire [97].

Liu et al. (2019) [20], after studying the global fire-climate response for the wildfires of 2003–2014, suggested that the positive warming response accounted for a decrease in ET, which lasted for 5 years post fire and the consecutive increase in albedo resulted in lowering the cooling effect. Tropical forests exhibit a net cooling effect as a result of their high ET rates. However, after undergoing extended wildfires, tropical areas exhibit lower but persistent positive surface warming response, driven by reduced evaporative cooling [20]. Dynamic interactions between ET and albedo at different ecosystems worldwide govern the surface warming and the radiative budget response after fires. According to the authors, the severity and frequency of fires will result in considerable changes in climate and the adiative budget especially for high latitudes [20]. Albedo values' offset between snow and non-snow periods allow the decreasing ET during the vegetation growth period to dictate the surface energy balance, resulting in warming over boreal forest areas which lasted for 5 years post fire. This positive feedback is a result of canopy loss. Liu et al. (2019) [20] deduced that these alterations in biophysical processes are not satisfactorily captured by satellite observations of burned areas.

High latitude biomes are found to be more sensitive to climatic change [20]. Wang et al. (2018) [98] put forward a mechanism potentially implemented to temperature-limited high latitude forests when there is a high diffuse of photosynthetic active radiation (PAR), given that increased longwave radiation is emitted from clouds. Successively, canopy temperature increases enhancing gross primary productivity and transpiration. Thus, diffuse solar radiation is another parameter considered to be critical regarding ET variations [52]. Moreover, this mechanism might be a reason that top-down models using remote sensing data to estimate ET are often biased towards clear sky conditions [98]. Hirano et al. (2015) [99] reported that the less-vegetated part of the burnt ex-peat swamp site was studded with open water which resulted in lower albedo in 2004–2005, while in 2006 El Niño drought dried off the burnt surface increasing the areal albedo. The latter parameter is critical especially for models employing satellite retrieved data.

The combined effects which LU/LC change along with climate change triggers on AET should be thoroughly studied since they cause alterations to variables such as albedo, LAI, and root depth which in turn lead to different ET rates [8]. Abatzoglou and Williams (2016) [100] after analyzing the consequences of several fire events in Western continental US forests over 1984–2015, found that anthropogenic climate change is responsible for 2/3 of the increase in RET which, along with VPD, is the most affected parameter by anthropogenic climate change. Therefore, they put forward RET as a metric of fuel aridity, interannually related to the burnt area. Häusler et al. (2019) [101] showed that using AET values acquired by remote sensing in drought indices would enhance the identification of fire risk areas by providing higher resolution. The main interactions between albedo and ET are displayed in Figure 2.

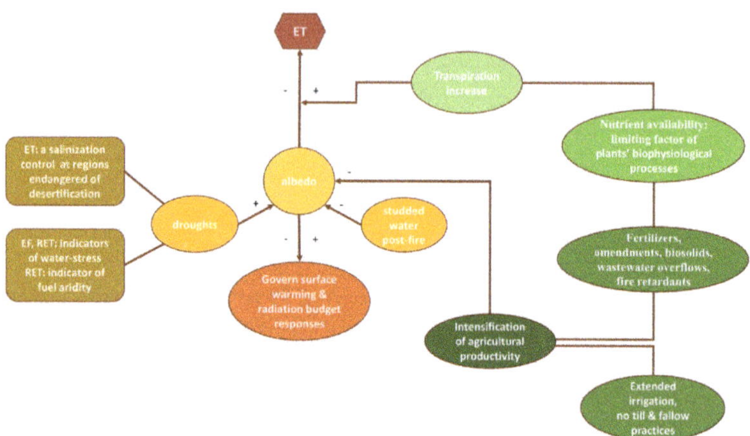

Figure 2. Contributions (positive (+) or negative (−)) of key environmental drivers to the fluctuation of actual ET rate.

5.2. Post-Fire ET and Groundwater

Poon and Kinoshita (2018) [102] underline the usefulness of evaporation time series, since pre-fire and post-fire biogeological processes would potentially substantially change due to fire disturbances. Alterations in vegetation and soil properties could create a water repellent topsoil layer which would increase surface runoff [102]. Thus, wildfires increase water repellency (or hydrophobicity) of soils, an attribute that substantially affects ET and water infiltration [23]. Johnk and Mays (2021) [103] reported a two-year post-fire reduction of groundwater level in Beaver County, Utah, USA, attributed to the wildfire of 1996. Bart and Tague (2017) [23] examined catchments in California (MCRs), where PET showed a statistically significant impact on baseflow recessions. An increase in the baseflow recession of 33.5% per mm of daily PET increasement has been predicted for eight catchments [23]. Hirano et al. (2015) [99] examined three sites of tropical peat swamp forest in SE Asia. Their results verified that in some settings ET appears strong relationship to groundwater level since the minimum mean value of monthly groundwater level appeared to be a robust predictor of annual ET for peatlands, showing statistically significant positive linearity for all sites despite their different disturbances (i.e., slight drainage, heavy drainage, fire) [99]. Specifically, according to the authors, a drawdown of 10 cm indicates decreases in annual ET between 19–33 mm for the three studied sites [99]. Kurylyk et al. (2015) [104] concluded that the decrease in ET due to canopy loss results in energy excess which warms the land surface. This warming can lead to successive warming of soil water and shallow aquifer water [104], thus ET may indirectly affect groundwater temperature.

These findings are in accordance with the research conducted by Menberg et al. (2014) [105] who underlined the vulnerability of shallow groundwater temperature to disturbances related to climate change.

5.3. Post-Fire ET and Streamflow

Wine and Cadol (2016) [106] suggested that there is a pattern between burn severity magnitude and overland flow in large catchments. Kinoshita and Hogue (2011, 2015) [107,108] found that reduced basin transpiration and infiltration after the wildfire in 2003 in California led to an increase in annual low flow by 118–1090%, which could potentially recharge water supplies in semi-arid areas (Figure 3). On the other hand, elevated flows deliver high loads of sediments. Streamflow increases are reported to sustain for longer than 7 years after the fire depending on the percentage of the burnt area and precipitation, and the types of vegetation loss and reestablishment [107]. Bart (2016) [109] attributed the intense fire effect during the first post-fire year in California to the larger effect on ET which is caused by postfire reductions in shrubland and other vegetation types of the watershed, compared to corresponding post-fire increases in herbaceous vegetation. Since both magnitude and sustainability of post-fire flow depend on scale, Wine and Cadol (2016) [106] highlighted that in large catchments, there is a threshold of 20% affected vegetation of the watershed in order for alterations in hydrogeological processes to have a measurable impact, a finding in line with Bosch and Hewlett's (1982) [110] research on catchment-scale post-fire ET and water yield.

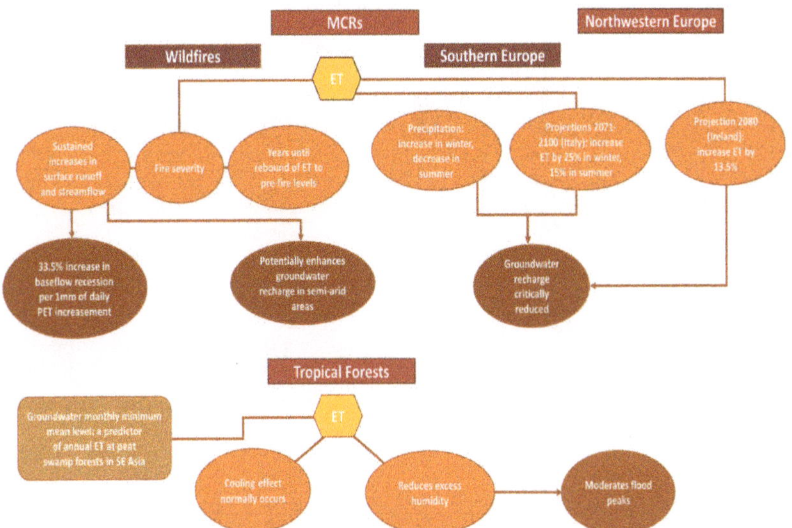

Figure 3. The main effects of ET variability on surface and ground water over Mediterranean Climate Regions, Northwestern Europe, and Tropical forests.

6. The Anthropogenic Footprint

6.1. Anthropogenic Impacts on ET

The effect of human activities on ET is twofold. One component is atmospheric pollution. The second component is LU/LC change. Global warming has decreased the differences between tropical and polar air temperatures leading to weak global atmospheric circulation, therefore, to a decrease in evaporative demand during past decades [16]. McVicar et al. (2012) [58] suggested that the elevated carbon dioxide concentrations in ambient air during past decades have enhanced vegetation growth. On the other hand, elevated carbon dioxide concentrations lead to a decrease in ET and an increase in soil moisture [111], a finding that may have partially offset the expected increases in ET [77]. Stomatal conductance, which constitutes a direct indicator of plant stress [112], is lower in elevated carbon dioxide environments, thereby decreasing transpiration sometimes by more than 20% [96]. As Liu et al. (2019) [20] demonstrated, an important service of forests is sequestering the carbon of the Earth's atmosphere. On the other hand, wildfires contribute to carbon dioxide and aerosol accumulation in the atmosphere [20]. However, aerosol is another factor that human activities are primarily accounted for. Wang and Yang (2014) [113] attributed the observed decrease in solar radiation, also referred to as dimming, at North China Plains to aerosols. They also suggested that the decreasing trend in PE in China (evaporation paradox) could be interpreted via surface solar radiation decrease (sunshine hours serve as a proxy of surface solar radiation) [113]. Sun et al. (2018) [9] attributed the decreasing trend in large-pan evaporation during 1985–2008 in North China Plain to the decrease in sunshine hours and VPD. The increasing trend during 2008–2014 was due to the increase in sunshine hours, VPD, and air temperature via heat waves and droughts during summer and spring [9]. Aerosols affect surface energy balance by absorbing and scattering solar radiation. Hallar et al. (2017) [114] refer that organic aerosol increases the optical depth of the atmosphere inhibiting radiation to reach the Earth's surface. Wildfires constitute significant sources of organic aerosols and are projected to increase optical depth in West US during summertime by 40% until 2050 [114]. As a result, net radiation is reduced by aerosol accumulation from anthropogenic emissions [17]. Jiang et al. (2019) [16] interpreted the larger magnitude of increase in minimum air temperature compared to maximum air temperature (also reported in Central Italy [115]) by the action of aerosols which absorb a part of solar radiation and emit longwave radiation at night. Industrialization of the recent decades in several regions in East Asia led to a significant increase in aerosol aggregation. However, in contradiction to East Asia, Central Mediterranean and NE America exhibited downward trends of aerosol optical depth in the 2000s as a result of the enforcement of emission-decreasing policies [116]. Urbanization density has been correlated with aerosol accumulation by Zhao et al. (2014) [117] in China, while it is asserted that high buildings in big cities inhibit aerosol diffuse by blocking wind flow, thus contributing to a decrease in solar radiation [17].

Human activities such as LU/LC change over the past decades account for ET variations by several researchers. Lv et al. (2019) [8] analyzed AET between 1986–2016 in the Yellow River basin. They found that the extended LU/LC change including conversion of the sloped terrain into terraced fields, dam constructions, forest, and vegetation implementation led to an increase in AET. They concluded that 90% of the AET increase was due to human activities and only 10% due to precipitation shift [8]. They attributed the former ratio for the thirty-year period to the reduction in surface runoff and to the increase in vegetation which increased the AET. The LU/LC changes affect the values of the physical parameter called "surface roughness". Human activities often increase the surface roughness. Even McVicar et al. (2012) [58] asserted that vegetation cover was due to agricultural abandonment of lands, surface roughness was shown to be increased due to agricultural land expansion [16]. The latter can be explained by the intensification of crop yielding the years following the former study, especially across specific areas (e.g., California). For example, Mueller et al. (2017) [77] referred that cropland expansion could affect climate by changing, among other parameters, the surface roughness. It seems that vegetation greenness along with agricultural land expansion affect AET variations by increasing surface roughness [8]. McVicar et al. (2012) [58], after analyzing numerous studies, concluded that terrestrial (wind) stilling has been observed in many regions globally during the last 30 years and led to a decline in evaporative demand reflected on PE and RET measurements.

6.2. Agricultural Practices Affect ET

The impacts of agricultural plantations on ET variations have attracted the interest of researchers especially at regions with high rates of crop yield and financial interest. Mueller et al. (2017) asserted that intensification of productivity, which incorporates extended irrigation, influences climate via increasing ET [77]. This fact results in regional cooling effect in accordance with the globally documented cooling trend in intensified croplands over the last decades, compared to adjacent regions. This could be attributed to the mediating role of vegetation in land–atmosphere coupling through controlling surface energy fluxes such as ET [118]. Irrigation directly affects the hydrological cycle of an area and the extra water on the soil enhances AET rates [1,119]. Uddin et al. (2016) studied the case of a cotton crop in Queensland, Australia (subtropical climate), during irrigation events, concluding that irrigation also changes the albedo value of the (wet) canopy [5]. They found that both during and following an irrigation event a considerable amount of the intercepted water evaporates: 11% of the applied water evaporates in highly advective conditions, while 8% in non-advective conditions [5]. Irrigation of croplands has been reported to increase ET in several regions globally such as the U.S., Asia, and Sudan [77]. The implementation of specific practices, such as multi-cropping, enhanced seasonal variation of ET at Brazilian Cerrado and North China Plain [77]. In addition, the availability of nitrogen at croplands (e.g., via land application of fertilizers, amendments, and biosolids [120]) has been correlated with an increase in ET since nitrogen availability is a limiting factor that governs plant biochemical processes (stomatal conductance, water uptake, photosynthesis) [121]. García-Llamas et al. (2019) underlined that PET is associated with the theoretical limit of the process of photosynthesis of plants [122]. French et al. (2016) asserted that a measure of relative ET such as the evaporative fraction (EF), which is a key factor in energy balance algorithms (equal to the ratio of latent heat flux to available land surface energy; [123]) could serve as a water stress indicator [124].

Irrigation decision making relies on the accuracy of ET estimations which in cases of remote sensing approaches depends on the overpass frequencies. Remote sensing is the usual approach for district-scale (regional) estimation, often used for fields too (local scale). French et al. (2016) [124] found significant benefit in ET accuracy when 8-day satellite ET products were used instead of 16-day products. The differences of seasonal water use estimation between 8-day and 16-day overpassing were up to 20%, suggesting considerable implications in regional water management in the second case.

6.3. ET Potentially Aggravates Soil Water and Groundwater Pollution

Nitrogen stress constitutes a control on ET rates by hindering the stomatal conductance, the leaf area, and the root development [77]. Post-fire ash is enriched in nutrients but is easily erodible and transported by wind and foremost by runoff. After intense rainfall events, nutrients are eluted from the topsoil layer. Recurrences of fires over the same site deteriorate soil deprivation in nutrients [125], jeopardizing stream water quality and, potentially, shallow aquifer or karst aquifer water quality which are vulnerable to pollution [126,127]. Tsypkin and Brevdo (1999) [128] after studying the evaporation impacts on groundwater quality, indicated a mechanism of pollutant deposition in groundwater caused by ET. They found that evaporation produces a gradient of the solute concentration with a vertical upward direction. For certain substances, such as NaCl, the maximum concentration at the evaporation front was greater than solubility, the latter defined as the critical concentration above which pollutant deposition begins [128]. Gran et al. (2009) [129] explained that the evaporation front divides soil into the upper dry area with salt content and the one below the front where salinity is low, suggesting that evaporation could serve as a moderating control on soil salinization since at least half of irrigated lands in arid and semi-arid regimes are subject to some degree of salinization). The salinization of soil has been correlated to erosion and desertification [130] and also to the salinization of rainfall that reaches the soil [131]. Considerable salt load is transported to freshwater ecosystems via runoff with potential salinization risk for shallow aquifers [131]. This conclusion becomes of major importance in the ongoing climate crisis. Chen et al. (2015) [132] reported that there has been a global increase in drought land since the late 1990s, with those in humid areas being the most concerning. They found that the ongoing air temperature rise accounts for 5% (humid areas) up to 45% (arid areas) of droughts in China [132]. On the other hand, high ET rates inhibit the infiltration of dissolved pollutants towards the aquifers. The evaporation enrichment concerns croplands' irrigation applied during periods when the evaporative demand is high, hence salt enrichment is considerable [133]. This water enriched in salts could potentially reach the aquifer triggering a cycle of successive enrichment if the enriched groundwater is pumped and used for irrigation, depending on the depth of the unsaturated zone, water fluxes in the saturated zone, and the evaporation rate [133]. The former process could lead to enhanced enrichment in cases where reclaimed water with high salt content is used for irrigation [134], a practice that has gained popularity not only for croplands but also for forests [135,136].

7. Discussion

This review has been carried out following the appropriate steps of a structured review [52,53,137] as shown in Figure 1, analyzing 141 studies, the majority of which were published during the last decade. The main findings are presented below:

ET is beneficial to tropical forests since it reduces excess humidity, moderating flood peaks [93] (see Figure 3). Moreover, ET could serve as a salinization control for regions endangered by desertification (see Figure 2).

A decrease in AET was found to be responsible for post-fire warming at a global scale after the extended wildfires during 2003–2014 [20]. The latter resulted in an increase in albedo which consequently reduced the net cooling effect [20]. The cooling effect is typical across tropical forests due to the high AET rates. However, the vegetation loss in tropical forests reduced the evaporative cooling resulting in a rise in the air temperature [20].

Moreover, droughts often increase albedo whereas studded open water after wildfires results in decreased albedo [99]. Overall, interactions between ET and albedo in different ecosystems govern surface warming and radiation budget responses [20] (see Figure 2). LE contributions of soil and canopy can be almost interchanged shortly after the fire [95].

Fire severity has been associated with AET differences in MCRs (see Figure 3) and with the number of years needed for AET to rebound to pre-fire levels [95–97].

ET rates affect groundwater recharge (see Figure 3). Projections for years 2071–2100 in Italy showed an increase in ET by 25% in winter, at least by 20% in fall, and by 15% in summer. This projection along with the anticipated alterations in precipitation patterns led to predicted groundwater recharge critically reduced in South Mediterranean [88]. However, this is not the case only for the endangered of high warming and drought in Southern Europe [60,138], which as a rule exhibits different precipitation patterns to North Europe. According to projections of ET for 2020 and 2080 in Ireland, North Europe is also anticipated to be critically affected [2]. Canopy removal reduces transpiration and can lead to increases in low baseflow and streamflow by 118–1090%. These alterations can sustain more than 7 years as reported in California (MCRs) [107,108]. This fact could be beneficial for groundwater recharge in semi-arid areas. The sustaining increases in low flow were also linked to the severity of the fire. Researchers concluded that 20% canopy removal from a catchment could be considered as a threshold above which alterations in hydrological processes occur [106,110]. Across MCRs, PET appeared statistically significant impact on baseflow recessions, quantified as a 33.5% increase in baseflow recession per 1 mm of daily PET increasement [23]. Furthermore, the minimum mean value of monthly groundwater level appeared to be a predictor of annual ET across tropical peat swamp forests in SE Asia, despite the different regional disturbances (see Figure 3). This relationship appears linearity: drawdown of 10 cm led to a decrease in annual ET by 19–33 mm [99].

Anthropogenic climate change has been accounted for 2/3 of the increase in RET [94]. The latter, along with VPD, appeared to be the most affected meteorological parameters, thus Abatzoglou and Williams (2016) [100] put forward RET as a metric of fuel aridity. It has been also shown by Häusler et al. (2019) [101] that AET values in fire risk indices would enhance spatial fire risk identification. Moreover, RET estimates constitute an indicator of water stress in areas prone to drought [139].

Wind stilling is the phenomenon responsible for the majority of decreasing ET trends during past decades on a global scale [58]. Wind stilling has been associated with the observed increase in surface roughness and led to a decrease in evaporative demand and successively to a decline of PE and RET values during past decades [8,58] (see Figure S1 on evaporation paradox in Supplementary Materials). According to McVicar et al. (2012) [58], the aforementioned drivers could explain the Evaporation paradox reported in several countries of the globe (see Figure S1 in Supplementary Materials). Roderick and Farquhar (2004) [85] shared the same opinion. However, in numerous countries, a rebound of PE occurred during the decades of 1980 and 1990. According to ET trends, benchmarks of ET rebound have been reported in the early to middle 1980s for Greece [64], Iran [1], and Nigeria [66], and in the 1990s for China [16,20] and Australia [87], while South Florida also exhibited an increasing trend at least since 1992 [79]. Air pollution is likely to affect ET rates: elevated carbon dioxide concentrations in the ambient air can decrease transpiration in some cases by more than 20% [77]. The latter may have offset the anticipated increase in ET trends [77]. Another phenomenon that has been accounted for decreasing trends in ET is global dimming, attributed to aerosol accumulation over past decades linked to industrialization and urbanization (see Figure S1). Global dimming was followed by global brightening [65,140,141]. The observed increase in solar and thermal radiative heating after the mid-1980s facilitated the intensification of the hydrological cycle and scattered aerosol loads [140,141]. For instance, the delay of ET rebound in China compared to Greece as reported above [9,64], is in line with the findings by Zerefos et al. (2009) [142] regarding the periods when the rebound in aerosol optical depth (brightening) was observed in those countries.

Several researchers have examined ET variations in light of the anthropogenic footprint. Lv et al. (2019) [8] deducted that 90% of the reported increase in AET was due to human activities and only 10% due to shifted and altered precipitation patterns which are associated with the climate crisis. AET was increased as a result of LU/LC shift towards large constructions such as dams, vegetation implementation (e.g., agriculture), and implemented reforestation. In addition, intensification of agricultural productivity involving extended irrigation and no-till and fallow practices increased ET via albedo, resulting in some cases (e.g., USA) in regional net cooling effect [8,77] (see Figure 2). Nitrogen availability has been also associated with elevated transpiration rates since nutrients constitute a limiting factor for plants' biophysiological processes. Nutrient sources are chemical fertilizers, amendments, raw sludge, or biosolids from wastewater treatment plants, and wastewater overflows which end up in natural receivers. In addition, some amount of nutrients also becomes available by the usage of fire retardants during fire distinguishing, since their composition resembles that of fertilizers [126,127] (see Figure 2).

It is apparent that the anthropogenic footprint in LU/LC, ambient air, and climate change have affected ET rates globally. Even if the evaporation paradox can be approached, the changeable climate patterns along with the significance of ET impacts on ecosystems, water supplies, and sustainability call for systematic research on PET, RET, PE, and AET to many more countries with different environmental conditions, aiming to quantify the relations between them [46,87,143]. Moreover, the mechanisms in which ET participates are related to cropland irrigation and practices, forest implementation, LU/LC management, fire management (preventing and mitigating), and water management (surface waterbodies and aquifers). Thus, it is recommended that stakeholders (scientists, engineers, water managers, and policymakers) consider the integrated role of ET in the elaborated mechanisms.

8. Conclusions

Climate change along with LU/LC change have aggravated or even triggered alterations to climatic variables and disturbances such as frequent droughts and extended wildfires which enlarge the effects of ET in the hydrological cycle and the surface energy balance. Thus, ET constitutes a major control of the prementioned processes. The specific hydrogeological, physical, and topographical characteristics along with the dynamics of regional climate conditions in the frame of the ongoing climate change practically turn each study area into a case study. Complex interactions and dynamics of several combinations of meteorological, lithological, hydrogeological, and physiological components with ET cannot lead to the formation of a general pattern of ET behavior or be reproduced, due to the nature of those relationships and the fact that numerous interfering factors such as weather conditions appearing temporal variability or even stochasticity govern the outcomes of the hydrological processes. Moreover, the review has a qualitative character and the deducted conclusions cannot be quantified. Consequently, the limitations of the present study are the impracticability of quantification and broad generalization. The latter could result in overgeneralization and oversimplification of mechanisms and interactions with many degrees of freedom. Furthermore, although all the main climate zones are covered, the review focuses on areas with considerable water-stress and desertification issues as the latter constitute severe impacts of climate change, bearing socio-economic consequences.

However, in some climate zones, some common behaviors of hydrological interactions could be detected, still need thorough future investigation aiming patterns to be extracted. This study does not aspire to discuss all possible interactions and impacts of ET in the environment. Instead, this study aims at setting the foundation for future research which addresses the integrated picture of ET as a major controlling factor of climate and sustainability. It provides a general overview of the main mechanisms in which ET participates, by pointing out the main interactions between ET, key environmental variables, and disturbances in different settings. Overall, since useful literature-based connections were made for specific areas under specific environmental conditions the former limitations are not considered critical for the validity of the conclusions.

Five broad conclusions can be deducted: First, MCRs appear to be vulnerable to the impacts of the ongoing increase in ET, especially during summertime, due to the ongoing precipitation shifting in winter and the air temperature warming (especially the rise of minimum air temperature values) which is expected to be more severe in MCRs such as Southern Europe, in the summertime. Air temperature is considered as a proxy of the energy state of the system. In water-limited areas, EF could serve as a water-stress indicator. Second, ET in tropical forests plays a rather beneficial role since it moderates the flooding risk during the wet season resulting in a net cooling effect. Third, in semi-arid to arid areas, an increase in ET and especially of evaporation constitutes an important problem due to sustained baseflow recessions which aggravate the limited water availability. In those drought-prone areas, ET exacerbates soil salinization. Fourth, the relationship between ET and wildfires is of major importance. The impacts are site-specific, climate, and fire-severity-dependent. The hydrological processes may be altered if a critical amount of canopy loss (e.g., 20% for semi-arid regions, 45% for tropical forests) occurs. Concurrently, RET could serve as a fuel aridity measure to assess forest fire risk. The case of Australia, with high rates of evaporation reported, may be a verification of the former deduction. Fifth, along with climate change, human activity consequences such as air pollution (aerosols, CO_2 emissions), LU/LC shifting to agricultural uses with intensive productivity practices, large reforestation implementation, and large constructions (e.g., dams, dense and high urban buildings) have substantially changed AET rates during last decades. Via the human footprint, the interpretation of the evaporation paradox has been made plausible.

In this context, future research is proposed to be designed towards two complementary axons. First, more refined, sophisticated ET modeling for global, regional, and local scales employing remote sensing techniques can be supported by eddy covariance, lysimeter, or pan evaporation measurements. Second, investigations that assess and quantify the dynamic anthropogenic impact on ET variability, aiming scientists, engineers as well as water managers to consider ET as a means to address the challenging environmental issues in two axons; either by finding methods to control it or by using it as an index of fire risk or water stress to help prevent or mediate the climate change impacts on water availability.

Supplementary Materials: The figure showing the evaporation paradox is available online at https://www.mdpi.com/article/10.3390/hydrology8040163/s1, Figure S1: Anthropogenic derived concepts which contribute to the interpretation of the evaporation paradox.

Author Contributions: Conceptualization, S.D. and K.G.N.; methodology, S.D. and K.G.N.; investigation, S.D.; data curation, S.D.; writing—original draft preparation, S.D.; writing—review and editing, S.D. and K.G.N.; supervision, K.G.N. All authors have read and agreed to the published version of the manuscript.

Funding: This research received no external funding.

Institutional Review Board Statement: Not applicable.

Informed Consent Statement: Not applicable.

Acknowledgments: Authors acknowledge the two anonymous reviewers for their contribution.

Conflicts of Interest: Authors declare no conflict of interest.

References

1. Xu, S.; Yu, Z.; Yang, C.; Ji, X.; Zhang, K. Trends in evapotranspiration and their responses to climate change and vegetation greening over the upper reaches of the Yellow River Basin. *Agric. For. Meteorol.* **2018**, *263*, 118–129. [CrossRef]
2. Gharbia, S.S.; Smullen, T.; Gill, L.; Johnston, P.; Pilla, F. Spatially distributed potential evapotranspiration modeling and climate projections. *Sci. Total Environ.* **2018**, *633*, 571–592. [CrossRef] [PubMed]
3. Ampas, V.; Baltas, E. Sensitivity analysis of different evapotranspiration methods using a new sensitivity coefficient. *Glob NEST J.* **2012**, *14*, 335–343. Available online: https://journal.gnest.org/sites/default/files/Journal%20Papers/335-343_882_Ambas_14_3.pdf (accessed on 29 October 2021).
4. Bulcock, H.H.; Jewitt, G.P.W. Modelling canopy and litter interception in commercial forest plantations in South Africa using the Variable Storage Gash model and idealised drying curves. *Hydrol. Earth Syst. Sci.* **2012**, *16*, 4693–4705. [CrossRef]
5. Uddin, J.; Foley, J.P.; Smith, R.J.; Hancock, N.H. A new approach to estimate canopy evaporation and canopy interception capacity from evapotranspiration and sap flow measurements during and following wetting. *Hydrol. Process.* **2016**, *30*, 1757–1767. [CrossRef]
6. Tegos, A.; Efstratiadis, A.; Koutsoyiannis, D. *A Parametric Model for Potential Evapotranspiration Estimation Based on a Simplified Formulation of the Penman-Monteith Equation, Evapotranspiration—An Overview*; Alexandris, S.G., Ed.; InTech: Rijeka, Croatia, 2013. Available online: https://www.intechopen.com/chapters/44363 (accessed on 29 October 2021). [CrossRef]
7. Yang, Y.; Chen, R.; Song, Y.; Han, C.; Liu, J.; Liu, Z. Sensitivity of potential evapotranspiration to meteorological factors and their elevational gradients in the Qilian Mountains, northwestern China. *J. Hydrol.* **2019**, *568*, 147–159. [CrossRef]
8. Lv, X.; Zuo, Z.; Sun, J.; Ni, Y.; Wang, Z. Climatic and human-related indicators and their implications for evapotranspiration management in a watershed of Loess Plateau, China. *Ecol. Indic.* **2019**, *101*, 143–149. [CrossRef]
9. Sun, Z.; Ouyang, Z.; Zhao, J.; Li, S.; Zhang, X.; Ren, W. Recent rebound in observational large-pan evaporation driven by heat wave and droughts by the Lower Yellow River. *J. Hydrol.* **2018**, *565*, 237–247. [CrossRef]
10. Allen, R.; Pereira, L.; Raes, D.; Smith, M. Crop evapotranspiration–Guidelines for computing crop water requirements. In *Irrigation and Drainage*; Paper No. 56; FAO: Rome, Italy, 1998; Volume 300.
11. Fu, G.; Liu, C.; Chen, S.; Hong, J. Investigating the conversion coefficients for free water surface evaporation of different evaporation pans. *Hydrol. Process.* **2014**, *18*, 2247–2262. [CrossRef]
12. Masoner, J.R.; Stannard, D.I.; Christenson, S.C. Differences in Evaporation Between a Floating Pan and Class A Pan on Land. *J. Am. Water Resour. Assoc.* **2018**, *44*, 552–561. [CrossRef]
13. Kitsara, G.; Papaioannou, G.; Retalis, A.; Paronis, D.; Kerkides, P. Estimation of air temperature and reference evapotranspiration using MODIS land surface temperature over Greece evapotranspiration using MODIS land surface temperature. *Int. J. Remote Sens.* **2018**, *39*, 924–948. [CrossRef]
14. Zamani Losgedaragh, S.; Rahimzadegan, M. Evaluation of SEBS, SEBAL, and METRIC models in estimation of the evaporation from the freshwater lakes (Case study: Amirkabir dam, Iran). *J. Hydrol.* **2018**, *561*, 523–531. [CrossRef]
15. Kim, S.; Kim, H.S. Neural networks and genetic algorithm approach for nonlinear evaporation and evapotranspiration modeling. *J. Hydrol.* **2008**, *351*, 299–317. [CrossRef]
16. Jiang, S.; Liang, C.; Cui, N.; Zhao, L.; Du, T.; Hu, X.; Feng, Y.; Guan, J.; Feng, Y. Impacts of climatic variables on reference evapotranspiration during growing season in Southwest China. *Agric. Water Manag.* **2019**, *216*, 365–378. [CrossRef]
17. Fan, J.; Wu, L.; Zhang, F.; Xiang, Y.; Zheng, J. Climate change effects on reference crop evapotranspiration across different climatic zones of China during 1956–2015. *J. Hydrol.* **2016**, *542*, 923–937. [CrossRef]
18. McCuen, H.R. A sensitivity and error analysis of procedures used for estimating evapotranspiration. *Water Resour. Bull.* **1974**, *10*, 486–498. [CrossRef]
19. Wang, Z.; Xie, P.; Lai, C.; Chen, X.; Wu, X.; Zeng, Z.; Li, J. Spatiotemporal variability of reference evapotranspiration and contributing climatic factors in China during 1961–2013. *J. Hydrol.* **2017**, *544*, 97–108. [CrossRef]
20. Liu, Z.; Ballantyne, A.P.; Cooper, L.A. Biophysical feedback of global forest fires on surface temperature. *Nat. Commun.* **2019**, *10*, 1–9. [CrossRef]
21. Tie, Q.; Hu, H.; Tian, F.; Holbrook, N.M. Comparing different methods for determining forest evapotranspiration and its components at multiple temporal scales. *Sci. Total Environ.* **2018**, *633*, 12–29. [CrossRef]
22. Xiao, Q.; McPherson, E.; Ustin, L.; Grismer, M.; Simpson, J. Winter rainfall interception by two mature open-grown trees in Davis, California. *Hydrol. Process.* **2000**, *14*, 763–784. [CrossRef]
23. Bart, R.R.; Tague, C.L. The impact of wildfire on baseflow recession rates in California. *Hydrol. Process.* **2017**, *31*, 1662–1673. [CrossRef]
24. Guevara-Escobar, A.; Gonzalez-Sosa, E.; Veliz-Chavez, C.; Ventura-Ramos, E.; Ramos-Salinas, M. Rainfall interception and distribution patterns of gross precipitation around an isolated Ficus benjamina tree in an urban area. *J. Hydrol.* **2007**, *333*, 532–541. [CrossRef]
25. Williams, D.G.; Cable, W.; Hultine, K.; Hoedjes, J.C.B.; Yepez, E.A.; Simonneaux, V.; Er-Raki, S.; Boulet, G.; de Bruin, H.A.R.; Chehbouni, A.; et al. Evapotranspiration components determined by stable isotope, sap flow and eddy covariance techniques. *Agric. For. Meteorol.* **2004**, *125*, 241–258. [CrossRef]

26. Qiu, X.; Zhang, M.; Wang, S.; Argiriou, A.A.; Chen, R.; Meng, H.; Guo, R. Water Stable Isotopes in an Alpine Setting of the Northeastern Tibetan Plateau. *Water* **2019**, *11*, 770. [CrossRef]
27. Shi, M.; Wang, S.; Argiriou, A.A.; Zhang, M.; Guo, R.; Jiao, R.; Kong, J.; Zhang, Y.; Qiu, X.; Zhou, S. Stable Isotope Composition in Surface Water in the Upper Yellow River in Northwest China. *Water* **2019**, *11*, 967. [CrossRef]
28. Gokmen, M.; Vekerdy, Z.; Verhoef, A.; Verhoef, W.; Batelaan, O.; van der Tol, C. Integration of soil moisture in SEBS for improving evapotranspiration estimation under water stress conditions. *Remote Sens. Environ.* **2012**, *121*, 261–274. [CrossRef]
29. Valipour, M.; Gholami Sefidkouhi, M.A.; Raeini−Sarjaz, M. Selecting the best model to estimate potential evapotranspiration with respect to climate change and magnitudes of extreme events. *Agric. Water Manag.* **2017**, *180*, 50–60. [CrossRef]
30. Mu, Q.; Zhao, M.; Running, S.W. Improvements to a MODIS global terrestrial evapotranspiration algorithm. *Remote Sens. Environ.* **2011**, *115*, 1781–1800. [CrossRef]
31. Allen, R.G.; Tasumi, M.; Morse, A.; Trezza, R. A Landsat-based energy balance and evapotranspiration model in Western US water rights regulation and planning. *Irrig Drain. Syst.* **2005**, *19*, 251–268. [CrossRef]
32. Anderson, M.C.; Kustas, W.P.; Norman, J.M.; Hain, C.R.; Mecikalski, J.R.; Schultz, L.; González-Dugo, M.P.; Cammalleri, C.; d'Urso, G.; Pimstein, A.; et al. Mapping daily evapotranspiration at field to continental scales using geostationary and polar orbiting satellite imagery. *Hydrol. Earth Syst. Sci.* **2015**, *15*, 223–239. [CrossRef]
33. Bastiaanssen, W.G.M.; Menenti, M.; Feddes, R.A.; Holtslag, A.A.M. A remote sensing surface energy balance algorithm for land (SEBAL). 1. Formulation. *J. Hydrol.* **1998**, *212–213*, 198–212. [CrossRef]
34. Sánchez, J.M.; Kustas, W.P.; Caselles, V.; Anderson, M.C. Modelling surface energy fluxes over maize using a two-source patch model and radiometric soil and canopy temperature observations. *Remote Sens. Environ.* **2008**, *112*, 1130–1143. [CrossRef]
35. Su, B.; Wang, L. Earth Observation of Water Resources (SEBS), Teaching Presentation, University of Twente The Netherlands. 2013. Available online: https://earth.esa.int/documents/10174/643007/D5P1c-1_SEBS_LTC2013.pdf (accessed on 9 September 2021).
36. Dimitriadou, S.; Nikolakopoulos, K.G. Remote sensing methods to estimate evapotranspiration incorporating MODIS derived data and applications over Greece: A review. In Proceedings of the SPIE 11524, Eighth International Conference on Remote Sensing and Geoinformation of the Environment (RSCy2020), Paphos, Cyprus, 26 August 2020.
37. Tabari, H.; Grismer, M.E.; Trajkovic, S. Comparative analysis of 31 reference evapotranspiration methods under humid conditions. *Irrig. Sci.* **2013**, *31*, 107–117. [CrossRef]
38. Malamos, N.; Tsirogiannis, I.L.; Tegos, A.; Efstratiadis, A.; Koutsoyiannis, D. Spatial interpolation of potential evapotranspiration for precision irrigation purposes. *Eur. Water* **2017**, *59*, 303–309. Available online: https://www.itia.ntua.gr/el/getfile/1776/1/documents/EW_2017_59_41_2HOxTxv.pdf (accessed on 9 September 2021).
39. Vasiliades, L.; Spiliotopoulos, M.; Tzabiras, J.; Loukas, A.; Mylopoulos, N. Estimation of crop water requirements using remote sensing for operational water resources management. In Proceedings of the Third International Conference on Remote Sensing and Geoinformation of the Environment (RSCy2015) SPIE 9535, Paphos, Cyprus, 19 June 2015; p. 95351B.
40. Demertzi, K.; Pisinaras, V.; Lekakis, E.; Tziritis, E.; Babakos, K.; Aschonitis, V. Assessing Annual Actual Evapotranspiration based on Climate, Topography and Soil in Natural and Agricultural Ecosystems. *Climate* **2021**, *9*, 20. [CrossRef]
41. Dimitriadou, S.; Nikolakopoulos, K.G. Annual Actual Evapotranspiration Estimation via GIS Models of Three Empirical Methods Employing Remotely Sensed Data for the Peloponnese, Greece, and Comparison with Annual MODIS ET and Pan Evaporation Measurements. *ISPRS Int. J. Geo-Inf.* **2021**, *10*, 522. [CrossRef]
42. Dimitriadou, S.; Nikolakopoulos, K.G. Reference evapotranspiration (ETo) methods implemented as ArcMap models with remote sensed and ground-based inputs, examined along with MODIS ET, for Peloponnese, Greece. *ISPRS Int. J. Geo-Inf.* **2021**, *10*, 390. [CrossRef]
43. Anderson, E. Calibration of Conceptual Hydrologic Models for Use in River Forecasting. NOAA Technical Report, NWS 45, Hydrology Laboratory, August 2002. Available online: https://www.semanticscholar.org/paper/Calibration-of-Conceptual-Hydrologic-Models-for-Use-Anderson/9ec4749a4064e6d41058c8c6fbcda108210e6865#paper-header (accessed on 10 September 2021).
44. Herman, M.R.; Nejadhashemi, A.P.; Abouali, M.; Hernandez-Suarez, J.S.; Daneshvar, F.; Zhang, Z.; Anderson, M.C.; Sadeghi, A.M.; Hain, C.R.; Sharifi, A. Evaluating the role of evapotranspiration remote sensing data in improving hydrological modeling predictability. *J. Hydrol.* **2018**, *556*, 39–49. [CrossRef]
45. Zhao, T.; Wang, Q.J.; Schepen, A.; Griffiths, M. Ensemble forecasting of monthly and seasonal reference crop evapotranspiration based on global climate model outputs. *Agric. For. Meteorol.* **2019**, *264*, 114–124. [CrossRef]
46. Johnson, F.; Sharma, A. A Comparison of Australian Open Water Body Evaporation Trends for Current and Future Climates Estimated from Class A Evaporation Pans and General Circulation Models. *J. Hydrometeor.* **2010**, *11*, 105–121. [CrossRef]
47. Sattari, M.T.; Apaydin, H.; Band, S.S.; Mosavi, A.; Prasad, R. Comparative analysis of kernel-based versus ANN and deep learning methods in monthly reference evapotranspiration estimation. *Hydrol. Earth Syst. Sci.* **2021**, *25*, 603–618. [CrossRef]
48. Tegos, A.; Efstratiadis, A.; Malamos, N.; Mamassis, N.; Koutsoyiannis, D. Evaluation of a Parametric Approach for Estimating Potential Evapotranspiration Across Different Climates. *Agric. Agric. Sci. Procedia* **2015**, *4*, 2–9. [CrossRef]
49. Batra, K.; Gandhi, P. Neural Network-Based Prediction Model for Evaporation Using Weather Data. *Agric. Res.* **2021**. [CrossRef]
50. Malik, A.; Kumar, A.; Kisi, O. Monthly pan-evaporation estimation in Indian central Himalayas using different heuristic approaches and climate based models. *Comput. Electron. Agric.* **2017**, *143*, 302–313. [CrossRef]
51. Güçlü, Y.S.; Subyani, A.M.; Şen, Z. Regional fuzzy chain model for evapotranspiration estimation. *J. Hydrol.* **2017**, *544*, 233–241. [CrossRef]

52. Boaz, A.; Ashby, D.; Young, K. Systematic Reviews: What Have They Got to Offer Evidence Based Policy and Practice? ESRC UK Centre for Evidence Based Policy and Practice, Queen Mary University of London, UK, 1–26. Available online: https://emilkirkegaard.dk/en/wp-content/uploads/Should-I-do-a-systematic-review.pdf (accessed on 10 September 2021).

53. Gill, J.C.; Malamud, B.D. Anthropogenic processes, natural hazards, and interactions in a multi-hazard framework. *Earth-Sci. Rev.* **2017**, *166*, 246–269. [CrossRef]

54. Hobbins, M.T.; Ramirez, J.A.; Brown, T.C. Trends in pan evaporation and actual evapotranspiration across the conterminous U.S.: Paradoxical or complementary? *Geophys. Res. Lett.* **2004**, *31*, 1–5. [CrossRef]

55. Qin, M.; Zhang, Y.; Wan, S.; Yue, Y.; Cheng, Y.; Zhang, B. Impact of climate change on "evaporation paradox" in province of Jiangsu in southeastern China. *PLoS ONE* **2021**, *16*, e0247278. [CrossRef]

56. Ohmura, A.; Wild, M. Is the hydrological cycle accelerating? *Science* **2002**, *298*, 1345–1346. [CrossRef]

57. Burn, D.H.; Hesch, N.M. Trends in evaporation for the Canadian Prairies. *J. Hydrol.* **2007**, *336*, 61–73. [CrossRef]

58. McVicar, T.R.; Roderick, M.L.; Donohue, R.J.; Li, L.T.; Van Niel, T.G.; Thomas, A.; Grieser, J.; Jhajharia, D.; Himri, Y.; Mahowald, N.M.; et al. Global review and synthesis of trends in observed terrestrial near-surface wind speeds: Implications for evaporation. *J. Hydrol.* **2012**, *416*, 182–205. [CrossRef]

59. Wang, M.; Zhang, Y.; Lu, Y.; Gong, X.; Gao, L. Detection and attribution of reference evapotranspiration change (1951–2020) in the Upper Yangtze River Basin of China. *J. Water Clim. Chang.* **2021**, *12*, 2624–2638. [CrossRef]

60. You, G.; Zhang, Y.; Liu, Y.; Song, Q.; Lu, Z.; Tan, Z.; Wu, C.; Xie, Y. On the attribution of changing pan evaporation in a nature reserve in SW China. *Hydrol. Process.* **2013**, *27*, 2676–2682. [CrossRef]

61. Liu, X.; Luo, Y.; Zhang, D.; Zhang, M.; Liu, C. Recent changes in pan-evaporation dynamics in China. *Geophys. Res. Lett.* **2011**, *38*, L13404. [CrossRef]

62. Jhajharia, D.; Shrivastava, S.K.; Sarkar, D.; Sarkar, S. Temporal characteristics of pan evaporation trends under the humid conditions of northeast India. *Agric. For. Meteorol.* **2009**, *149*, 763–770. [CrossRef]

63. Maček, U.; Bezak, N.; Šraj, M. Reference evapotranspiration changes in Slovenia, Europe. *Agric. For. Meteorol.* **2018**, *260*, 183–192. [CrossRef]

64. Papaioannou, G.; Kitsara, G.; Athanasatos, S. Impact of global dimming and brightening on reference evapotranspiration in Greece. *J. Geophys. Res.* **2011**, *116*, D09107. [CrossRef]

65. Kitsara, G.; Papaioannou, G.; Papathanasiou, A.; Retalis, A. Dimming/brightening in Athens: Trends in Sunshine Duration, Cloud Cover and Reference Evapotranspiration. *Water Resour. Manag.* **2013**, *27*, 1623–1633. [CrossRef]

66. Ogolo, E.O. Regional trend analysis of pan evaporation in Nigeria (1970 to 2000). *J. Geogr. Reg. Plan.* **2011**, *4*, 566–577. Available online: https://pdfs.semanticscholar.org/57a9/ce191a35968d7e6ab147416b24ce3e64f852.pdf (accessed on 8 September 2021).

67. Cohen, S.; Ianetz, A.; Stanhill, G. Evaporative climate changes at Bet Dagan, Israel, 1964–1998. *Agric. For. Meteorol.* **2002**, *111*, 83–91. [CrossRef]

68. Dadaser-Celik, F.; Cengiz, E.; Guzel, O. Trends in reference evapotranspiration in Turkey: 1975–2006. *Int. J. Climatol.* **2016**, *36*, 1733–1743. [CrossRef]

69. Yeşilırmak, E. Temporal changes of warm-season pan evaporation in a semi-arid basin in Western Turkey. *Stoch Environ. Res. Risk Assess.* **2013**, *27*, 311–321. [CrossRef]

70. Lionello, P.; Scarascia, L. The relation between climate change in the Mediterranean region and global warming. *Reg. Environ. Chang.* **2018**, *18*, 1481–1493. [CrossRef]

71. Vicente-Serrano, S.M.; Bidegain, M.; Tomas-Burguera, M.; Dominguez-Castro, F.; El Kenawy, A.; McVicar, T.R.; Azorin-Molina, C.; López-Moreno, J.I.; Nieto, R.; Gimeno, L.; et al. A comparison of temporal variability of observed and model-based pan evaporation over Uruguay (1973–2014). *Int. J. Climatol.* **2018**, *38*, 337–350. [CrossRef]

72. Breña-Naranjo, J.A.; Laverde-Barajas, M.Á.; Pedrozo-Acuña, A. Changes in pan evaporation in Mexico from 1961 to 2010. *Int. J. Climatol.* **2017**, *37*, 204–213. [CrossRef]

73. Aschonitis, V.; Miliaresis, G.; Demertzi, K.; Papamichail, D. Terrain Segmentation of Greece Using the Spatial and Seasonal Variation of Reference Crop Evapotranspiration. *Adv. Meteorol.* **2016**, 1–14. [CrossRef]

74. Estévez, J.; Gavilán, P.; Berengena, J. Sensitivity analysis of a Penman-Monteith type equation to estimate reference evapotranspiration in southern Spain. *Hydrol. Process.* **2009**, *23*, 3342–3353. [CrossRef]

75. Azizzadeh, M.; Javan, K. Analyzing trends in Reference Evapotranspiration in northwest part of Iran. *J. Ecol. Eng.* **2015**, *16*, 1–12. [CrossRef]

76. Dinpashoh, Y.; Jhajharia, D.; Fakheri-Fard, A.; Singh, V.P.; Kahya, E. Trends in reference crop evapotranspiration over Iran. *J. Hydrol.* **2011**, *399*, 422–433. [CrossRef]

77. Mueller, N.D.; Rhines, A.; Butler, E.E.; Ray, D.K.; Siebert, S.; Holbrook, N.M.; Huybers, P. Global Relationships between Cropland Intensification and Summer Temperature Extremes over the Last 50 Years. *J. Clim.* **2017**, *30*, 7505–7528. [CrossRef]

78. Limjirakan, S.; Limsakul, A. Trends in Thailand pan evaporation from 1970 to 2007. *Atmos. Res.* **2012**, *108*, 2859–2876. [CrossRef]

79. Abtew, W.; Obeysekera, J.; Iricanin, N. Pan evaporation and potential evapotranspiration trends in South Florida. *Hydrol. Process.* **2011**, *25*, 958–969. [CrossRef]

80. Stanhill, G. Is the class A evaporation pan still the most practical and accurate meteorological method for determining irrigation water requirements? *Agric. For. Meteorol.* **2002**, *112*, 233–236. [CrossRef]

81. Doorenbos, J.; Pruitt, W.O. *Guidelines for Prediction of Crop Water Requirements in Irrigation and Drainage*; Paper No. 24 (revised); FAO: Rome, Italy, 1977. Available online: http://www.fao.org/3/a-f2430e.pdf (accessed on 8 September 2021).

82. Hobbins, M. What Drives the Variability of Evaporative Demand across the Conterminous United States? *J. Hydrometeorol.* **2012**, *13*, 1195–1214. [CrossRef]

83. Kim, S.; Anabalón, A.; Sharma, A. An Assessment of Concurrency in Evapotranspiration Trends across Multiple Global Dataset. *J. Hydrometeorol.* **2021**, *22*, 231–244. [CrossRef]

84. Craig, I.P. Loss of Storage Water Due to Evaporation. National Centre for Engineering in Agriculture University of Southern Queensland: Toowoomba, 2005. Available online: https://core.ac.uk/download/pdf/11036429.pdf (accessed on 8 September 2021).

85. Roderick, M.L.; Farquhar, G.D. Changes in Australian pan evaporation from 1970 to 2002. *Int. J. Climatol.* **2004**, *24*, 1077–1090. [CrossRef]

86. Rayner, D.P. Wind Run Changes: The Dominant Factor Affecting Pan Evaporation Trends in Australia. *J. Clim.* **2007**, *20*, 3379–3394. [CrossRef]

87. Stephens, C.M.; McVicar, T.R.; Johnson, F.M.; Marshall, L.A. Revisiting Pan Evaporation Trends in Australia a Decade on. *Geophys. Res. Lett.* **2018**, *45*, 164–172. [CrossRef]

88. Baruffi, F.; Cisotto, A.; Cimolino, A.; Ferri, M.; Monego, M.; Norbiato, D.; Cappelletto, M.; Bisaglia, M.; Pretner, A.; Galli, A.; et al. Climate change impact assessment on Veneto and Friuli plain groundwater. Part I: An integrated modeling approach for hazard scenario construction. *Sci. Total Environ.* **2012**, *440*, 154–166. [CrossRef]

89. Lipczynska-Kochany, E. Effect of climate change on humic substances and associated impacts on the quality of surface water and groundwater: A review. *Sci. Total Environ.* **2018**, *640*, 1548–1565. [CrossRef]

90. Pitz, C.F. *Predicted Impacts of Climate Change on Groundwater Resources of Washington State*; Environmental Assessment Program, Publication No. 16-03-006; Washington State Department of Ecology: Olympia, WA, USA, 2016. Available online: https://apps.ecology.wa.gov/publications/documents/1603006.pdf (accessed on 9 September 2021).

91. Nolin, A.W. Perspectives on Climate Change, Mountain Hydrology, and Water Resources in the Oregon Cascades, USA. *Mt. Res. Dev.* **2010**, *32*, 35–46. [CrossRef]

92. Collins, M.; Knutti, R.; Arblaster, J.; Dufresne, J.L.; Fichefet, T.; Friedlingstein, P.; Gao, X.; Gutowski, W.J.; Johns, T.; Krinner, G.; et al. Long-term Climate Change: Projections, Commitments and Irreversibility. In *Climate Change 2013: The Physical Science Basis. Contribution of Working Group I to the Fifth Assessment Report of the Intergovernmental Panel on Climate Change*; Stocker, T.F., Qin, D., Plattner, G.-K., Tignor, M., Allen, S.K., Boschung, J., Nauels, A., Xia, Y., Bex, V., Midgley, P.M., Eds.; Cambridge University Press: Cambridge, UK, 2013. Available online: https://www.ipcc.ch/site/assets/uploads/2018/02/WG1AR5_Chapter12_FINAL.pdf (accessed on 9 September 2021).

93. Krishnaswamy, J.; Bonell, M.; Venkatesh, B.; Purandara, B.K.; Rakesh, K.N.; Lele, S.; Kiran, M.C.; Reddy, W.; Badiger, S. The groundwater recharge response and hydrologic services of tropical humid forest ecosystems to use and reforestation: Support for the "infiltration-evapotranspiration trade-off hypothesis. *J. Hydrol.* **2013**, *498*, 191–209. [CrossRef]

94. Rodrigues, E.L.; Jacobi, C.M.; Figueira, J.E.C. Wildfires and their impact on the water supply of a large neotropical metropolis: A simulation approach. *Sci. Total Environ.* **2019**, *651*, 1261–1271. [CrossRef]

95. Häusler, M.; Nunes, J.P.; Soares, P.; Sánchez, J.M.; Silva, J.M.N.; Warneke, T.; Keizer, J.J.; Pereira, J.M.C. Assessment of the indirect impact of wildfire (severity) on actual evapotranspiration in eucalyptus forest based on the surface energy balance estimated from remote-sensing techniques. *Int. J. Remote Sens.* **2018**, *39*, 195–209. [CrossRef]

96. Nolan, R.H.; Lane, P.N.J.; Benyon, R.G.; Bradstock, R.A.; Mitchell, P.J. Trends in Evapotranspiration and Streamflow following Wildfire in Resprouting Eucalypt Forests. *J. Hydrol.* **2015**, *524*, 614–624. [CrossRef]

97. Sánchez, J.M.M.; Bisquert, E.R.; Caselles, V. Impact of Land Cover Change Induced by a Fire Event on the Surface Energy Fluxes Derived from Remote Sensing. *Remote Sens.* **2015**, *7*, 14899–14915. [CrossRef]

98. Wang, S.; Ibrom, A.; Bauer-Gottwein, P.; Garcia, M. Incorporating diffuse radiation into a light use efficiency and evapotranspiration model: An 11-year study in a high latitude deciduous forest. *Agric. For. Meteorol.* **2018**, *248*, 479–493. [CrossRef]

99. Hirano, T.; Kusin, K.; Limin, S.; Osaki, M. Evapotranspiration of tropical peat swamp forests. *Glob. Chang. Biol.* **2015**, *21*, 1914–1927. [CrossRef]

100. Abatzoglou, J.T.; Williams, A.P. Impact of anthropogenic climate change on wildfire across western US forests. *Proc. Natl. Acad. Sci. USA* **2016**, *113*, 11770–11775. [CrossRef]

101. Häusler, M.; Nunes, J.P.; Silva, J.M.N.; Keizer, J.J.; Warneke, T.; Pereira, J.M.C. A promising new approach to estimate drought indices for fire danger assessment using remotely sensed data. *Agric. For. Meteorol.* **2019**, *274*, 195–209. [CrossRef]

102. Poon, P.; Kinoshita, A. Estimating Evapotranspiration in a Post-Fire Environment Using Remote Sensing and Machine Learning. *Remote Sens.* **2018**, *10*, 1728. [CrossRef]

103. Johnk, B.T.; Mays, D.C. Wildfire Impacts on Groundwater Aquifers: A Case Study of the 1996 Honey Boy Fire in Beaver County, Utah, USA. *Water* **2021**, *13*, 2279. [CrossRef]

104. Kurylyk, B.L.; MacQuarrie, K.T.B.; Caissie, D.; McKenzie, J.M. Shallow groundwater thermal sensitivity to climate change and land cover disturbances: Derivation of analytical expressions and implications for stream temperature modeling. *Hydrol. Earth Syst. Sci.* **2015**, *19*, 2469–2489. [CrossRef]

105. Menberg, K.; Blum, P.; Kurylyk, B.L.; Bayer, P. Observed groundwater temperature response to recent climate change. *Hydrol. Earth Syst. Sci.* **2014**, *18*, 4453–4466. [CrossRef]

106. Wine, M.L.; Cadol, D. Hydrologic effects of large southwestern USA wildfires significantly increase regional water supply: Fact or fiction? *Environ. Res. Lett.* **2016**, *11*, 1–13. [CrossRef]
107. Kinoshita, A.M.; Hogue, T.S. Catena Spatial and temporal controls on post-fire hydrologic recovery in Southern California watersheds. *Catena* **2011**, *87*, 240–252. [CrossRef]
108. Kinoshita, A.M.; Hogue, T.S. Increased dry season water yield in burned watersheds in Southern California Increased dry season water yield in burned watersheds in Southern California. *Environ. Res. Lett.* **2015**, *10*, 14003. [CrossRef]
109. Bart, R.R. A regional estimate of postfire streamflow change in California: A regional estimate of postfire streamflow change. *Water Resour. Res.* **2016**, *52*, 1465–1478. [CrossRef]
110. Bosch, J.M.; Hewlett, J.D. A review of catchment experiments to determine the effect of vegetation changes on water yield and evapotranspiration. *J. Hydrol.* **1982**, *55*, 3–23. [CrossRef]
111. Leakey, A.D.B.; Ainsworth, E.A.; Bernacchi, C.J.; Rogers, A.; Long, S.P.; Ort, D.R. Elevated CO_2 effects on plant carbon, nitrogen, and water relations: Six important lessons from FACE. *J. Exp. Bot.* **2009**, *60*, 2859–2876. [CrossRef]
112. Blonquist, J.; Norman, J.; Bugbee, B. Automated measurement of canopy stomatal conductance based on infrared temperature. *Agric. For. Meteorol.* **2009**, *149*, 2183–2197. [CrossRef]
113. Wang, Y.W.; Yang, Y.H. China's dimming and brightening: Evidence, causes and hydrological implications. *Ann. Geophys.* **2014**, *32*, 41–55. [CrossRef]
114. Hallar, A.G.; Molotch, N.P.; Hand, J.L.; Livneh, B.; McCubbin, I.B.; Petersen, R.; Michalsky, J.; Lowenthal, D.; Kunkel, K.E. Impacts of increasing aridity and wildfires on aerosol loading in the intermountain Western US. *Environ. Res. Lett.* **2017**, *12*, 1–8. [CrossRef]
115. Vergni, L.; Todiso, F. Spatio-temporal variability of precipitation, temperature and agricultural drought indices in Central Italy. *Agric. For. Meteorol.* **2011**, *151*, 301–313. [CrossRef]
116. De Meij, A.; Pozzer, A.; Lelieveld, J. Trend analysis in aerosol optical depths and pollutant emission estimates between 2000 and 2009. *Atmos. Environ.* **2012**, *51*, 75–85. [CrossRef]
117. Zhao, L.; Xia, J.; Sobkowiak, L.; Li, Z. Climatic Characteristics of Reference Evapotranspiration in the Hai River Basin and Their Attribution. *Water* **2014**, *6*, 1482–1499. [CrossRef]
118. Williams, I.N.; Torn, M.S. Vegetation controls on surface heat flux partitioning, and land-atmosphere coupling: Vegetation and Land-Atmosphere coupling. *Geophys. Res. Lett.* **2015**, *42*, 9416–9424. [CrossRef]
119. Lu, Y.; Jin, J.; Kueppers, L.M. Crop growth and irrigation interact to influence surface fluxes in a regional climate-cropland model (WRF3.3-CLM4crop). *Clim. Dyn.* **2015**, *45*, 3347–3363. [CrossRef]
120. Giannakopoulos, E.; Svarnas, P.; Dimitriadou, S.; Kalavrouziotis, I.; Papadopoulos, P.K.; Georga, S.; Krontiras, C. Emerging Sanitary Engineering of Biosolids: Elimination of *Salmonella*, *Escherichia coli*, and Coliforms by means of Atmospheric Pressure Air Cold Plasma. *J. Hazard. Toxic Radioact. Waste* **2021**, *25*, 6021001. [CrossRef]
121. Rudnick, D.R.; Irmak, S. Impact of Nitrogen Fertilizer on Maize Evapotranspiration Crop Coefficients under Fully Irrigated, Limited Irrigation, and Rainfed Settings. *J. Irrig. Drain. Eng.* **2014**, *140*, 4014039. [CrossRef]
122. García-Llamas, P.; Suárez-Seoane, S.; Taboada, A.; Fernández-Manso, A.; Quintano, C.; Fernández-García, V.; Fernández-Guisuraga, J.M.; Marcos, E.; Calvo, L. Environmental drivers of fire severity in extreme fire events that affect Mediterranean pine forest ecosystems. *For. Ecol. Manag.* **2019**, *433*, 24–32. [CrossRef]
123. Gentine, P.; Entekhabi, D.; Polcher, J. The Diurnal Behavior of Evaporative Fraction in the Soil–Vegetation–Atmospheric Boundary Layer Continuum. *J. Hydrometeorol.* **2011**, *12*, 1530–1546. [CrossRef]
124. French, A.N.; Hunsaker, D.J.; Throp, K.R. Remote sensing of evapotranspiration over cotton using the TSEB and METRIC energy balance models. *Remote Sens. Environ.* **2016**, *158*, 281–294. [CrossRef]
125. Hosseini, M.; Geissen, V.; González-Pelayo, O.; Serpa, D.; Machado, A.I.; Ritsema, C.; Keizer, J.J. Effects of fire occurrence and recurrence on nitrogen and phosphorus losses by overland flow in maritime pine plantations in north-central Portugal. *Geoderma* **2017**, *289*, 97–106. [CrossRef]
126. Dimitriadou, S.; Katsanou, K.; Stratikopoulos, K.; Lambrakis, N. Investigation of the chemical processes controlling the groundwater quality of Ilia Prefecture. *Environ. Earth Sci.* **2019**, *78*, 401. [CrossRef]
127. Dimitriadou, S.; Katsanou, K.; Charalabopoulos, S.; Lambrakis, N. Interpretation of the Factors Defining Groundwater Quality of the Site Subjected to the Wildfire of 2007 in Ilia Prefecture, South-Western Greece. *Geosciences* **2018**, *8*, 108. [CrossRef]
128. Tsypkin, G.G.; Brevdo, L. A Phenomenological Model of the Increase in Solute Concentration in Ground Water Due to Evaporation. *Transp. Porous Media* **1999**, *37*, 129–151. [CrossRef]
129. Gran, M.; Carrera, J.; Olivella, S.; Massana, J.; Saaltink, M.W.; Ayora, C.; Lloret, A. Salinity is reduced below the evaporation front during soil salinization. *Estud. Zona No Saturada Suelo* **2009**, *9*, 12678.
130. Amezketa, E. An integrated methodology for assessing soil salinization, a pre-condition for land desertification. *J. Arid. Environ.* **2006**, *67*, 594–606. [CrossRef]
131. Neave, M.; Rayburg, F. Salinity and erosion: A preliminary investigation of soil erosion on a salinized hillslope. In Proceedings of the Symposium Sediment Dynamics and the Hydromorphology of Fluvial Systems, Dundee, UK, 2–7 July 2006; IAHS Publ.: Oxfordshire, UK, 2006; Volume 306, pp. 531–539. Available online: https://iahs.info/uploads/dms/13587.68-531-539-03-306-Neave.pdf (accessed on 9 September 2021).
132. Chen, H.; Sun, J. Changes in Drought Characteristics over China Using the Standardized Precipitation Evapotranspiration Index. *J. Clim.* **2015**, *28*, 5430–5447. [CrossRef]

133. California State Water Resources Control Board Division of Water Quality GAMA Program. *Groundwater Information Sheet, Salinity*; Water Boards: Sacramento, CA, USA, 2017. Available online: https://www.waterboards.ca.gov/gama/docs/coc_salinity.pdf (accessed on 9 September 2021).
134. Guo, W.; Andersen, M.N.; Qi, X.; Li, P.; Li, Z.; Fan, X.; Zhou, Y. Effects of reclaimed water irrigation and nitrogen fertilization on the chemical properties and microbial community of soil. *J. Integr. Agric.* **2017**, *16*, 679–690. [CrossRef]
135. Kalavrouziotis, I.K. The reuse of Municipal Wastewater in soils. *Glob. Nest J.* **2015**, *17*, 474–486.
136. Kalavrouziotis, I.K.; Kokkinos, P.; Oron, G.; Fatone, F.; Bolzonella, D.; Vatyliotou, M.; Fatta-Kassinos, D.; Koukoulakis, P.H.; Varnavas, S.P. Current status in wastewater treatment, reuse and research in some mediterranean countries. *Desalination Water Treat.* **2015**, *53*, 2015–2030. [CrossRef]
137. Tavares., P.; Beltrão, N.; Guimarães, U.; Teodoro, A.C.; Gonçalves, P. Urban ecosystem services quantification through remote sensing approach: A systematic review. *Environments* **2019**, *6*, 51. [CrossRef]
138. Almeida, C.R.D.; Teodoro, A.C.; Gonçalves, A. Study of the Urban Heat Island (UHI) Using Remote Sensing Data/Techniques: A Systematic Review. *Environments* **2021**, *8*, 105. [CrossRef]
139. Zipper, S.; Loheide, S. Using evapotranspiration to assess drought sensitivity on a subfield scale with HRMET, a high resolution surface energy balance model. *Agric. For. Meteorol.* **2014**, *123*, 91–102. [CrossRef]
140. Wild, M.; Gilgen, H.; Roesch, A.; Ohmura, A.; Long, C.N.; Dutton, E.G.; Forgan, B.; Kallis, A.; Rusak, V.; Tsvetkov, A. From dimming to brightening: Decadal changes in solar radiation at Earth's surface. *Science* **2005**, *308*, 847–850. [CrossRef]
141. Wild, M.; Grieser, G.; Schar, C. Combined surface solar brightening and increasing greenhouse effect support recent intensification of the global land-based hydrological cycle. *Geophys. Res. Lett.* **2008**, *35*, 1–5. [CrossRef]
142. Zerefos, C.S.; Eleftheratos, K.; Meleti, C.; Kazadzis, S.; Romanou, A.; Ichoku, C.; Tselioudis, G.; Bais, A. Solar dimming and brightening over Thessaloniki, Greece, and Beijing, China. *Tellus B* **2009**, *61*, 657–665. [CrossRef]
143. Dimitriadou, S.; Nikolakopoulos, K.G. Development of GIS models via optical programming and python scripts to implement four empirical methods of reference and actual evapotranspiration (ETo, ETa) incorporating MODIS LST inputs. In Proceedings of the SPIE 11856, Remote Sensing for Agriculture, Ecosystems, and Hydrology XXIII, 118560K, Madrid, Spain, 12 September 2021. [CrossRef]

 *hydrology*

Article

Precipitation and Potential Evapotranspiration Temporal Variability and Their Relationship in Two Forest Ecosystems in Greece

Stefanos Stefanidis [1,*] and Vasileios Alexandridis [2]

1 Laboratory of Mountainous Water Management and Control, Faculty of Forestry and Natural Environment, Aristotle University of Thessaloniki, 54124 Thessaloniki, Greece
2 Independent Researcher, 54621 Thessaloniki, Greece; alexandridisvasileios@gmail.com
* Correspondence: ststefanid@gmail.com

Abstract: The assessment of drought conditions is important in forestry because it affects forest growth and species diversity. In this study, temporal variability and trends of precipitation (P), potential evapotranspiration (PET), and their relationship (P/PET) were examined in two selected forest ecosystems that present different climatic conditions and vegetation types due to their location and hypsometric zone. The study area includes the forests of Pertouli and Taxiarchis, which are managed by the Aristotle University Forest Administration and Management Fund. The Pertouli is a coniferous forest in Central Greece with a maximum elevation of 2073 m a.s.l, and Taxiarchis is a broadleaved forest in Northern Greece with a maximum elevation of 1200 m a.s.l. To accomplish the goals of the current research, long–term (1974–2016) monthly precipitation and air temperature data from two mountainous meteorological were collected and processed. The PET was estimated using a parametric model based on simplified formulation of the Penman–Monteith equation rather than the commonly used Thornthwaite approach. Seasonal and annual precipitation, potential evapotranspiration (PET), and their ratio (P/PET) values were subjected to Mann–Kendall tests to assess the possible upward or downward trends, and Sen's slope method was used to estimate the trends magnitude. The results indicated that the examined climatic variables vary greatly between seasons. In general, negative trends were detected for the precipitation time series of Pertouli, whereas positive trends were found in Taxiarchis; both were statistically insignificant. In contrast, statistically significant positive trends were reported for PET in both forest ecosystems. These circumstances led to different drought conditions between the two forests due to the differences of their elevation. Regarding Pertouli, drought trend analysis indicated downward trends for annual, winter, spring, and summer values, whereas autumn showed a slight upward trend. In addition, the average magnitude trend per decade was approximately −2.5%, −3.5%, +4.8%, −0.8%, and +3.3% for annual, winter, autumn, spring, and summer seasons, respectively. On the contrary, the drought trend and the associated magnitude per decade for the Taxiarchis forest were found to be as follows: annual (+2.2%), winter (+6.2%), autumn (+9.2%), spring (+1.0%), and summer (−5.0%). The performed statistical test showed that the reported trend was statistically insignificant at a 5% significance level. These results may be a useful tool as a forest management practice and can enhance the adaptation and resilience of forest ecosystems to climate change.

Keywords: precipitation; evapotranspiration; drought; Mann–Kendall; trend analysis

Citation: Stefanidis, S.; Alexandridis, V. Precipitation and Potential Evapotranspiration Temporal Variability and Their Relationship in Two Forest Ecosystems in Greece. *Hydrology* **2021**, *8*, 160. https://doi.org/10.3390/hydrology8040160

Academic Editors: Aristoteles Tegos and Nikolaos Malamos

Received: 28 September 2021
Accepted: 16 October 2021
Published: 18 October 2021

Publisher's Note: MDPI stays neutral with regard to jurisdictional claims in published maps and institutional affiliations.

1. Introduction

Global warming has increasingly raised the concerns of both governments and the scientific community in recent decades. The Mediterranean basin has been identified as one of the most sensitive regions to climate change, with future warming potentially exceeding the global average [1,2]. According to the Intergovernmental Panel on Climate Change (IPCC) Fifth Assessment Report, the Mediterranean is facing the possibility of experiencing warmer and drier conditions in the near future [3]. Therefore, most countries in the southeastern Mediterranean region are likely to address water scarcity issues [4–6]. It is widely acknowledged that the ongoing changing patterns of precipitation and temperature have revealed an increased frequency of drought and the growing impact of aridity severity [7–9]. The aforementioned climatic conditions have profound effects on streamflow [10,11], lake levels [12], crop production [13], desertification [14], and forest growth [5,15,16].

There is a fundamental difference in terminology between aridity and drought, as noted by the World Meteorological Organization (WMO) [17]. Aridity is a long–term (climatic) phenomenon and concerns a period of at least 30 years. It is usually defined by low–average precipitation, available water, or humidity, and is a permanent climatic feature. Drought, by comparison, is a short–term (meteorological) phenomenon, which can vary from year to year. Assessing and monitoring the drought and aridity conditions prevailing in a certain area is a key element in climate research and provide important information for sustainable ecosystem management.

Numerous indices have been proposed to quantify the degree of dryness in a particular location [18,19]. The most common indicators are defined by the ratio of precipitation (P) to potential evapotranspiration (PET). Thereby, the indices express the water availability concept in a single number. The Thornthwaite method is the easiest way of calculating PET in data–scarce areas [20]. However, PET is preferably estimated, in terms of accuracy, using the Penman formula [21].

There has been a large amount of research on the trends and fluctuations of precipitation in Europe and the Mediterranean regions. Regarding annual precipitation in Europe, a positive trend in the north and a negative trend in the south have been noticed throughout the last century [22]. These findings revealed a remarkable negative tendency in eastern Mediterranean, whereas a positive trend was detected in central and northern Europe [23]. Similar results have been demonstrated for western Europe, indicating a downward trend in annual precipitation, prolongation of the dry period, and increase in the number of rainy days [24]. Moreover, the negative trend of annual precipitation was found to be statistically significant (95% confidence level) for the majority of Mediterranean regions, except north Africa, southern Italy, and western Iberian Peninsula, where a slight positive trend was recorded [25]. Concerning the Greek territory, a decrease in precipitation was recorded at the end of the 19th century, followed by a positive trend in the first three decades of the 20th century, then a smaller fluctuation, before returning to a decreasing trend [26–28]. In terms of PET tendency, fewer studies have been undertaken compared to those examining precipitation [29]. In Greece, trend analysis of weekly time series of PET shows an increasing trend in spring, and particularly in summer, no trend in winter (almost stable), and downward trends in winter; annual trends are also increasing [30]. Finally, in a survey conducted by Myronidis and Theophanous [31], the trend analysis of the P/PET ratio in the South Aegean shows insignificant negative trends, indicating that drought phenomena have slightly intensified [32].

Valuable Mediterranean–type ecosystems, which have remarkable forest species biodiversity, will be highly influenced by drought conditions. Moreover, the uneven precipitation and temperature distribution makes the assessment of climate tendency and variability a necessity for dealing with water scarcity problems. This is particularly important considering that dry conditions will continue to dominate and intensify in the forthcoming period [33,34]. Additionally, the spatial and temporal trends of precipitation and temperature conditions have been identified as the main factor affecting tree growth [16,35,36]. Thus, it is crucial to evaluate trends in the climatic variables in mountainous forested areas. Seasonal patterns can be identified and recorded to improve forest and water management. To the best of the author's knowledge, there are limited studies in Greece analyzing the seasonal trend of precipitation (P) and potential evapotranspiration (PET) based on long–term time series from mountainous meteorological stations [5,37,38]. In addition, these studies examine a short period of meteorological records. This is due to the difficulties of installation and maintenance of meteorological instruments, especially at the high elevations of the mountainous regions. In most cases, studies of the Greek territory take into account stations, including those in the network of The Hellenic National Meteorologic Service, which are usually installed at elevations less than 100 m a.s.l. Thus, the results cannot be exploited for high–elevation forestland and for areas with complex terrain.

The object of this study was to investigate temporal variability and detect trends in drought conditions in two different types of forest ecosystems using long–term time series meteorological data from mountainous meteorological stations. For this purpose, the P/PET ratio was used as a proxy indicator for the evaluation of drought conditions at different timescales (annual/seasonal). The Mann–Kendal and Sen's slope methods were applied in order to evaluate the significance and magnitude of the tendency, and to identify the time of abrupt changes.

2. Materials and Methods

2.1. Study Area

The study was conducted in two selected forest ecosystems in Greece with different forest types (broadleaved, coniferous) and climatic conditions (Figure 1). These are the university forests of Taxiarchis and Pertouli. The management rights of these two forests have been assigned to the Aristotle University of Thessaloniki since 1934 for research and educational purposes.

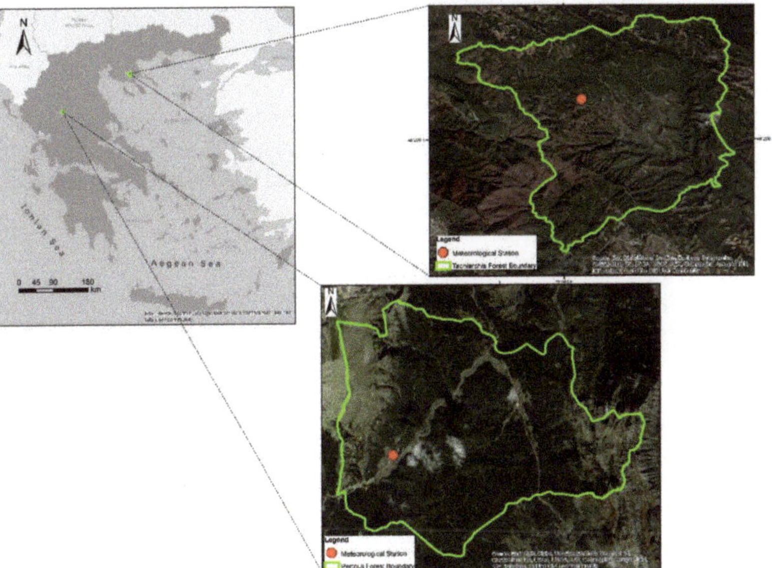

Figure 1. Location map of the selected forest ecosystems.

The Pertouli University Forest is located in the mountainous range of Pindus (Trikala Prefecture, Central Greece) and covers an area of 3290 ha. It consists mainly of pure fir stands (*Abies borisii regis*) and the elevation ranges between 1100 and 2073 m above sea level (a.s.l). The climate is transitional, Mediterranean–Mid–European, with cold, rainy winters and warm, dry summers. The average annual precipitation is 1542.2 mm and the average mean annual air temperature is 8.9 °C according to the existing meteorological station (1180 m a.s.l), which has operated since 1961 and is located at latitude 39°32′35.8″ and longitude 21°28′8.5″. Specifically, the total precipitation was found to be 525.7 mm in winter, 432.6 mm in autumn, 337.9 mm in spring, and 128.1 mm in summer. The average air temperature was found to be 0.8 °C in winter, 9.9 °C in autumn, 7.7 °C in spring, and

17.9 °C in summer. Furthermore, the region is a member of the European environmental protection network, Natura 2000, and is particularly designated as a Site of Community Importance (SCI) with the code GR1440002, namely: *"Kerketio Oros (Koziakas)"*.

The Taxiarchis University Forest is located in the mountainous range of Cholomontas (Chalkidiki Prefecture, Northern Greece) and covers an area of approximately 5800 ha. The area is a coppice oak forest (*Quercus frainetto Ten.*) and the elevation ranges from 320 to 1200 m a.s.l. The climate is characterized as sub–humid Mediterranean, expressed by short periods of drought, hot summers, and mild winters. A meteorological station (860 m a.s.l) has operated in the area since 1974, and is located at latitude 40°25′54.7″ and longitude 23°30′20.1″. According to the station long–term data the average annual precipitation is 808.3 mm and the average mean annual air temperature is 11.5 °C. Precipitation and temperature data show significant seasonal variations. Specifically, the total precipitation was found to be 242.9 mm in winter, 206.7 mm in autumn, 196.4 mm in spring, and 162.3 mm in summer. The average air temperature was found to be 2.7 °C in winter, 12.3 °C in autumn, 10.3 °C in spring, and 20.7 °C in summer. The area is also a part of the Natura 2000 network, including the SCI site with code GR1270001 known as *"Oros Cholomontas"*.

Monthly precipitation and temperature data, for the common operating period (1974–2016) of the two meteorological stations, were collected and processed. These stations are operated by the Aristotle University Forest Administration and Management Fund. The data series are complete without missing values. Moreover, the equipment and observation techniques were common and remained consistent throughout the examined period. The stations' data were checked for homogeneity on a monthly time step using the double mass method and two parametric statistical tests (Student's t–test and chi–squared test), as detailed by WMO [39]. The results verified that the data are homogeneous and can be further processed.

2.2. Potential Evapotranspiration (PET)

In the current approach, a parsimonious regional parametric evapotranspiration model (PET) based on a simplification of the Penman–Monteith formula [29,30] was applied, as proposed by Tegos et al. [40,41]:

$$PET = \frac{\alpha R_a + b}{1 - cT} \tag{1}$$

where R_a (kJ m^{-2}) is the extraterrestrial shortwave radiation calculated without measurements and T (°C) is the air temperature. The model has three additional parameters a (kg kJ^{-1}), b (kg m^{-2}), and c (°C) that should be inferred from either measurement or modeled calibration. The a, b, and c factors were set equal to 0.0000976, 0.83, and 0.02, respectively, in the case of the Pertouli area. Respectively, in the case of Taxiarchis, the values of 0.0000485, −0.19, and 0.03 was given for the parameters a, b, and c.

It should be noted that the aforementioned parameterizations have some physical similarities with the original Penman–Monteith approach. This is because the overall energy term (incoming minus outgoing solar radiation) is represented by αR_a, the missing aerodynamic term is represented by b, and ($1-cT$) is a rough approximation of the formula's denominator term. This approach is characterized as a radiation–based method because it uses two explanatory variables: extraterrestrial radiation, R_a; and temperature, T. The model variables are highly connected with location characteristics and can be predicted from global spatial interpolation maps [42].

2.3. Trend Analysis

The non–parametric Mann–Kendall (M–K) test was applied in order to analyze the aridity tendency in the study areas and investigate the statistical significance of the tracked trends at the 95% confidence level. This is the most commonly used trend analysis test in climatological time series, and better fits non–normally distributed data with extreme and missing values, which are frequently encountered in environmental time series [43]. The M–K test was conducted as proposed by Sneyers [44] in order to investigate both annual and seasonal trends, and to detect the turning point using the data series at different timescales. Therefore, for each x_i ($i = 1 \dots$, n) of the time series, the number n_i of lower elements x_j ($x_j < x_i$) preceding it ($j < i$) was calculated, and the test statistic t was given by:

$$t = \sum_i n_i \tag{2}$$

In the absence of any trend (null hypothesis), t is asymptotically normal and independent from the distribution function of the data:

$$u(t) = \frac{(t - \overline{t})}{\sqrt{var(t)}} \tag{3}$$

and has a standard normal distribution, with t and var(t) given by:

$$t = \frac{n(n-1)}{4} \tag{4}$$

$$var(t) = \frac{n(n-1)(2n+5)}{72} \tag{5}$$

Therefore, the null hypothesis can be rejected for high values of $|u(t)|$, with the probability α_1 of rejecting the null hypothesis when it is derived by a standard normal distribution table:

$$\alpha_1 = P(|u| > |u(t)|) \tag{6}$$

The Mann–Kendall test in its sequential form was also utilized for a progressive study of the series. This consists of applying the test to all of the series, beginning with the first term and concluding with the ith term (and the reverse).

In the absence of a trend, the graphical depiction of the direct (u_t) and backward ($u_{t'}$) series created curves that overlapped multiple times. Nevertheless, in the case of a significant trend (5% level $|u_t| > 1.96$), the intersection of the curves made it possible to approximately detect the time of occurrence [44].

Although the M–K test is a useful nonparametric test for temporal trend, it is premised on the assumption that observations are independent. This is due to the fact that a positive or negative autocorrelation may led to overestimation or underestimation of the trend's significance. Thus, before applying the M–K test, all datasets have to be checked for the presence of autocorrelation. Herein, the lag–1 autocorrelation coefficient (r_1) at a 5% significance level was calculated.

Additionally, trend magnitudes were calculated using the Theil–Sen technique (TSA) [45,46]. This method is based on slope and is often referred to as Sen's slope. It is preferred to linear regression because it minimizes the influence of outliers on the slope [47].

3. Results

The precipitation and potential evapotranspiration present great inter–annual and intra–annual variability in the selected forest ecosystems of the study areas. The lag–1 autocorrelation coefficient was computed for the examined variables and it was found that the r_1 value does not exceed the confidence interval bounds. Thus, the latter variables are considered to be serially independent and therefore the M–K test can be applied.

Trend analysis of the annual and seasonal precipitations indicated negative trends in Pertouli, except during autumn, whereas positive trends were identified in Taxiarchis in each examined period. However, these trends were found to be statistically insignificant (at the 0.05 significance level). The trend magnitudes were −12.7, −15.8, +11.6, −3.6, and −2.9 mm per decade for annual, winter, autumn, spring, and summer precipitation, respectively, of the Pertouli station. In contrast, the positive trends in Taxiarchis were found to be 45.9, 16.8, 27.2, 1.4, and 0.9 mm for annual, winter, autumn, spring, and summer precipitations, respectively. The graphical representations of the M–K test for the precipitation time series of the Pertouli and Taxiarchis forests are shown in Figures 2 and 3, respectively.

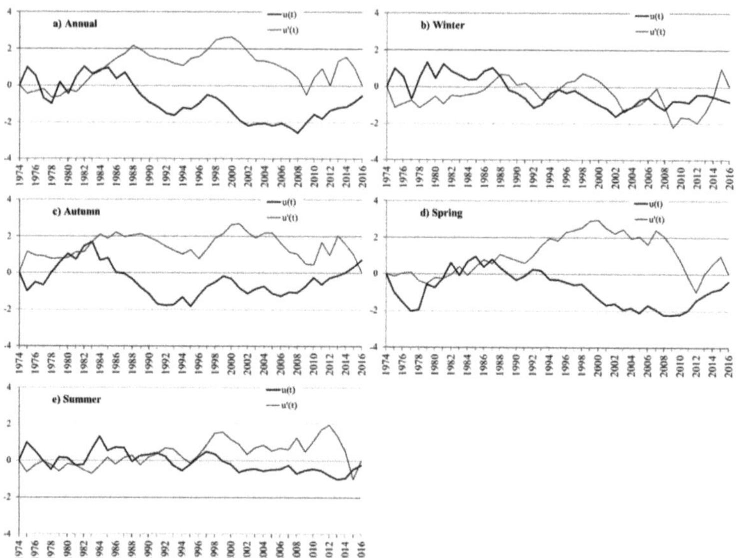

Figure 2. Graphical representation of the series $u_{(t)}$ and the retrograde series $u'_{(t)}$ of the sequential version of the Mann–Kendall test for (**a**) annual, (**b**) winter, (**c**) autumn, (**d**) spring, and (**e**) summer precipitations in the University Forest of Pertouli.

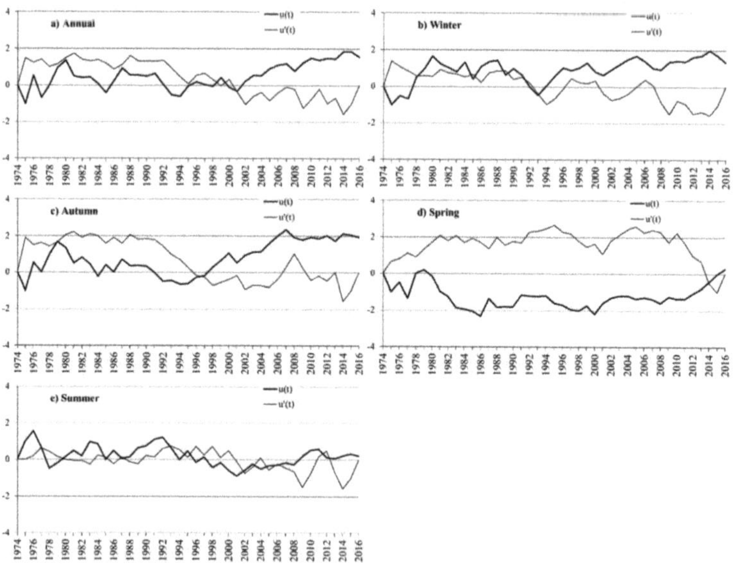

Figure 3. Graphical representation of the series $u_{(t)}$ and the retrograde series $u'_{(t)}$ of the sequential version of the Mann–Kendall test for (**a**) annual, (**b**) winter, (**c**) autumn, (**d**) spring, and (**e**) summer precipitations in the University Forest of Taxiarchis.

Regarding the potential evapotranspiration (PET), upward trends were detected in Pertouli. These trends were found to be statistically significant (at the 0.05 significance level), except in winter and autumn. The timing of this abrupt change was identified as 1992 for the annual and summer time series, and 1997 for summer. Moreover, the increase per decade was 6.8 mm (annual), 0.3 mm (winter), 0.8 mm (autumn), 1.7 mm (spring), and 3.9 mm (summer). The same pattern was also found in Taxiarchis. The trend analysis shows statistically significant trends (at the 0.05 significance level) in all seasons. The abrupt change was identified in 1997 for the annual and summer precipitation, 1996 for winter, 2005 for autumn, and 1999 for spring. The graphs representing the results of the M–K test for the potential evapotranspiration (PET) in the study areas are shown in Figures 4 and 5. Furthermore, the magnitude of the increase in PET in Taxiarchis was higher than that in Pertouli; specifically, it was estimated to be 24.2, 1.3, 2.4, 2.5, and 17.8 mm per decade for annual, winter, autumn, spring, and summer precipitations, respectively. Subsequently, in order to track the effect of precipitation (P) and potential evapotranspiration (PET) on drought conditions, the same analysis was performed for the P/PET ratio.

The results of annual and seasonal ratio P/PET estimation indicate that excess water is available (P > PET) in most cases, with the exception of summer and the annual values in the Taxiarchis forest. This is typical for mountainous areas with complex topography, and uneven precipitation and temperature regime [5,36]. The analytical results of the average P/PET values for the reference period (1974–2016) are presented in Table 1.

Table 1. Analysis of the P/PET ratio for the selected forest ecosystems.

Study Area	Annual	Winter	Autumn	Spring	Summer
Pertouli	1.6	7.5	2.7	1.3	0.3
Taxiarchis	0.9	2.6	1.2	1.0	0.4

Seasonal variability shows winter as the most humid season in the two forest ecosystems, followed by autumn and spring. The variation in P/PET values in the two forests ecosystems can be justified by the difference in their elevation.

Results of the graphical representation of the M–K test for the Pertouli forest are illustrated in Figure 6. It is shown that an insignificant downward trend was exhibited for annual, winter, spring, and summer drought conditions, whereas autumn showed a slight upward trend.

The results from the application of the M–K test based on the data of Taxiarchis forest are presented in Figure 7. In contrast to the Pertouli forest, insignificant positive trends were noted in all seasons, except summer, where a slight downward trend was found. The timing of the abrupt change related to drought was not a consideration for either forest, because the upward and downward trends were not found to be statistically significant [32].

Additionally, concerning the output of the Sen's slope estimation, the magnitude of the trends was determined. In Pertouli, the magnitudes of the P/PET values were equal to −0.04, −0.236, +0.13, −0.013, and −0.01 per decade for annual, winter, autumn, spring, and summer, respectively. On the contrary, the magnitudes of the P/PET values per decade were +0.02 for annual, +0.15 in winter, +0.11 in autumn, +0.01 in spring, and +0.02 in summer for the Taxiarchis forest. These trends are considered negligible, because the increase was found to be greater than 5% per decade compared to the corresponding average AI value for winter and autumn. Moreover, a shift in the climate zone classification did not occur in any of the seasons. The detailed results from the implemented method, showing the percentage of influence on the average annual and seasonal AI, are presented in Table 2.

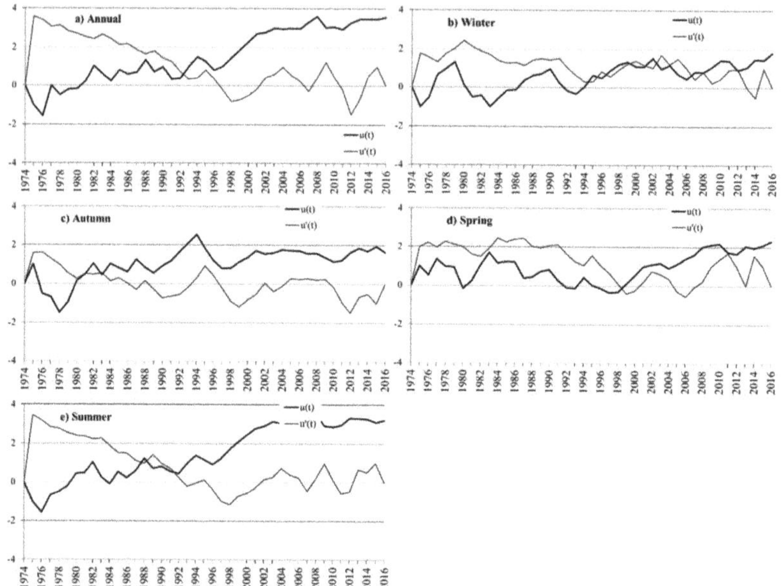

Figure 4. Graphical representation of the series $u_{(t)}$ and the retrograde series $u'_{(t)}$ of the sequential version of the Mann–Kendall test for (**a**) annual, (**b**) winter, (**c**) autumn, (**d**) spring, and (**e**) summer PET in the University Forest of Pertouli.

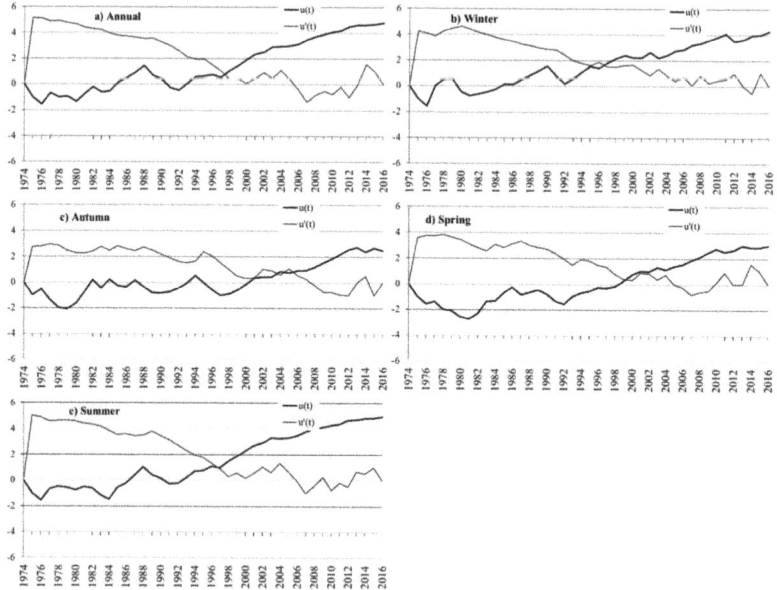

Figure 5. Graphical representation of the series $u_{(t)}$ and the retrograde series $u'_{(t)}$ of the sequential version of the Mann–Kendall test for (**a**) annual, (**b**) winter, (**c**) autumn, (**d**) spring, and (**e**) summer PET in the University Forest of Taxiarchis.

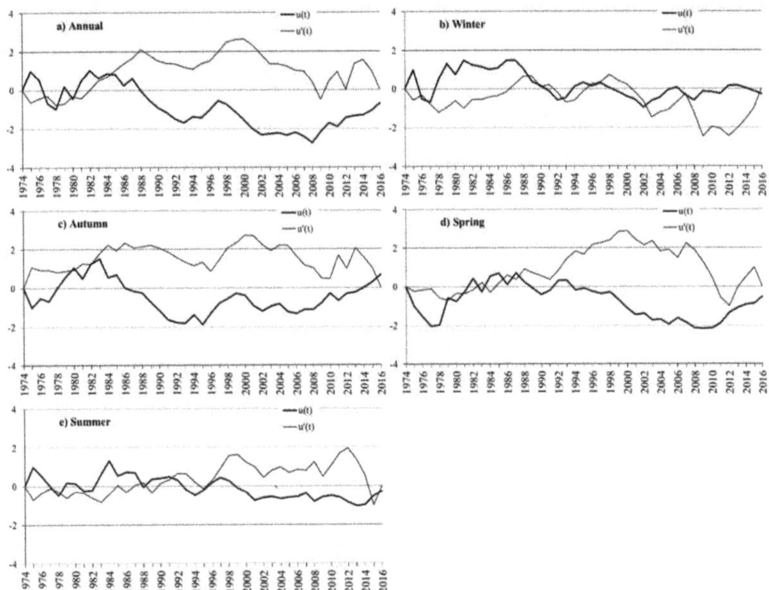

Figure 6. Graphical representation of the series $u_{(t)}$ and the retrograde series $u'_{(t)}$ of the sequential version of the Mann–Kendall test for (**a**) annual, (**b**) winter, (**c**) autumn, (**d**) spring, and (**e**) summer P/PET in the University Forest of Pertouli.

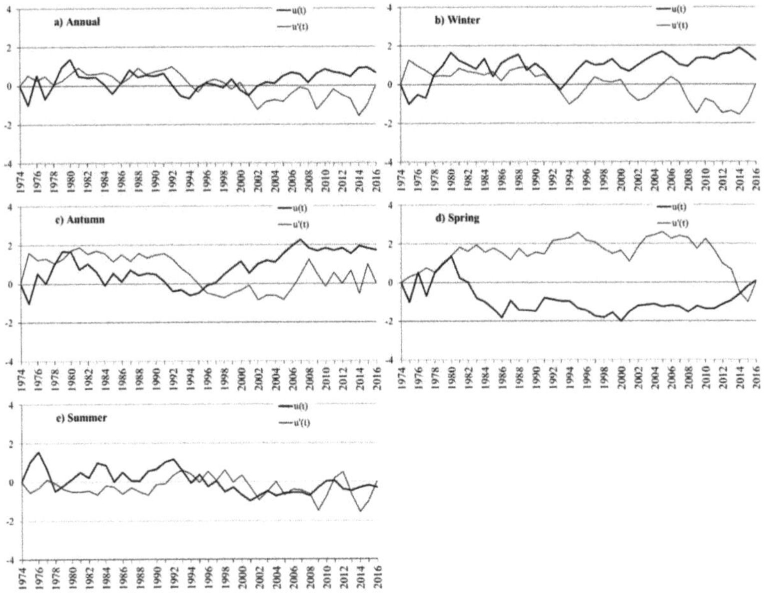

Figure 7. Graphical representation of the series $u_{(t)}$ and the retrograde series $u'_{(t)}$ of the sequential version of the Mann–Kendall test for (**a**) annual, (**b**) winter, (**c**) autumn, (**d**) spring, and (**e**) summer P/PET in the University Forest of Taxiarchis.

Table 2. Sen slope β values and trend magnitude (%) per decade for the P/PET ratio.

Temporal Coverage	Pertouli		Taxiarchis	
	Slope β	% Magnitude per Decade	Slope β	% Magnitude per Decade
Annual	−0.004	−2.5	0.002	+2.2
Winter	−0.026	−3.5	0.016	+6.2
Autumn	0.013	+4.8	0.011	+9.2
Spring	−0.001	−0.8	0.001	+1.0
Summer	−0.001	+3.3	−0.002	−5.0

4. Discussion

In recent decades, concerns have increased about climate variability and changes. Therefore, in the context of climate crisis, the tendency in climatic variables has been examined in many studies. However, in the majority of these cases, data from lowland meteorological stations have been analyzed. In forest research, the usage of these meteorological data does not fulfill the conditions for studying the weather and climate conditions in mountainous areas of forest growth.

Reviewing the results of related studies in Europe and Greece revealed both similarities and differences. Caloiero et al. [23] evaluated the tendency of precipitation in continental Europe and the Mediterranean basin for the period 1901–2009 using gridded reanalysis data. The results for annual precipitation demonstrated a negative trend of about −20 mm per decade in eastern Mediterranean (including Greece), −16 mm per decade in North Africa, and a positive trend (+20 mm per decade) in central and north Europe. Moreover, focusing on the results of the aforementioned study [23] for the seasonal precipitation in the Mediterranean, the magnitudes of the negative trends were approximately −10, −6, −4, and −2 mm per decade for winter, autumn, spring, and summer, respectively. Additionally, Philandras et al. [25] highlighted statistically significant negative trends in the Mediterranean region concerning annual precipitation data from stations for the period 1951–2010. Specifically, the magnitude of the trend was found to be 36.1 mm per decade for West Mediterranean, 30.1 mm per decade for central Mediterranean, and 15 mm per decade in east Mediterranean. The results of these two studies [23,25] showed a relative convergence. Several authors mentioned that the reported trends are related to the teleconnection pattern, which has been extensively described in the literature [47,48]. In comparison with our results, it seems that Pertouli station followed the general negative trend in the Mediterranean region, except during winter, whereas Taxiarchis showed an increase in all seasons. Nevertheless, the percentage decrease in precipitation is negligible because the study area is considered one of the rainiest regions in Greece.

Precipitation in Greece has undergone significant changes over the last century, as represented in a national study taking into account stations with different conditions [5]. An average increase of about 4% was reported by comparing the periods 1900–1929 and 1930–1960. On the contrary, a considerable decrease (−15%) was found between the periods 1961–1997 and 1900–1929 [5]. Moreover, in the previous referenced work [5], the findings indicated that the respective average changes in PET were negligible. Furthermore, another study [49] showed a decreasing trend in PET in annual and warm periods (1979–1999) based on data from low–elevation stations. However, in the present study, the trends in PET were found to be statistically significant (95% confidence level). Despite the differences in PET among the examined forest ecosystems, the variability and fluctuation of drought are considered to be almost stable at different timescales. A more recent study [30] analyzed meteorological and hydrological variables from 17 stations located across Greece during 1961–2006. The results generally indicated statistically significant downward trends for seasonal and hydrological year precipitation. Regarding PET, negative trends were found only in winter.

The assessment of aridity and drought is challenging and may have a profound effect in forest ecosystems. A large–scale survey conducted for the entire Greek territory mentioned a progressive shift from the humid aridity class, which previously characterized the wider area of Greece, towards the sub–humid and semi–arid class, especially in southeastern Greece [50]. It should be noted, however, that for a number of areas, mostly located in the western coastal region, the AI class (humid) remains unchanged [5]. In addition, the drought conditions were evaluated and a tool was presented for water resource management [36] and recently linked to drinking water availability [32].

Although aridity is important in forests, custom modification of the basic indices has been undertaken [51,52]. It has also been stated that the climate variability affects forest growth, species adjustability [16], and shifts in the tree line [53]. Nevertheless, trends in AI values and drought in mountainous ecosystems have not been extensively examined.

5. Conclusions

In the present study, the precipitation and potential evapotranspiration, and their relationship, were analyzed in two forest ecosystems with different climate conditions and forest types. Variability and trend analysis of these parameters was performed at seasonal and annual timescales using the M–K and Sen's slope statistical tests. Potential evapotranspiration (PET) is a key parameter in the hydrological cycle and, related to precipitation, has been used as a proxy indicator of drought. In contrast to several studies that commonly used Thornthwaite's PET method, a parametric model for PET based on a simplified formulation of the Penman–Monteith equation was applied in the proposed approach.

The results indicated that humid conditions prevail in both forest areas and that dry conditions occur in summer. The examined parameters present significant variability between seasons, following the Mediterranean climate pattern. The trend analysis showed that the reported upward and downward trends in AI are, in general, statistically insignificant, and the magnitude of the trend is considered negligible.

Southern Europe will be significantly affected by climate changes due to an increase in temperature, a decrease in precipitation, and more frequent extreme weather events. These changes will have an impact on vegetation phenology and the flowering season, in addition to the growth rate and productivity of forest ecosystems. Forest species will face increased competition, whereas those species more resistant to dry–thermal conditions will survive. These conditions are expected to impact the regeneration potential and diversity of forests. The monitoring of precipitation and potential evapotranspiration is recommended for forest ecosystems to contribute to adaptive and resilient forest management in the context of climate change.

A target of future research can be the evaluation of different evapotranspiration methods in forest environments, because PET is considered the most important component of the hydrologic cycle. Moreover, the drought conditions under future climate conditions and different emission scenarios can be evaluated with the use of high–resolution regional climate models (RCMs). The evaluation of the aridity index (AI) and investigating variability by dividing the time series into two climatic periods in the two forests is a particularly interesting subject of study.

Author Contributions: Conceptualization, S.S and V.A.; methodology, S.S.; investigation, S.S.; data curation, V.A.; writing—original draft preparation, S.S and V.A.; writing—review and editing, S.S. and V.A; visualization, S.S. and V.A. All authors have read and agreed to the published version of the manuscript.

Funding: This research received no external funding.

Acknowledgments: The authors would like to express their gratitude to the Aristotle University Forest Administration and Management Fund for providing the meteorological data. Also, the authors express their sincere gratitude to the editor, and the two anonymous reviewers for sharing their valuable and constructive comments. Their feedback resulted in many significant improvements in this work.

Conflicts of Interest: The authors declare no conflict of interest.

References

1. Giorgi, F.; Lionello, P. Climate change projections for the Mediterranean region. *Glob. Planet. Chang.* **2008**, *63*, 90–104. [CrossRef]
2. Diffenbaugh, N.S.; Giorgi, F. Climate change hotspots in the CMIP5 global climate model ensemble. *Clim. Chang.* **2012**, *114*, 813–822. [CrossRef] [PubMed]
3. IPCC. Climate change 2013: The physical science basis. In *Contribution of Working Group I to the Fifth Assessment Report of the Intergovernmental Panel on Climate Change*; Cambridge University Press: Cambridge, UK, 2013.
4. Sofroniou, A.; Bishop, S. Water scarcity in Cyprus: A review and call for integrated policy. *Water* **2014**, *6*, 2898–2928. [CrossRef]
5. Tsiros, I.X.; Nastos, P.; Proutsos, N.D.; Tsaousidis, A. Variability of the aridity index and related drought parameters in Greece using climatological data over the last century (1900–1997). *Atmos. Res.* **2020**, *240*, 104914. [CrossRef]
6. Tramblay, Y.; Koutroulis, A.; Samaniego, L.; Vicente-Serrano, S.M.; Volaire, F.; Boone, A.; Le Page, M.; Llasat, M.C.; Albergel, C.; Burak, S.; et al. Challenges for drought assessment in the Mediterranean region under future climate scenarios. *Earth-Sci. Rev.* **2020**, *210*, 103348. [CrossRef]
7. Spinoni, J.; Naumann, G.; Vogt, J.; Barbosa, P. European drought climatologies and trends based on a multi-indicator approach. *Glob. Planet. Chang.* **2015**, *127*, 50–57. [CrossRef]
8. Caloiero, T.; Veltri, S.; Caloiero, P.; Frustaci, F. Drought analysis in Europe and in the Mediterranean basin using the standardized precipitation index. *Water* **2018**, *10*, 1043. [CrossRef]
9. Greve, P.; Roderick, M.L.; Ukkola, A.M.; Wada, Y. The aridity index under global warming. *Environ. Res. Lett.* **2019**, *14*, 124006. [CrossRef]
10. Myronidis, D.; Ioannou, K.; Fotakis, D.; Dörflinger, G. Streamflow and hydrological drought trend analysis and forecasting in Cyprus. *Water Resour. Manag.* **2018**, *32*, 1759–1776. [CrossRef]
11. Sarailidis, G.; Vasiliades, L.; Loukas, A. Analysis of streamflow droughts using fixed and variable thresholds. *Hydrol. Process.* **2019**, *33*, 414–431. [CrossRef]
12. Myronidis, D.; Stathis, D.; Ioannou, K.; Fotakis, D. An integration of statistics temporal methods to track the effect of drought in a shallow Mediterranean Lake. *Water Resour. Manag.* **2012**, *26*, 4587–4605. [CrossRef]
13. Tigkas, D.; Vangelis, H.; Tsakiris, G. Drought characterisation based on an agriculture-oriented standardised precipitation index. *Theor. Appl. Climatol.* **2019**, *135*, 1435–1447. [CrossRef]
14. Sidiropoulos, P.; Dalezios, N.R.; Loukas, A.; Mylopoulos, N.; Spiliotopoulos, M.; Faraslis, I.N.; Alpanakis, N.; Sakellariou, S. Quantitative classification of desertification severity for degraded aquifer based on remotely sensed drought assessment. *Hydrology* **2021**, *8*, 47. [CrossRef]
15. Fyllas, N.M.; Christopoulou, A.; Galanidis, A.; Michelaki, C.Z.; Giannakopoulos, C.; Dimitrakopoulos, P.G.; Arianoutsou, M.; Gloor, M. Predicting species dominance shifts across elevation gradients in mountain forests in Greece under a warmer and drier climate. *Reg. Environ. Chang.* **2017**, *17*, 1165–1177. [CrossRef]
16. Proutsos, N.; Tigkas, D. Growth Response of Endemic Black Pine Trees to Meteorological Variations and Drought Episodes in a Mediterranean Region. *Atmosphere* **2020**, *11*, 554. [CrossRef]
17. World Meteorological Organization (WMO). *Drought*; SER-5; WMO: Geneva, Switzerland, 1975.
18. Maliva, R.; Missimer, T. Aridity and drought. In *Arid Lands Water Evaluation and Management*; Springer: Berlin/Heidelberg, Germany, 2012; Volume 1, pp. 21–39.
19. Myronidis, D.; Fotakis, D.; Ioannou, K.; Sgouropoulou, K. Comparison of ten notable meteorological drought indices on tracking the effect of drought on streamflow. *Hydrol. Sci. J.* **2018**, *63*, 2005–2019. [CrossRef]
20. Thornthwaite, C.W. An approach toward a rational classification of climate. *Geogr. Rev.* **1948**, *38*, 55–94. [CrossRef]
21. Penman, H.L. Natural evaporation from open water, bare soil and grass. *Proc. R. Soc. Lond. Ser. A Math. Phys. Sci.* **1948**, *193*, 120–145.
22. Fleig, A.K.; Tallaksen, L.M.; James, P.; Hisdal, H.; Stahl, K. Attribution of European precipitation and temperature trends to changes in synoptic circulation. *Hydrol. Earth Syst. Sci.* **2015**, *19*, 3093–3107. [CrossRef]
23. Caloiero, T.; Caloiero, P.; Frustaci, F. Long-term precipitation trend analysis in Europe and in the Mediterranean basin. *Water Environ. J.* **2018**, *32*, 433–445. [CrossRef]
24. Valdes-Abellan, J.; Pardo, M.A.; Tenza-Abril, A.J. Observed precipitation trend changes in the western Mediterranean region. *Int. J. Climatol.* **2017**, *37*, 1285–1296. [CrossRef]
25. Philandras, C.M.; Nastos, P.T.; Kapsomenakis, J.; Douvis, K.C.; Tselioudis, G.; Zerefos, C.S. Long term precipitation trends and variability within the Mediterranean region. *Nat. Hazards Earth Syst. Sci.* **2011**, *11*, 3235–3250. [CrossRef]
26. Katsoulis, B.D.; Kambetzidis, H.D. Analysis of the long-term precipitation series at Athens, Greece. *Clim. Chang.* **1989**, *14*, 263–290. [CrossRef]
27. Amanatidis, G.T.; Repapis, C.C.; Paliatsos, A.G. Precipitation trends and periodicities in Greece. *Fresenius Environ. Bull.* **1997**, *6*, 314–319.
28. Feidas, H.; Noulopoulou, C.; Makrogiannis, T.; Bora-Senta, E. Trend analysis of precipitation time series in Greece and their relationship with circulation using surface and satellite data: 1955–2001. *Theor. Appl. Climatol.* **2007**, *87*, 155–177. [CrossRef]
29. Chaouche, K.; Neppel, L.; Dieulin, C.; Pujol, N.; Ladouche, B.; Martin, E.; Salas, D.; Caballero, Y. Analyses of precipitation, temperature and evapotranspiration in a French Mediterranean region in the context of climate change. *Comptes. Rendus Geosci.* **2010**, *342*, 234–243. [CrossRef]

30. Mavromatis, T.; Stathis, D. Response of the water balance in Greece to temperature and precipitation trends. *Theor. Appl. Climatol.* **2011**, *104*, 13–24. [CrossRef]

31. Cheval, S.; Dumitrescu, A.; Birsan, M.V. Variability of the aridity in the South-Eastern Europe over 1961–2050. *Catena* **2017**, *151*, 74–86. [CrossRef]

32. Myronidis, D.; Theofanous, N. Changes in climatic patterns and tourism and their concomitant effect on drinking water transfers into the region of South Aegean, Greece. *Stoch. Environ. Res. Risk Assess.* **2021**, *35*, 1725–1739. [CrossRef] [PubMed]

33. Zarch, M.A.A.; Sivakumar, B.; Malekinezhad, H.; Sharma, A. Future aridity under conditions of global climate change. *J. Hydrol.* **2017**, *554*, 451–469. [CrossRef]

34. Fyllas, N.M.; Christopoulou, A.; Galanidis, A.; Michelaki, C.Z.; Dimitrakopoulos, P.G.; Fulé, P.Z.; Arianoutsou, M. Tree growth-climate relationships in a forest-plot network on Mediterranean mountains. *Sci. Total Environ.* **2017**, *598*, 393–403. [CrossRef] [PubMed]

35. Camarero, J.J.; Sánchez-Salguero, R.; Ribas, M.; Touchan, R.; Andreu-Hayles, L.; Dorado-Liñán, I.; Meko, D.M.; Gutiérrez, E. Biogeographic, atmospheric and climatic factors influencing tree growth in Mediterranean Aleppo pine forests. *Forests* **2020**, *11*, 736. [CrossRef]

36. Paparrizos, S.; Maris, F.; Matzarakis, A. Integrated analysis and mapping of aridity over Greek areas with different climate conditions. *Glob. NEST J.* **2016**, *18*, 131–145.

37. Proutsos, N.D.; Tsiros, I.X.; Nastos, P.; Tsaousidis, A. A note on some uncertainties associated with Thornthwaite's aridity index introduced by using different potential evapotranspiration methods. *Atmos. Res.* **2021**, *260*, 105727. [CrossRef]

38. World Meteorological Organization (WMO). *Guidelines on the Quality Control of Surface Climatological Data*; WCP-85; WMO: Geneva, Switzerland, 1986.

39. Monteith, J.L. Evaporation and the environment in the state and movement of water in living organisms. In Proceedings of the Society for Experimental Biology, Symposium No. 19, Cambridge, UK, 1 January 1965; Cambridge University Press: Cambridge, UK, 1965; pp. 205–234.

40. Tegos, A.; Malamos, N.; Efstratiadis, A.; Tsoukalas, I.; Karanasios, A.; Koutsoyiannis, D. Parametric modelling of potential evapotranspiration: A global survey. *Water* **2017**, *9*, 795. [CrossRef]

41. Tegos, A.; Malamos, N.; Koutsoyiannis, D. A parsimonious regional parametric evapotranspiration model based on a simplification of the Penman–Monteith formula. *J. Hydrol.* **2015**, *524*, 708–717. [CrossRef]

42. Goossens, C.; Berger, A. Annual and seasonal climatic variations over the northern hemisphere and Europe during the last century. *Ann. Geophys.* **1986**, *4*, 385–400.

43. Sneyers, R. *Technical Note No 143 on the Statistical Analysis of Series of Observations*; World Meteorological Organization: Geneva, Switzerland, 1990.

44. Thiel, H. A rank-invariant method of linear and polynomial regression analysis, part 3. In *Advanced Studies in Theoretical and Applied Econometrics*; Springer: Berlin, Germany, 1992; pp. 345–381.

45. Sen, P.K. Estimates of the regression coefficients based on Kendall's tau. *J. Am. Stat. Assoc.* **1968**, *63*, 1379–1389. [CrossRef]

46. Hirsch, R.M.; Slack, J.R.; Smith, R.A. Techniques of trend analysis for monthly water quality data. *Water Resour. Res.* **1982**, *18*, 107–121. [CrossRef]

47. Hatzaki, M.; Flocas, H.A.; Asimakopoulos, D.N.; Maheras, P. The eastern Mediterranean teleconnection pattern: Identification and definition. *Int. J. Climatol. A J. R. Meteorol. Soc.* **2007**, *27*, 727–737. [CrossRef]

48. Mathbout, S.; Lopez-Bustins, J.A.; Royé, D.; Martin-Vide, J.; Benhamrouche, A. Spatiotemporal variability of daily precipitation concentration and its relationship to teleconnection patterns over the Mediterranean during 1975–2015. *Int. J. Climatol.* **2020**, *40*, 1435–1455. [CrossRef]

49. Kitsara, G.; Floros, J.; Papaioannou, G.; Kerkides, P. Spatial and Temporal Analysis of Pan Evaporation in Greece. In Proceedings of the 7th International Conference of European Water Resources Association (EWRA): Resources Conservation and Risk Reduction under Climatic Instability, Limassol, Cyprus, 25 June 2009.

50. Nastos, P.T.; Politi, N.; Kapsomenakis, J. Spatial and temporal variability of the Aridity Index in Greece. *Atmos. Res.* **2013**, *119*, 140–152. [CrossRef]

51. Führer, E.; Horváth, L.; Jagodics, A.; Machon, A.; Szabados, I. Application of a new aridity index in Hungarian forestry practice. *Időjárás* **2011**, *115*, 205–216.

52. Gavrilov, M.B.; Lukić, T.; Janc, N.; Basarin, B.; Marković, S.B. Forestry Aridity Index in Vojvodina, North Serbia. *Open Geosci.* **2019**, *11*, 367–377. [CrossRef]

53. Zindros, A.; Radoglou, K.; Milios, E.; Kitikidou, K. Tree Line Shift in the Olympus Mountain (Greece) and Climate Change. *Forests* **2020**, *11*, 985. [CrossRef]

Article

Evaluation of Evaporation from Water Reservoirs in Local Conditions at Czech Republic

Eva Mellišová [1,2,*,†], Adam Vizina [1,2,†], Martin Hanel [1,2,†], Petr Pavlík [1,2] and Petra Šuhájková [1,2]

1 Department of Hydrology, T. G. Masaryk Water Research Institute, Podbabská 2582/30, 160 00 Prague, Czech Republic; adam.vizina@vuv.cz (A.V.); hanel@fzp.czu.cz (M.H.); petr.pavlik@vuv.cz (P.P.); petra.suhajkova@vuv.cz (P.Š.)
2 Faculty of Environmental Sciences, Czech University of Life Sciences Prague, Kamýcká 129, 165 00 Prague, Czech Republic
* Correspondence: eva.melisova@vuv.cz; Tel.: +420-220-197-263
† These authors contributed equally to this work.

Abstract: Evaporation is an important factor in the overall hydrological balance. It is usually derived as the difference between runoff, precipitation and the change in water storage in a catchment. The magnitude of actual evaporation is determined by the quantity of available water and heavily influenced by climatic and meteorological factors. Currently, there are statistical methods such as linear regression, random forest regression or machine learning methods to calculate evaporation. However, in order to derive these relationships, it is necessary to have observations of evaporation from evaporation stations. In the present study, the statistical methods of linear regression and random forest regression were used to calculate evaporation, with part of the models being designed manually and the other part using stepwise regression. Observed data from 24 evaporation stations and ERA5-Land climate reanalysis data were used to create the regression models. The proposed regression formulas were tested on 33 water reservoirs. The results show that manual regression is a more appropriate method for calculating evaporation than stepwise regression, with the caveat that it is more time consuming. The difference between linear and random forest regression is the variance of the data; random forest regression is better able to fit the observed data. On the other hand, the interpretation of the result for linear regression is simpler. The study introduced that the use of reanalyzed data, ERA5-Land products using the random forest regression method is suitable for the calculation of evaporation from water reservoirs in the conditions of the Czech Republic.

Keywords: evaporation; water reservoir; regression; observed data; ERA5-Land data; R language

Citation: Melišová, E.; Vizina, A.; Hanel, M.; Pavlík, P.; Šuhájková P. Evaluation of Evaporation from Water Reservoirs in Local Conditions Evaluation of Evaporation from Water Reservoirs in Local Conditions at Czech Republic. *Hydrology* 2021, 8, 153. https://doi.org/10.3390/hydrology8040153

Academic Editors: Aristoteles Tegos and Nikolaos Malamos

Received: 31 August 2021
Accepted: 29 September 2021
Published: 12 Ocotober 2021

Publisher's Note: MDPI stays neutral with regard to jurisdictional claims in published maps and institutional affiliations.

1. Introduction

Water management, changes in natural water regime and sustainable landscape became an important topic of social discussions and policies not only in the Czech Republic, but also around the world [1]. It is clear that global and local climatic conditions are changing and will have an impact on the water management sector and therefore they should be given the highest attention. The evaporation in the Czech Republic also changes [2].

However, not only the climatic conditions change, but also the technology and knowledge that can be used in water management and specifically in hydrology. With the rapid development of remote sensing tools through recent decades an onset of easy-to-use high quality products supplied both professionals and public in water resources.

In recent years, there has been a significant development in the supply of information from remote sensing of the Earth utilizable in water management, not only for the professional public [3–5]. Another option is, for example, the use of globally available climate reanalyses or other available data sources. Despite the development of data availability and modelling tools, a question arises: How significant is the impact of the ongoing climate change on hydrological balance components and the consequent impact on water management [6]?

The hydrological balance is tied to rainfall-runoff processes, which are driven by climatic, geographical and geomorphological factors. The climatic factors include meteorological factors affecting the evaporation and evapotranspiration from the catchment, such as: precipitation, humidity, soil moisture, evaporation, air temperature, wind speed and direction and atmospheric pressure [7].

Recently, a number of studies pointed out that evapotranspiration significantly affects the hydrological balance. The key role of evapotranspiration in hydrological balance was the subject of many recent studies, e.g., [8–11]. And it is nowadays widely recognized, that on the most of the Earth's surface evaporation plays crucial role in the hydrological cycle.

The study [12] illustrates the impacts of climate change on the water cycle, which may impact from total evaporation, precipitation, atmospheric humidity and horizontal moisture transport at the global scale.

There are many methods to calculate evaporation, which can be calculated from free water, from the soil surface or from vegetation over a period of time. The evaluation of evaporation can be done by direct methods namely measurement or by indirect methods: empirical methods, remote sensing of the Earth on regional or global scales [13,14], the use of models that are classified as fully physically-based combination models, semi-physically based models or black-box models [15].

The total evaporation can be divided into actual, potential or reference evapotranspiration. The potential evaporation can be determined by empirical relationships or by measurement, the empirical relationships may differ in the input data or in the time step [8,16]. The calculation of the reference evapotranspiration is defined according to the FAO methodology, with the reference area being devided in [17].

The studies [18,19] evaluated evapotranspiration calculated on the base of empirical equations, which were divided into categories: mass-transfer, radiation based method and temperature-based method. The best equations from each category were then selected and compared based on the FAO and Penman–Monteith equations [20].

The estimation of reference evapotranspiration was used in the study [21], where the Penman–Monteith temperature-based equation achieved the best rating for the evaluation of reference evapotranspiration because it preserves the physical philosophy of the Penman–Monteith equation method. The method was applied at a global scale using the Köppen climate classification system with respect to the world dataset under different climate conditions. Calculation of reference evapotranspiration based on indirect methods can provide acceptable results when direct measurements of are not available [15].

Since most of the empirical formulas are based on geographical location, it is straightforward that the empirical calculation of evapotranspiration is not the same for different regions, due to the different climatic conditions [17]. National standards, legislation and expertise also takes place resulting that different methods are preferred in different countries, e.g., Netherlands—Makkink's method [22], Slovakia—Budyko's method [23], Bulgaria—Delibaltov–Hristov–Tsonev method [24].

The Penman–Monteith method is considered the sole standard for calculating reference evapotranspiration. The inputs to the equation are climatic data, solar radiation, air temperature, humidity and wind speed. It allows the calculation of evapotranspiration at different times of the year and in different regions, yet a precise measurement at a given location can easily replace the simplified Penman–Monteith equation [17].

Other methods of calculating evapotranspiration include the use of empirical relationships, e.g., the relationship between observed evaporation from evaporation stations and meteorological quantities, these relationships can be calculated either linearly or non-linearly [25,26] using machine learning algorithms [27,28] linear regression or random forest regression.

The assessment of long-term climate variables can be based on time series. The time series is a sequence of measurements recorded over time, that can be analysed using, e.g., Least-Squares Spectral Analysis, Least-Squares Wavelet Analysis, Least-Squares Cross Wavelet Analysis [29].

Other methods for evaluation may include parametric and non-parametric trend tests, which are used in machine learning [30,31]. The parametric method (logistic regression, linear discriminant analysis and simple neural network) use a fixed number of parameters to build models, require fewer variables and the result may be affected by outliers. The non-paramtric method (the Mann–Kendall, Spearman's Rho and k-Nearest neighbors) use a flexible number of parameters, both variable and attribute can be used in the models, the result is not affected by outliers.

In this paper, we explore the relationships for the calculation of evaporation from water surface in the Czech Republic using reanalyzed climate data and the constructed linear models (LM) and random forest models (RFM) for the calculation of evaporation. Evaporation estimated from the derived models was compared with observed evaporation from evaporation stations. Finally, the derived relationships were applied to the selected water reservoirs.

Specifically, we aim to answer the following questions: Which statistical method for calculating evaporation achieves better linear regression or random forest regression? How many variables are important for determining the formula for calculating evaporation? How important is the geomorphological information (elevation and location) for calculating evaporation using linear and non-linear models? The main objective of the evaporation estimation from water surface was to derive a universal relationship for the whole territory of the Czech Republic.

This paper is structured as follows: Section 2 introduces the area of interest and input data. The statistical method for evaluation evaporation with respect to goodness-of-fit (GOF) is evaluated in the R environment [32] and described Section 3. The results and discussion are in Section 4 along with a detailed evaluation of the goodness-of-fit (GOF) regression for evaporation stations and subsequently for water reservoirs. The paper is concluded in Section 5.

2. Study Area and Data

The study area is defined by the state border of the Czech Republic. Within the region (51°03′ N to 48°33′ N latitude and 12°05′ E to E 18° 51′ longitude) the long–term (1981–2010) mean annual precipitation totals at 709.5 mm, mean annual air temperature is 7.9 °C, mean runoff is 205.5 mm [33] and long-term runoff coefficient is thus 0.29 (29% of precipitation totals runs off).

Figure 1 describe long-term temperature, evaporation trend at evaporation station Hlasivo. The Hlasivo evaporation measuring station provides a consistent time series of 58 years, the evaporation values are measured by a 20 $[m^{-2}]$ benchmark evaporator. Other observed variables are: air temperature at 2 m [°C], water surface temperature in the evaporimeter [°C], relative humidity [%], global solar radiation $[W \cdot m^{-2}]$ and wind speed at 2 m $[m \cdot s^{-1}]$ [34].

Figure 2 shows the selected 24 evaporation stations and 33 water reservoirs. The evaporation stations were assigned to water reservoirs based on the Quitt classification and the elevation [35]. The elevation differences between the evaporation stations and water reservoirs do not exceed 100 m a.s.l. The Quitt classification divides the Czech Republic into three climatic regions (cold, moderately warm and warm regions), with an evaporation station in the same climatic region always assigned to a reservoir. The observed evaporation from the evaporation station was recorded between 1957 and 2019 (most evaporation station was recorded from 2005).

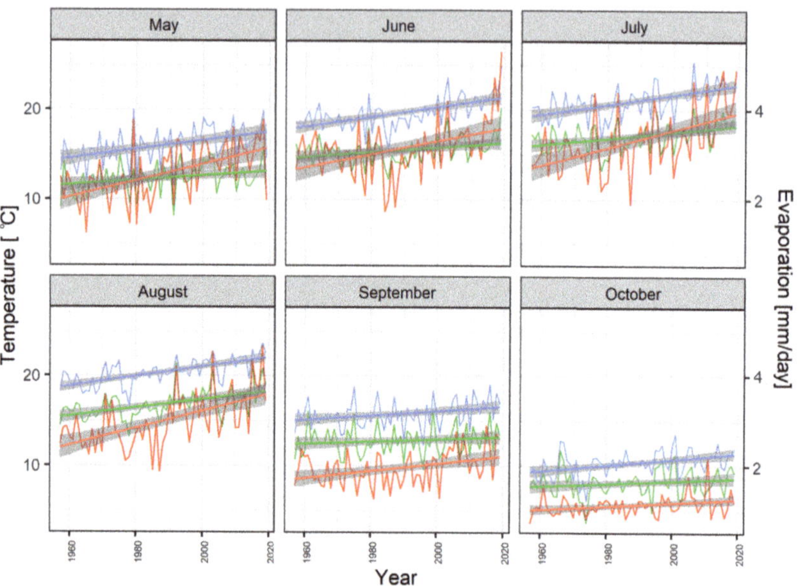

Figure 1. Mean monthly temperature (green lines), water temperature (blue line) and evaporation (red line) at Hlasivo evaporation station for period 1957–2019.

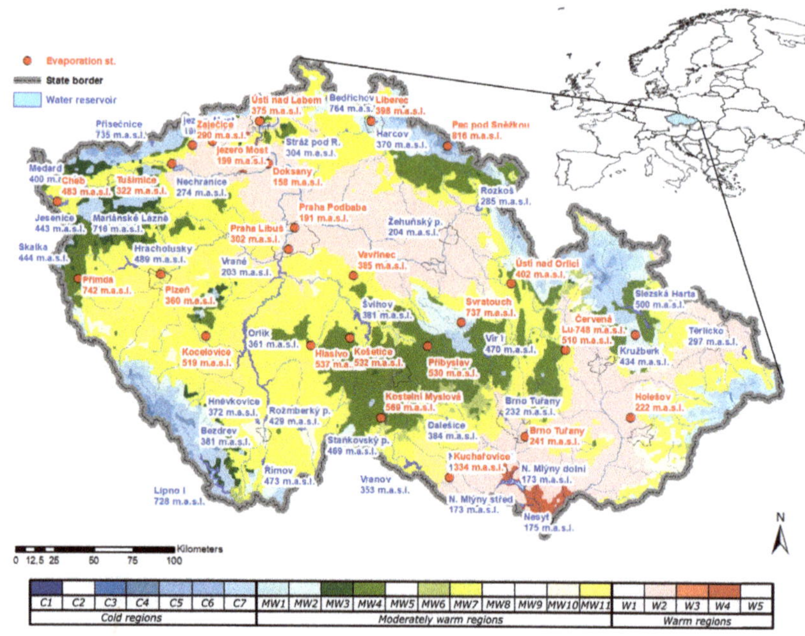

Figure 2. Study area: Czech Republic with climatic regions [35], blue color: water reservoirs with altitude of dam and red color: evaporation stations with altitude.

The data from the evaporimeter (EWM) were provided by the Czech Hydrometeorological Institute, Palivový kombinát Ústí, state-owned enterprise. The T. G. Masaryk Water

Research Institute, public research institution (TGM WRI, p.r.i.) provided data from the floating evaporator and data from the evaporation station Hlasivo.

Observed data from evaporation stations were aggregated into monthly step which were then used to evaluate evaporation from water reservoir surface, because the measured daily values are affected by random error [36]. The observed evaporation (may–october) is from 459 [mm·year^{-1}] (Pec pod Sněžkou) to 760 [mm·year^{-1}] (Holešov), mean evaporation (from evaporation stations) 627 [mm·year^{-1}], minimum mean daily rate (1.38 [mm·year^{-1}]) is in October and maximum mean daily rate is in July (4.53 [mm·year^{-1}]), with maximum in June 2017 (5 stations exceeded 6.5 [mm·year^{-1}]).

The relationships for calculating evaporation from the water surface were developed using linear and nonlinear regression. Measured evaporation from evaporation stations serves as the dependent variable. ERA5-Land climate reanalysis data were used for the non-dependent variables from 1981 to 2019.

Climate Reanalysis

The purpose of the reanalysis is to provide an estimate of quantities describing atmospheric, climatic and hydropedological processes and behavior of oceans with global coverage and relatively high spatiotemporal resolution.

The reanalyses are outputs of various models, usually including a hydrological, atmospheric and ocean model and a model of the Earth's surface. The advantage is the provision of multidimensional spatially complete and coherent information about the global circulation and hydroclimatic quantities. Climate reanalyses are generated in a similar manner as in numerical weather forecasts, where the prediction models based on the development of the climate system from the initial state are used to predict the future state of the atmosphere. The initial state of the climate is a key input into the forecast determining the future development of the model simulation. Data assimilation is used to estimate the initial state that best matches the available data, while taking into account model errors. The climate reanalysis is performed as the only version of data assimilation that includes the use of the prediction model [37].

The reanalysis uses a combination of modeled data and observed data with emphasis on the laws of physics. The data are stored in the ECMWF archive and copied to the COPERNICUS Climate Data Store archive, from where they are freely downloadable using the CDS catalog or the CDS API application in the GRIB or NetCDF format.

The data was downloaded in NetCDF, which is a common format in drought or flood forecasting [38]. The spatial resolution is 0.1° × 0.1°, which represents approximately a grid of 9 km × 9 km.

The data set consisting of 2 m temperature [K], skin temperature [K], 2 m dew-point temperature [K], 10 m v-component of wind [m·s^{-1}], surface pressure [Pa], surface net solar radiation [J·m^{-2}] was selected to calculate evaporation from water reservoir. Temperature units [K] were converted to [°C] and energy units from [J·m^{-2}] were converted to [W·m^{-2}], values divided by the accumulation time expressed in seconds. Relative humidity [%] was calculated using the August–Roche–Magnus approximation [39], where the input data were dew point and temperature.

In the final dataset preparation, evaporation data from evaporation stations and geomorphological variables (elevation, latitude and longitude) were added to the reanalyzed data.

3. Methods

Statistical methods of linear and non-linear regression (random forest regression) were used to evaluate evaporation from the water reservoir. In this case, the main objective of the regression is to determine the best fit between the observed values from the evaporation stations and the variables from the ERA5-Land project. The resulting linear and non-linear models were evaluated based on cross validation and goodness-of-fit (GOF): mean absolute error (MAE), root mean squared error (RMSE), coefficient of determination (R^2)

and relative error (RERR). This section introduced building linear and non-linear models and their evaluating.

3.1. Linear Regression

Linear regression attempts to explain the values of a dependent variable through other quantities. In our case, an attempt was made to explain the dependent variable (evaporation value or evaporation rate from evaporimeter stations and evaporimeters EWM) using other variables (air temperature, surface temperature, wind, surface net solar radiation, dew point, pressure, latitude and altitude, evaporation type distribution) using 18 linear models created by sequential testing manually (8 models) and on the basis of stepwise regression (10 models).

The first set of models (built manually) was evaluated based on the Akaike Information Criterion (AIC) [40] value and the QQ plot was used for visual diagnostics [41]. The value of AIC is the sum of two terms, the first is proportional to the logarithm of the residual sum of squares, the second term is proportional to the complexity of the model (number of its members). When building the LM models, it can often happen that more independent variables reduce the sum of residues (improves the fit of the model with the observed data), however, this can result in an overfitted LM. The part of the AIC that penalizes the complexity of the model should prevent overfitting. When verifying the assumptions of the model (normality of residues), the QQ plot of residues can help. In the QQ plot of residues, two quantiles are plotted against each other—the theoretical quantile from distribution and the quantile with the actual residues of the model.

The second part of the linear models was developed using stepwise regression. R-packages caret, leaps, MASS [42] were used for this regression. The R-package caret uses the principle of machine learning and the R-package leaps are used to calculate the stepwise regression. The R-package caret has a function $train()$, which allows the implementation of a sequential selection of predictors, where the linear regression selection is selected:

- leapBackward,
- leapForward,
- leapSeq.

In this work, a method with backward selection was selected. The hyperparameter nvmax corresponds to the maximum number of predictors that are included in the model. In this work, 11 predictors were used. Furthermore, it is also possible to set the parameters of the validation method, in this work it was cross validation with 500 iterations.

3.2. Random Forest Regression

Random forest (RF) is a combined learning method for classification and regression that creates multiple decision trees during learning and then outputs the modus (most frequent value) of the classes returned by each tree to form a regression forest. The resulting regression function is defined as a weighted average of the regression functions of multiple trees. Regression forests belong to the so-called committee or ensemble methods, the main idea of which is to combine several separate models into a single ensemble. Thus, it uses the so-called collective decision [26,43]. A random forest consists of a set of trees $T_1,\ldots,T_N$ whose classification or regression functions can be expressed as follows:

$$h(\mathbf{X},\mathbf{O}_1),\ldots,h(\mathbf{X},\mathbf{O}_N), \tag{1}$$

where h is a function, $\mathbf{X}$ is a predictor and $\mathbf{O}_1,\ldots,\mathbf{O}_N$ are independent equally distributed random vectors. For the Random forests method, binary trees of type CART [44] are used. Similar to the creation of individual trees or other calibrations, a split into test and training sets is used. The R-package randomforest [27] was used in this work.

Random forest is an approach to build predictive models for both classification and regression tasks. It is a way to combine poorer performing baseline models to obtain better predictive models. Due to their simple nature, low assumptions and high performance,

RF models have been widely used in machine learning. The term "forest" refers to a set of decision trees that are themselves "weak" classifiers. A regression forest does not have the same predictive power as a stand-alone regression tree. If a single tree splits into a single criterion, it is very sensitive to changes. RF models classify variables based on their importance to achieve the best RF model [45].

3.3. Evaluation of Regression

Cross validation is used to improve the quality of regression models [46]. Depending on the method chosen, cross-validation is divided into k-fold cross validation, k-fold cross validation and leave-one-out. In our experiment, the method selected was leave-one-out validation. The dataset was split into training and test data, with one subset of data removed for the training data. The dataset consisted of the selected stations and in the training data the subset consisted of one sampled station, for a total of 24 stations, resulting in 24 iterations. Goodness-of-fit (GOF) criteria were used for further evaluation.

3.4. Evaluation of Regression by Goodness-of-Fit (GOF)

The linear regression and random forest regression set were evaluated based on their GOF (R^2 [47], RMSE [48], MAE [48] and RERR [49]). This means that we would like to identify the best model which is the most suitable for the calculation of evaporation in the Czech Republic.

(i)   The R^2 is given by:

$$R^2 = 1 - \frac{RSS}{TSS},\tag{2}$$

where RSS is the residual sum of squares and TSS the total sum of squares from predicted evaporation values Ep and of tested data of cross validation Et.

- R^2 indicates a measure of the quality of the regression model and explains the proportion of variability in the dependent variable of the model R^2, it may attain maximum value of 1, which means perfect prediction of the dependent variable. Conversely, value of 0 means that the model provides no information for understanding the dependent variable and is useless.

(ii)   RMSE is given by:

$$RMSE = \sqrt{\left(\frac{1}{n}\right) \sum_{i=1}^{n} (Ep_i - Et_i)^2},\tag{3}$$

where Ep_i is predicted evaporation values i-th case, Et_i tested data from cross validation and N is the total number of simulated values.

- It was used as the standard statistical metric providing a relatively high weight to large errors.

(iii)   MAE is given by:

$$MAE = \frac{1}{n} \sum_{i=1}^{n} |Ep_i - Et_i|,\tag{4}$$

The mean absolute error (MAE) is calculated as the average of the absolute differences between the predicted evaporation values Ep_i and tested data from cross validation Et_i.

- MAE is used to measure how close the predictions or forecasts are to the final results. 'Absolute' means that negative values are converted to positive values. The error is less sensitive to occasional very large errors because it does not amplify calculation errors.

(iv)   RERR is given by:

$$\delta = \frac{|Ep - Et|}{Et}.\tag{5}$$

is the ratio of the absolute error between Ep-predicted evaporation values and Et-tested data to the true of the value Ep-predicted evaporation values.

- It is a dimensionless quantity and can be given in percentages, it may attain both positive and negative values. Relative error can be used to compare quantities with different dimensions.

3.5. Final Evaluation of Regression Models

The last step of the evaluation was to create a scoring matrix and consecutively remove the models from the end (order of removal was from the worst models to the best). In order for the removal to occur, the individual models had to be ranked (from best to worst) or standardized using a GOF. Based on this procedure, the final evaluation was performed.

4. Results and Discussion

In this section, a detailed evaluation of linear and random forest regression with respect to GOF (R^2, RMSE, MAE and RERR) is presented. After evaluating all GOFs, RMSE was selected. Then, the best evaporation formulas are selected from the group of linear models (LM) and random forest models (RFM). Selected models were used to calculate evaporation from the water reservoirs.

4.1. Evaluation of Regression Models

Regression models LM and RFM were evaluated by cross-validation. The cross-validation procedure was as follows:

(i)   In the training data, one station out of 24 stations was selected and validation of the inferred patterns from 23 stations was performed for this station.
(ii)   Validation was carried out successively for all stations and models.
(iii)   For validations, the goodness-of-fit R^2, RMSE, MAE and RERR were calculated.
(iv)   Based on the RMSE, the function of R [32] `rank()` was used, which lists the order of individual values corresponding in an ascending order to the sorted vector. After creating a unique identifier, a matrix was created where the models were on the x-axis and on the stations on the y-axis were. Based on this matrix, the best models were selected.

The models were evaluated and compared using GOF (see Figure 3). The results show that RF models can fit the data better than LM models. RF models are more consistent than LM models for all criterion functions. It can also be seen from the graph and results that for some stations the models do not achieve a good fit.

Outliers (the worst 10% GOF values) are present in all LM models, which also happens in RF models, but on a smaller scale. The outliers corresponded to 70% of the maximum value, thus setting the limit value for selected GOF. Table 1 shows evaporative stations that have exceeded the limit values for the selected GOF.

Table 1. Evaluation of evaporation stations based on GOF.

Goodness-of-Fit	Limit Values	Station above the Limit Value
R^2	<0.2	Hlasivo, reservoir Most
RMSE	<1.5	Hlasivo, reservoir Most, Praha Podbaba
MAE	<1	Hlasivo, reservoir Most, Praha Podbaba, Praha Libuš, Dukovany
RERR	<1.3	Hlasivo, reservoir Most, Praha Podbaba, Praha Libuš

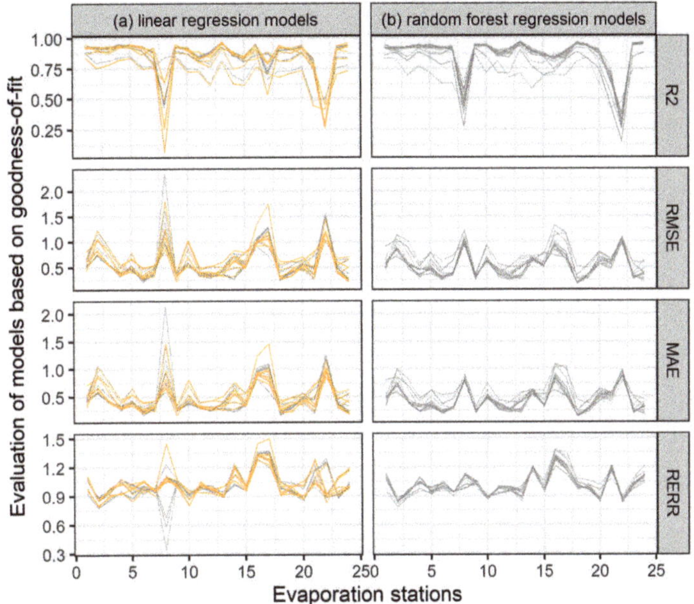

Figure 3. Model evaluation using GOF (R^2, RMSE, MAE, RERR). The lines in the plot represent the LM and RFM models. Part (**a**) linear regression models (18 models) is divided into two parts: orange line: models created manually (8 models), grey part: stepwise regression was used (10 models). Part (**b**) random forest regression models (15 models).

By the method of sorting, using the R function `rank()`, 3 linear and 3 random forest models (RFM) were selected. The selected regression models with average values of GOF are presented in Table 2.

Table 2. Average values of goodness-of-fit of selected models.

ID	R^2	RMSE	MAE	RERR
LM1	**0.85**	**0.58**	**0.47**	**1.04**
LM7	0.84	0.56	0.45	1.01
LM8	0.84	0.56	0.46	1.02
RFM4	0.86	0.51	0.42	1.02
RFM5	0.86	0.51	0.42	1.02
RFM15	**0.86**	**0.51**	**0.42**	**1.01**

The top 3 linear models according to all criterion functions are LM1, LM7 and LM8 and the top 3 RFM are RFM4, RFM5 and RFM15. The selected models are shown in Figure 4, green line represents linear models and blue line represents random forest models. The average value of RMSE for the selected linear models is 0.57, the minimum value is 0.22. The selected RFM had an average RMSE value of 0.51 and a minimum value of 0.18. The models that were designed based on stepwise regression achieved worse results than the models that were built manually based on data analysis. Models designed using manual regression achieved better results; however, some models designed using stepwise regression achieved good results in some cases, with less demanding inputs. The linear models were further supplemented with LM12, which also showed good results and the derived equation is more useful for practice due to its simplicity. All regression models are presented in the Sect. Appendix A in Table A1.

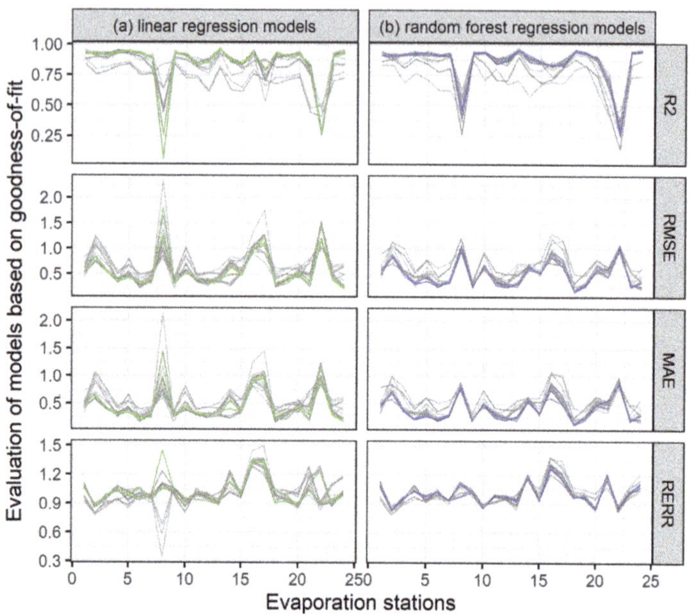

Figure 4. Evaluation of models using GOF, where the best models are selected. Part (**a**) linear regression models, green lines are LM1, LM7 and LM8. Part (**b**) random forest regression models, blue lines are RFM4, RFM5 and RFM15.

Selected regression formulas based on linear models:

$$LM1 = 45.84 + (0.173T \cdot (-0.004R))^{0.0008} - 0.183D - 0.0002P - 0.0002asl - 0.475Y + 0.063X, \tag{6}$$

$$LM7 = 16.97 + 0.082W + (0.235T \cdot (-0.263D))^{0.007} + 0.008R \quad 0.0003asl \quad 0.368Y \mid 0.063X, \tag{7}$$

$$LM8 = 17.33 + 0.055X - 0.367Y - 0.0003asl + (0.2134T * (-0.277D))^{0.009} + 0.008R, \tag{8}$$

$$LM12 = 19.82 + 0.302ST + 0.006R - 0.170D - 0.419Y. \tag{9}$$

Selected variables for best random forest regression models:

$$RFM4 : W, (T * D), R, asl, Y, X, \tag{10}$$

$$RFM5 : X, Y, asl, (T * D), R, \tag{11}$$

$$RFM15 : W, T, ST, R, D, P, H, asl, Y, X. \tag{12}$$

where:

- LM1, LM7, LM8, LM12, RFM4, RFM5, RFM15 are formula identifiers for evaporation,
- T ... temperature (2 m) [°C],
- ST ... surface temperature [°C],
- P ... surface pressure [Pa],
- W ... wind speed [m·s^{-1}],
- R ... surface net solar radiation [W·m^{-2}],
- D ... dew point [°C],
- H ... relative humidity [%],
- asl ... elevation above sea level [m],
- X ... longitude,
- Y ... latitude.

4.2. Model Application to Water Reservoirs

For testing, the best LM models (LM1, LM7, LM8 and LM12), RF models (RFM4, RFM5, RFM15) already described above were applied to selected reservoirs in the Czech Republic for the period May–October. The selection of the May–October period is because the evaporation from the observed data in the winter months is not measured due to freezing. The calculated evaporation values for the water reservoirs are introduced in the Sect. Appendix A in Table A2.

The difference and seasonality in evaporation between the water reservoirs is described in Figure 5 where green lines represent linear models (LM), blue lines random forest models (RFM) and red lines introduce observed data. The average across all data is represented by the bold lines. The mean value from LM models and RFM models over the period (1981–2020) for reservoirs for May–October is 546.54 [mm·year^{-1}] and for RFM is 546.02 [mm·year^{-1}]. The mean value of the evaporation stations (2005–2019) is 497.26 [mm·year^{-1}]. The highest increase in evaporation is observed in the month of July, however, in the summer months (June–August) a significant increase in evaporation can be observed for all models.

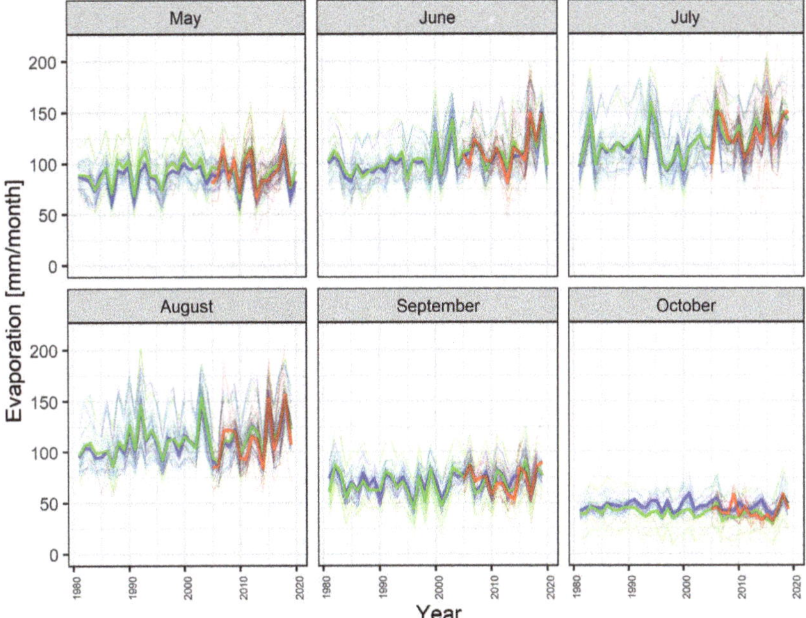

Figure 5. Monthly evaporation for the water reservoirs throughout 1981–2019 (green lines represent linear regression, blue lines represent random forest regression and red lines observed data) and from evaporation stations throughout 2005–2019. Bold lines represent means for water reservoirs and evaporation's stations.

Top models LM1 and RFM12 are compared with elevation for the whole water reservoir. The following Figure 6 shows the relationship between elevation and evaporation, where the green line represents linear regression model and blue line represents random forest model. The elevation of water reservoirs is 170.54–781.91 m a.s.l. The evaporation decreases with the elevation above sea level. Both models are influenced by local conditions because both models have input geographic coordinates and elevation.

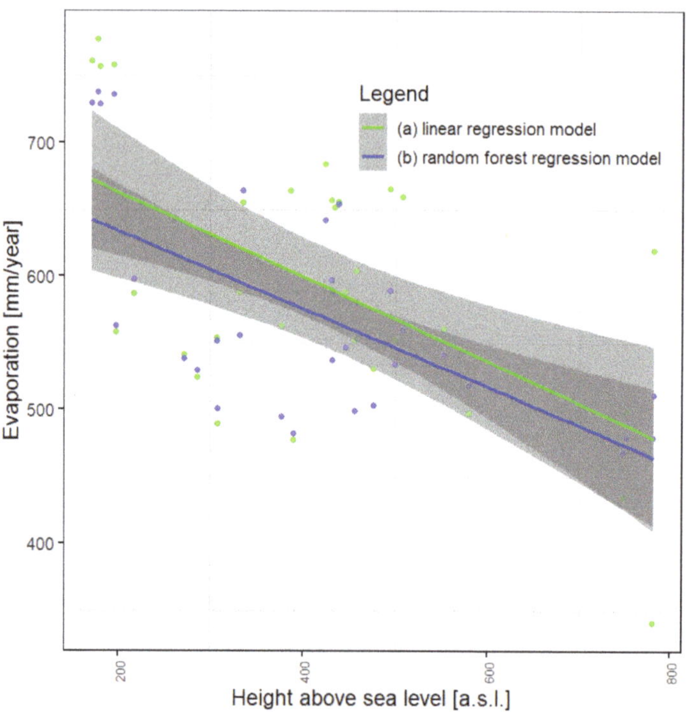

Figure 6. The lines (a) and (b) represent the relationship between yearly total evaporation and elevation based on the derived formulas. The points represent yearly evaporation and altitude for both models.

The results of the study will be implemented to the hydrological model Bilan [50,51] and for assessing climate change studies in the Czech Republic [52].

5. Concluding Remarks

The main objective of the estimation of evaporation from the water reservoirs was to derive a universal relationship for the whole territory of the Czech Republic.

The estimation of evaporation from water reservoirs is complicated because a large number of water reservoirs do not have observed evaporation data. In this work, Quitt's climate classification was used to assign a evaporimeter station that is not near a reservoir to a given reservoir based on climate region and elevation. Within the Czech Republic, the evaporation value from water reservoirs is determined on the basis of a handling order, which is established according to a Czech technical standard which is based on old climatic data and does not deal with climate change. For this reason, the determination of the evaporation from water reservoirs is based on estimation using statistical methods rather than exact measurement.

The ERA5-Land climate reanalysis data were used for derivation and were chosen for their comprehensiveness, availability, high spatial resolution, long time series and advantageous management. Relative humidity was included into the results based on the calculated August–Roche–Magnus approximation. The climate reanalysis data were exported for stations and water reservoirs.

The derivation of the relationship for evaporation was based on the multiple linear regression method, where the values of the dependent variable (evaporation) were sought, based on two or more variables (predictors: air temperature, surface temperature, wind speed, surface net solar radiation, dew point, surface pressure, dew point, altitude, latitude, longitude and calculated humidity). The construction of the models was done (i) manually, where the evaluation was done using the AIC parameter and the quantile–quantile (plot-QQ) was used for visual diagnostics, this method was time consuming, (ii) using stepwise regression, where the predictors are entered sequentially and models from one to X-selected variables were generated, this method is not time consuming. Random forest regression was used to account for non-linear relationships. Linear and random forest regression models were cross-validated and evaluated using criterion functions (R^2, RMSE, MAE and RERR). Finally, 3(+1) LM models and 3 RF models were selected. The models contained a large number of independent variables (6–7), possibly leading to model overfitting and therefore another model was selected which performed best for the RMSE criterion function and is based only on 4 independent variables and is therefore more user friendly.

It turned out that geomorphological information (elevation, location) appeared more in the manually derived models as opposed to models constructed using the stepwise regression method. When comparing linear models (LM) and random forest models (RFM), LM was found to have much more variability in the outcome compared to the RFM. The advantage of RFM is their adaptability, but the subsequent interpretation of the results can be a problem. This has been shown in the design of LM and RFM as well as when applying the proposed models to water reservoirs.

Evaporation values for the period 1981–2019 were calculated for the selected water reservoirs and selected formulas based on ERA5-Land climate reanalysis data.

For the evaluation of evaporation, models from LM and RFM models were used. Among the best models that were evaluated by linear regression, models LM1 from the manual linear regression group and LM12 from the stepwise regression group were used. Model LM1 was selected as the best model among the six predictors. The LM1 model can be replaced by an alternative model LM12 with which also performed satisfactorily with four predictors.

Author Contributions: Conceptualization, A.V., E.M. and M.H.; methodology, A.V., E.M. and M.H.; validation, A.V., E.M., M.H., P.P. and P.Š.; formal analysis, A.V., E.M., M.H. and P.P.; investigation, A.V.; resources, A.V.; data curation, A.V., E.M. and M.H.; writing—original draft preparation, A.V., P.P. and P.Š.; writing—review and editing, A.V., E.M., M.H. and P.P.; visualization, A.V., E.M., M.H. and P.P.; supervision, A.V. and M.H; project administration, P.Š.; funding acquisition, A.V. All authors have read and agreed to the published version of the manuscript.

Funding: This research was funded by T. G. Masaryk Water Research Institute, public research institution (VÚV TGM, v.v.i.) grant number 3600.52.26.

Institutional Review Board Statement: Not applicable.

Informed Consent Statement: Not applicable.

Data Availability Statement: The data underlying the analyses in this paper are available from the lead authors E.M., A.V., M.H. and P.P.

Conflicts of Interest: The authors declare no conflict of interest.

Abbreviations

The following abbreviations are used in this manuscript:

LM	Linear Model
RFM	Random forest Model
GOF	goodness-of-fit
EWM	evaporimeter
ERA5-Land	climate reanalysis product
ECMWF	European Centre for Medium-Range Weather Forecasts
COPERNICUS	European Union's Earth Observation Programme
CDS	Climate Data Store

API   Application Program Interfaces
GRIB   General Regularly-distributed Information in Binary form
NetCDF   Network Common Data Form
MAE   Mean Absolute Error
RMSE   Root Mean Squared Error
R^2 Coefficient of Determination
RERR   Relative Error
AIC   Akaike Information Criterion
QQ   Quantile-Quantile Plot

Appendix A

Table A1. Review of linear models and random forest models, where evaporation E [mm·month^{-1}], temperature T (2 m) [°C], surface temperature ST [°C], wind speed W [m·s^{-1}], surface net solar radiation R [W·m^{-2}], dew point D [°C], relative humidity H [%], surface pressure P [Pa], elevation above sea level asl [m], longitude X, latitude Y.

ID	Manual Linear Regression
LM1	**E ~ (T * R) + D + P + asl + Y**
LM2	E ~ T + (ST^2) + R + Y
LM3	E ~ H + W + T + ST + asl
LM4	E ~ W + T + asl + (X * Y)
LM5	E ~ W + T + (X * Y) + asl
LM6	E ~ W + T + R + (X * Y) + asl
LM7	**E ~ W + (T * D) + R + asl + Y + X**
LM8	**E ~ X + Y + asl + (T * D) + R)**

ID	Stepwise regression
LM9	E ~ ST
LM10	E ~ ST + R
LM11	E ~ ST + R + Y
LM12	**E ~ ST + R + D + Y**
LM13	E ~ ST + R + D + asl + Y
LM14	E ~ ST + R + D + P + asl + Y
LM15	E ~ ST + R + D + P + asl + Y + X
LM16	E ~ W + ST + R + D + P + asl + Y + X
LM17	E ~ W + T + ST + R + D + P + asl + Y + X
LM18	E ~ W + T + ST + R + D + P + H + asl + Y + X

ID	Random forest regression
RFM1	E ~ (T * R) + D + P + asl + Y
RFM2	E ~ H + W + T + ST + asl
RFM3	E ~ W + T + asl + (X * Y)
RFM4	**E ~ W + (T * D) + R + asl + Y + X)**
RFM5	**E ~ X + Y + asl + (T * D) + R**
RFM6	E ~ ST
RFM7	E ~ ST + R
RFM8	E ~ ST + R + Y
RFM9	E ~ ST + R + D + Y
RFM10	E ~ ST + R + D + asl + Y
RFM11	E ~ ST + R + D + P + asl + Y
RFM12	E ~ ST + R + D + P + asl + Y + X
RFM13	E ~ W + ST + R + D + P + asl + Y + X
RFM14	E ~ W + T + ST + R + D + P + asl + Y + X
RFM15	**E ~ W + T + ST + R + D + P + H + asl + Y + X**

Table A2. Average year evaporation [mm·year^{-1}] values for selected reservoirs by model.

	Water Reservoir	LM 1	LM 7	LM 8	LM 12	RF 4	RF 5	**RF 15**
1	Mariánské Lázně	456.57	406.37	407.53	460.76	448.25	441.19	440.32
2	Medard	517.42	419.89	420.58	445.50	464.99	461.08	452.22
3	Nesyt	734.47	738.63	738.35	720.95	693.63	694.18	693.05
4	Rožmberk	621.84	577.57	579.15	581.44	581.58	576.09	555.99
5	Staňkovský pond	629.94	570.13	570.85	586.79	566.87	562.56	550.49
6	Bezdrev	629.52	568.50	568.80	564.59	571.94	567.83	564.59
7	jezero Most	507.42	470.84	468.07	449.22	540.34	546.20	510.86
8	Bedřichov	302.96	379.47	381.93	398.94	442.42	435.92	442.27
9	Brno	612.95	625.37	626.29	606.97	627.63	625.78	609.18
10	Dalešice	617.56	606.14	606.28	607.93	608.56	607.86	597.58
11	Harcov	437.17	405.10	408.10	404.30	452.33	447.10	445.74
12	Hněvkovice	616.43	565.34	566.31	570.69	569.44	565.61	546.01
13	Nové Mlýny dolní	715.26	723.92	724.69	703.13	690.91	692.57	689.31
14	Nové Mlýny horní	716.69	718.96	719.29	696.68	684.96	685.64	682.96
15	Nové Mlýny střed	719.17	718.65	719.25	693.89	685.19	685.79	682.40
16	Orlík	566.86	536.74	536.87	551.54	557.05	557.89	539.68
17	Přísečnice	392.88	364.87	363.56	404.01	448.05	445.16	429.48
18	Rozkoš	483.40	480.53	481.91	453.28	514.23	513.06	489.23
19	Skalka	485.34	421.82	425.00	460.64	472.03	467.27	459.67
20	Slezská Harta	455.25	487.71	483.71	463.05	507.73	511.28	480.48
21	Stráž pod Ralskem	447.56	439.12	441.68	430.19	464.72	459.26	462.60
22	Těrlicko	510.93	547.67	546.09	492.15	538.75	537.80	507.56
23	Vranov	646.21	617.51	616.28	621.22	602.25	599.90	590.68
24	Vrané	547.91	543.23	540.54	543.30	512.07	509.25	520.30
25	Vír I	521.65	513.87	512.16	510.30	528.49	531.97	504.17
26	Hracholusky	543.98	491.44	490.70	513.37	500.41	497.08	491.13
27	Jesenice	507.08	421.91	424.18	456.80	468.90	464.60	456.41
28	Kružberk	508.67	503.72	499.90	480.90	512.99	515.47	489.92
29	Lipno I	586.06	508.19	510.88	541.07	516.80	512.38	476.28
30	Nechranice	493.24	479.92	476.84	465.56	502.13	500.34	490.44
31	Římov	625.90	558.38	559.47	564.04	560.89	557.42	524.97
32	Švihov	546.86	527.77	526.51	533.85	513.68	512.63	506.66
33	Žehuňský pond	536.57	564.87	563.07	548.24	543.89	539.64	551.57

References

1. Beran, A.; Hanel, M.; Nesládková, M.; Vizina, A. Increasing water resources availability under climate change. *Procedia Eng.* **2016**, *162*, 448–454. [CrossRef]
2. Možný, M.; Trnka, M.; Vlach, V.; Vizina, A.; Potopová, V.; Zahradníček, P.; Štepánek, P.; Hájková, L.; Staponites, L.; Žalud, Z. Past (1971–2018) and future (2021–2100) pan evaporation rates in the Czech Republic. *J. Hydrol.* **2020**, *590*, 125390. [CrossRef]
3. Tang, Q.; Gao, H.; Lu, H.; Lettenmaier, D.P. Remote sensing: Hydrology. *Prog. Phys. Geogr.* **2009**, *33*, 490–509. [CrossRef]
4. Schultz, G.A.; Engman, E.T. *Remote Sensing in Hydrology and Water Management*; Springer Science & Business Media: Berlin/Heidelberg, Germany, 2012.
5. Calera, A.; Campos, I.; Osann, A.; D'Urso, G.; Menenti, M. Remote sensing for crop water management: From ET modelling to services for the end users. *Sensors* **2017**, *17*, 1104. [CrossRef]
6. Beran, A.; Hanel, M.; Nesládková, M.; Vizina, A.; Kožín, R.; Vyskoč, P. Climate change impacts on water balance in Western Bohemia and options for adaptation. *Water Supply* **2019**, *19*, 323–335. [CrossRef]
7. Chahine, M.T. The hydrological cycle and its influence on climate. *Nature* **1992**, *359*, 373–380. [CrossRef]
8. Zhao, L.; Xia, J.; Xu, C.; Wang, Z.; Sobkowiak, L.; Long, C. Evapotranspiration estimation methods in hydrological models. *J. Geogr. Sci.* **2013**, *23*, 359–369. [CrossRef]
9. Beran, A.; Kašpárek, L.; Vizina, A.; Šuhájková, P. Water loss by evaporation from free water surface. *Vodohospod. Tech.-Ekon. Inf.* **2019**, *61*, 12–18.
10. Šuhájková, P. Evaporation from VÚV TGM evaporation stations. *Vodohospod. Tech.-Ekon. Inf.* **2020**, *5*, 16–27.
11. Markonis, Y.; Kumar, R.; Hanel, M.; Rakovec, O.; Máca, P.; AghaKouchak, A. The rise of compound warm-season droughts in Europe. *Sci. Adv.* **2021**, *7*, eabb9668. [CrossRef]
12. Rodell, M.; Beaudoing, H.K.; L'ecuyer, T.; Olson, W.S.; Famiglietti, J.S.; Houser, P.R.; Adler, R.; Bosilovich, M.G.; Clayson, C.A.; Chambers, D.; et al. The observed state of the water cycle in the early twenty-first century. *J. Clim.* **2015**, *28*, 8289–8318. [CrossRef]

13. Glenn, E.P.; Huete, A.R.; Nagler, P.L.; Hirschboeck, K.K.; Brown, P. Integrating remote sensing and ground methods to estimate evapotranspiration. *Crit. Rev. Plant Sci.* **2007**, *26*, 139–168. [CrossRef]

14. Hersbach, H.; Bell, B.; Berrisford, P.; Hirahara, S.; Horányi, A.; Muñoz-Sabater, J.; Nicolas, J.; Peubey, C.; Radu, R.; Schepers, D.; others. The ERA5 global reanalysis. *Q. J. R. Meteorol. Soc.* **2020**, *146*, 1999–2049. [CrossRef]

15. Srivastava, A.; Sahoo, B.; Raghuwanshi, N.S.; Singh, R. Evaluation of Variable-Infiltration Capacity Model and MODIS-Terra Satellite-Derived Grid-Scale Evapotranspiration Estimates in a River Basin with Tropical Monsoon-Type Climatology. *J. Irrig. Drain. Eng.* **2017**, *143*, 04017028. [CrossRef]

16. Singh, V.P.; Xu, C.Y. Evaluation and Generalization of 13 Equations for Determining Free Water Evaporation. *Hydrol. Process.* **1997**, *11*, 311–323. [CrossRef]

17. Allen, R.G.; Pereira, L.S.; Raes, D.; Smith, M. *Crop Evapotranspiration-Guidelines for Computing Crop Water Requirements-FAO Irrigation and Drainage Paper 56;* FAO: Rome, Italy, 1998; Volume 300, p. D05109.

18. Xu, C.Y.; Singh, V.P. Evaluation and generalization of radiation-based methods for calculating evaporation. *Hydrol. Process.* **2000**, *14*, 339–349. [CrossRef]

19. Xu, C.Y.; Singh, V.P. Evaluation and generalization of temperature-based methods for calculating evaporation. *Hydrol. Process.* **2001**, *15*, 305–319. [CrossRef]

20. Xu, C.Y.; Singh, V.P. Cross comparison of empirical equations for calculating potential evapotranspiration with data from Switzerland. *Water Resour. Manag.* **2002**, *16*, 197–219. [CrossRef]

21. Almorox, J.; Senatore, A.; Quej, V.H.; Mendicino, G. Worldwide assessment of the Penman–Monteith temperature approach for the estimation of monthly reference evapotranspiration. *Theor. Appl. Climatol.* **2018**, *131*, 693–703. [CrossRef]

22. Jacobs, A.F.G.; De Bruin, H.A.R. Makkink's equation for evapotranspiration applied to unstressed maize. *Hydrol. Process.* **1998**, *12*, 1063–1066. [CrossRef]

23. Budyko, M.I. *Climate and Life;* Academic Press: Cambridge, MA, USA, 1974.

24. Moteva, M.; Kazandjiev, V.; Zhivkov, Z.; Kireva, R.; Mladenova, B.; Matev, A.; Kalaydzhieva, R. Estimation of Crop Evapotranspiration in Bulgaria Climate Conditions. *Sci. Pap.-Ser. A Agron.* **2014**, *57*, 255–263.

25. Beran, A.; Vizina, A. Derivation of regression equations for calculation of evaporation from a free water surface and identification of trends in measured variables in Hlasivo station. *Vodohospod. Tech.-Ekon. Inf.* **2013**, *4*, 4–8.

26. Breiman, L. Ranfom forests. *Mach. Learn.* **2001**, *45*, 5–32. [CrossRef]

27. Ishwaran, H.; Kogalur, U.B. *Fast Unified Random Forests for Survival, Regression, and Classification (RF-SRC);* R Package Version 2.10.1; 2021. Available online: https://rdrr.io/cran/randomForestSRC/man/rfsrc.html (accessed on 25 August 2021).

28. Kumar, M.; Bandyopadhyay, A.; Raghuwanshi, N.S.; Singh, R. Comparative study of conventional and artificial neural network-based ETo estimation models. *Irrig. Sci.* **2008**, *26*, 531–545. [CrossRef]

29. Ghaderpour, E.; Vujadinovic, T.; Hassan, Q.K. Application of the Least-Squares Wavelet software in hydrology: Athabasca River Basin. *J. Hydrol. Reg. Stud.* **2021**, *36*, 100847. [CrossRef]

30. Shadmani, M.; Marofi, S.; Roknian, M. Trend Analysis in Reference Evapotranspiration Using Mann-Kendall and Spearman's Rho Tests in Arid Regions of Iran. *Water Resour. Manag. Int. J. Publ. Eur. Water Resour. Assoc. (EWRA)* **2012**, *26*, 211–224. [CrossRef]

31. Chen, Y.; Guan, Y.S.G.; Zhang, D. Investigating trends in streamflow and precipitation in Huangfuchuan Basin with wavelet analysis and the Mann-Kendall test. *Water* **2016**, *8*, 77. [CrossRef]

32. R Core Team. *R: A Language and Environment for Statistical Computing;* R Foundation for Statistical Computing: Vienna, Austria, 2020.

33. Tolasz, R.; Míková, T.; Valeriánová, A.; Voženílek, V. Atlas Podneb í Česka. Palacký University Olomouc, Czech Hydrometeorological Institute. 2007. Available online: https://www.geoinformatics.upol.cz/publikace/atlas-podnebi-ceska (accessed on 25 August 2021).

34. Šuhájková, P.; Kožín, R.; Beran, A.; Melišová, E.; Vizina, A.; Hanel, M. Update of empirical relationships for calculation of free water surface evaporation based on observation at Hlasivo station. *Vodohospod. Tech.-Ekon. Inf.* **2019**, *61*, 4–11.

35. Vondráková, A.; Vávra, A.; Voženílek, V. Climatic regions of the Czech Republic. *J. Maps* **2013**, *9*, 425–430. [CrossRef]

36. Mrkvičková, M. Evaluation of the measurements at the Hlasivo evaporation station. *Vodohospod. Tech.-Ekon. Inf.* **2007**, *49*, 9–11.

37. Dee, D.P.; Uppala, S.M.; Simmons, A.J.; Berrisford, P.; Poli, P.; Kobayashi, S.; Andrae, U.; Balmaseda, M.A.; Balsamo, G.; Bauer, P.; et al. The ERA-Interim reanalysis: Configuration and performance of the data assimilation system. *Q. J. R. Meteorol. Soc.* **2011**, *137*, 553–597. [CrossRef]

38. Muñoz-Sabater, J. ERA5-Land monthly averaged data from 1981 to present. *Copernic. Clim. Chang. Serv. (C3s) Clim. Data Store (CDS)* **2019**, *10*, 24381.

39. Alduchov, O.A.; Eskridge, R.E. Improved Magnus form approximation of saturation vapor pressure. *J. Appl. Meteorol.* **1996**, *35*, 601–609. [CrossRef]

40. Akaike, H. Factor analysis and AIC. In *Selected Papers of Hirotugu Akaike;* Springer: Berlin/Heidelberg, Germany, 1987; pp. 371–386.

41. Hadley, W. *ggplot2: Elegant Graphics for Data Analysis;* Springer: New York, NY, USA, 2016.

42. Bruce, P.; Bruce, A. Practical Statistics for Data Scientists: 50 Essential Concepts. 2017. Available online: http://www.sthda.com/english/articles/37-model-selection-essentials-in-r/154-stepwise-regression-essentials-in-r/ (accessed on 11 March 2018).

43. Aleh, M. Regresní Stromy. Master's Thesis, Univerzita Karlova v Praze, Staré Město, Czech, 2013.

44. Antipov, E.A.; Pokryshevskaya, E.B. Mass appraisal of residential apartments: An application of Random forest for valuation and a CART-based approach for model diagnostics. *Expert Syst. Appl.* **2012**, *39*, 1772–1778. [CrossRef]

45. Nhu, V.H.; Shahabi, H.; Nohani, E.; Shirzadi, A.; Al-Ansari, N.; Bahrami, S.; Miraki, S.; Geertsema, M.; Nguyen, H. Daily Water Level Prediction of Zrebar Lake (Iran): A Comparison between M5P, Random Forest, Random Tree and Reduced Error Pruning Trees Algorithms. *ISPRS Int. J. Geo-Inf.* **2020**, *9*, 479. [CrossRef]
46. Westerhuis, J.A.; Hoefsloot, H.C.; Smit, S.; Vis, D.J.; Smilde, A.K.; van Velzen, E.J.; van Duijnhoven, J.P.; van Dorsten, F.A. Assessment of PLSDA cross validation. *Metabolomics* **2008**, *4*, 81–89. [CrossRef]
47. Nagelkerke, N.J. A note on a general definition of the coefficient of determination. *Biometrika* **1991**, *78*, 691–692. [CrossRef]
48. Chai, T.; Draxler, R.R. Root mean square error (RMSE) or mean absolute error (MAE)?–Arguments against avoiding RMSE in the literature. *Geosci. Model Dev.* **2014**, *7*, 1247–1250. [CrossRef]
49. Park, H.; Stefanski, L. Relative-error prediction. *Stat. Probab. Lett.* **1998**, *40*, 227–236. [CrossRef]
50. Vizina, A.; Horáček, S.; Hanel, M. Recent Developments of the Bilan Model. *Vodohospod. Tech.-Ekon. Inf.* **2015**, *57*, 7–10.
51. Melišová, E.; Vizina, A.; Staponites, L.R.; Hanel, M. The Role of Hydrological Signatures in Calibration of Conceptual Hydrological Model. *Water* **2020**, *12*, 3401. [CrossRef]
52. Vyskoč, P.; Beran, A.; Peláková, M.; Kožín, R.; Vizina, A. Preservation of drinking water demand from water reservoirs in climate change conditions. *Vodohospod. Tech.-Ekon. Inf.* **2021**, *63*, 4–18.

 *hydrology*

Article

Integrating Drone Technology into an Innovative Agrometeorological Methodology for the Precise and Real-Time Estimation of Crop Water Requirements

Stavros Alexandris [1], Emmanouil Psomiadis [1,*], Nikolaos Proutsos [2], Panos Philippopoulos [3], Ioannis Charalampopoulos [4], George Kakaletris [5], Eleni-Magda Papoutsi [1], Stylianos Vassilakis [1] and Antonios Paraskevopoulos [6]

1   Department of Natural Resources Management and Agricultural Engineering, School of Environment and Agricultural Engineering, Agricultural University of Athens, 75 Iera Odos Str., Votanikos, 11855 Athens, Greece; stalex@aua.gr (S.A.); stud611010@aua.gr (E.-M.P.); stud611059@aua.gr (S.V.)
2   Institute of Mediterranean Forest Ecosystems, Hellenic Agricultural Organization "Demeter", Terma Alkmanos, 11528 Athens, Greece; np@fria.gr
3   Department of Digital Systems, University of the Peloponnese, Kladas, 23100 Sparta, Greece; p.filippopoulos@uop.gr
4   Department of Crop Science, School of Plant Sciences, Agricultural University of Athens, 75 Iera Odos Str., Votanikos, 11855 Athens, Greece; icharalamp@aua.gr
5   Communication & Information Technologies Experts S.A., Omiriou 22, Kessariani, 16122 Athens, Greece; gkakas@cite.gr
6   Department of Agriculture Economy and Veterinary of Trifilia, 29 Eleftheriou Venizelou Str., 24500 Kyparissia, Greece; paraskmd@otenet.gr
*   Correspondence: mpsomiadis@aua.gr; Tel.: +30-2105294078

Citation: Alexandris, S.; Psomiadis, E.; Proutsos, N.; Philippopoulos, P.; Charalampopoulos, I.; Kakaletris, G.; Papoutsi, E.-M.; Vassilakis, S.; Paraskevopoulos, A. Integrating Drone Technology into an Innovative Agrometeorological Methodology for the Precise and Real-Time Estimation of Crop Water Requirements. *Hydrology* **2021**, *8*, 131. https://doi.org/10.3390/hydrology8030131

Academic Editors: Aristoteles Tegos and Nikolaos Malamos

Received: 18 June 2021
Accepted: 25 August 2021
Published: 1 September 2021

Publisher's Note: MDPI stays neutral with regard to jurisdictional claims in published maps and institutional affiliations.

Abstract: Precision agriculture has been at the cutting edge of research during the recent decade, aiming to reduce water consumption and ensure sustainability in agriculture. The proposed methodology was based on the crop water stress index (CWSI) and was applied in Greece within the ongoing research project GreenWaterDrone. The innovative approach combines real spatial data, such as infrared canopy temperature, air temperature, air relative humidity, and thermal infrared image data, taken above the crop field using an aerial micrometeorological station (AMMS) and a thermal (IR) camera installed on an unmanned aerial vehicle (UAV). Following an initial calibration phase, where the ground micrometeorological station (GMMS) was installed in the crop, no equipment needed to be maintained in the field. Aerial and ground measurements were transferred in real time to sophisticated databases and applications over existing mobile networks for further processing and estimation of the actual water requirements of a specific crop at the field level, dynamically alerting/informing local farmers/agronomists of the irrigation necessity and additionally for potential risks concerning their fields. The supported services address farmers', agricultural scientists', and local stakeholders' needs to conform to regional water management and sustainable agriculture policies. As preliminary results of this study, we present indicative original illustrations and data from applying the methodology to assess UAV functionality while aiming to evaluate and standardize all system processes.

Keywords: CWSI; UAV; remote sensing; micrometeorological data; spatial IRT measurements; crop irrigation scheduling and management; infrared radiometer sensors; real-time data analysis

1. Introduction

As a result of the environmental impact factors and the growing demands on food production and consumption, combined with the global market demand to keep merchandise prices low, the modern agricultural industry faces a major challenge [1,2]. There is greater urgency than ever before for producers, farmers, and agronomists around the world to improve the management of their agricultural practices on farms in response to the reduction in their budgets through the optimization in water-saving and farm inputs and, at the same time, maintain high product quality [3]. Future water requirements, combined with limited or bad quality water resources and requirements to adequately meet a growing population's increasing nutritional needs, require improved crop yields and productivity and increased irrigated land, even in the planet's arid areas.

Globally irrigated agriculture delivers 40% of the food production and consumes 70% of the available water [1,4]. Therefore, irrigation scheduling is an important tool for improving the efficient use of irrigation water. Recently, several new sensors and emerging technologies, such as the Internet of things (IoT), have been developed and applied in precision agriculture for the management of available water resources and the biometeorological monitoring of plants and soil [2,5]. These provide significant potential in precision agriculture (PA) and smart farming since it has a direct impact on improving the management of irrigation systems and enabling a long-term increase in productivity [6]. Nowadays, the scientific community focuses not only on estimating the duration of irrigation but also on the accurate calculation of actual needs (actual evapotranspiration (Eta)) regarding the water required by each crop [7]. This trend is contrary to the general tendency of determining the maximum rate of evapotranspiration of the crop (ET_C) in large estimation time steps. Additionally, the effect of the vegetation surface (temperature, physical characteristics, etc.) on the precise estimation of actual evapotranspiration values is also one of the main issues of research [7].

Earth observation data with variations in spectral, spatial, and temporal characteristics have been widely used for vegetation mapping and crop water stress monitoring [8–11]. Traditional remote sensing methods place remote sensors over crop fields or use aircraft and satellites where the appropriate fixed position or the temporal and spatial resolution significantly bounds their utility for PA [12–14]. However, freely available optical-sensors-based imagery is often unsuitable for monitoring crop water stress at the farm level due to the poor revisiting times and coarse spatial resolutions. At the same time, the higher accuracy data are too expensive [15,16]. Furthermore, these data are typically unavailable or not useful on cloudy days [2,17]. Nevertheless, for decades, a lot of studies were undertaken to monitor the crop water stress index using satellite- and aerial-based instruments in several crops, such as those of Rud et al. [18] and Cucho-Padin et al. [19] in potato fields, Veysi et al. [20] and Lebourgeois et al. [21] regarding sugarcane, da Silva et al. [22] regarding melon, Sepulcre-Cantó et al. [23] in olive orchards, and Gutiérrez et al. [24] regarding vineyards.

The synergistic utilization of innovative UAV advanced high-precision thermal cameras and other infrared sensors have enhanced the usefulness of these systems to monitor water statuses [25,26]. As a result, numerous studies have researched several different crops to evaluate the crop water stress conditions, such as those of Zhang et al. [16,27], Yang et al. [28], Gago et al. [12], Gonzalo-Dugo et al. [29], Bellvert et al. [30], Berni et al. [31], Matese et al. [32], and Santesteban et al. [33]. However, the process of calibrating and processing thermal images takes a long time to give the final decision on whether to irrigate. In addition, this process involves more empiricism. Therefore, even a high-precision thermal camera could not be described as a useful tool for everyday use on extensive arable land. The CWSI can be determined using at least two different methodologies: the theoretical model proposed by Jackson et al. [34], which is based on meteorological models, and the empirical model proposed by Idso et al. [35], which has achieved considerably more popularity due to the limited data requirements [8,16]. Due to its simplicity and requiring only three variables to be measured, this practical approach has received much attention in the literature. However, it received criticism concerning its inability to account for temperature changes due to radiation and wind speed [36]. The theoretical method is more complicated because it requires these two additional variables to be measured to evaluate the aerodynamic resistance. Moreover, given the assumed net radiation and wind speed, the theoretical approach to calculating the CWSI promises to improve the estimation of plant water trends [36]. Considering the above, the methodology for estimating the CWSI coefficients is based mainly on the theoretical approach. For this purpose, all the necessary sensor equipment was installed for the CWSI calibration stage for three crops independently.

The GreenWaterDrone (GWD) project proposed a pilot implementation of an innovative and autonomous system (onboard micrometeorological station) that identifies a farmer's real-time irrigation needs through the CWSI to estimate each crop's water status. The direct estimation of the CWSI is based on spatial data that is remote and directly from the field and is based mainly on the infrared temperature of the crop canopy. Even in the recent (2021) literature, but also in previous years, the calculation of the CWSI with direct spatial measurements of the temperature of the crop canopy has never been proposed or adapted [37–43]. Such measurements are constantly achieved indirectly through thermal images. The raw data and thermal images from the infrared camera were also used for comparison and calibration purposes. The use of the spatial measurements of canopy temperature, air temperature, and relative humidity from sensors incorporated into an unmanned aerial vehicle (UAV) may be a global novelty. This study aimed to present the new methodology and the equipment used in the assessment of crop water stress by presenting preliminary results from the initial stages of a pilot implementation in a potato cultivation field at the farm scale.

2. Description of the GWD Concept

2.1. CWSI—Empirical and Theoretical Approaches

CWSI is a measure of the relative transpiration rate that occurs from a crop and is found by using the canopy temperature and the vapor pressure deficit as relevant variables. The latter measure is related to the atmospheric dryness over the crop. The CWSI approach [34] utilizes the energy balance theory that separates the net radiation over the canopy from the sensible heat (thermal content of the air) and latent heat that is consumed for transpiration. The energy balance considerations show how the difference between the canopy and air temperatures ($T_c - T_a$) is related to the vapor pressure deficit (VPD), and the flux density of the net radiation (R_{net}) presents a theoretical basis for the CWSI.

When the plant canopy is fully transpiring, the leaf temperature is some degrees below the overlying air layer temperature and the CWSI is equal to 0 (stomata are closed). Conversely, as the transpiration decreases, the leaf temperature rises and can reach some degrees above the overlying air temperature. When the canopy is no longer transpiring, the CWSI is equal to 1. Because of the scatter in the measured canopy minus air temperature vs. the vapor pressure deficit, the crop does not need to be watered until the CWSI reaches 0.1 to 0.15. For example, for eggplant crops, it could be possible to use values of CWSI between 0.18–0.20 for high and good quality yields. Irmak et al.'s [44] work showed that when the CWSI value exceeds more than 0.22, this resulted in a decreased corn grain yield. Determining the temperature upper line (maximum stress) is not a simple process for any crop. However, this empirical approach has attracted much interest due to its simplicity. It requires measuring the canopy temperature (T_c), air temperature (T_a), and the vapor pressure of the atmosphere's deficit (VPD).

After the empirical approach suggested by Idso et al. [35], a theoretical method for calculating the CWSI was presented. The index's practical estimation is calculated by determining the relative distance between the lower baseline representing the soil water adequacy conditions in the rhizosphere (no stress) and the upper baseline representing the crop's maximum water stress (no transpiration by plants; Figure 1).

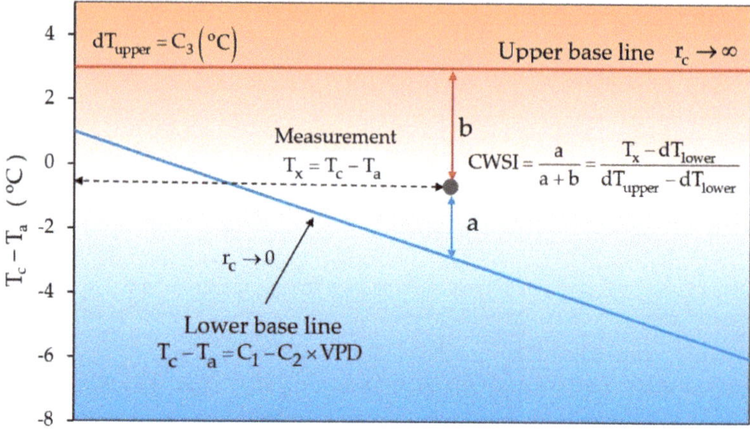

Figure 1. A schematic representation of the CWSI, which equals the relative distance of the measurement between the lower and the upper baselines.

The two lines (blue and red) presented in Figure 1 are suitable for assessing a crop's water status during the warm seasons of the year when irrigation is required. These lines are the result of local calibration and concern the specific crop that is studied each time. The lower baseline is determined for the crop by plotting $(T_c - T_a)$ against the local VPD after a significant watering. A sufficient number of measurements (10 min time step) around midday were taken for the precise estimation of the canopy temperature (T_c). The canopy's temperature sampling must be repeated several times, including measurements under dry and wet atmospheric conditions. This results in a full range of measures representing the lower regression line for a wide range of value pairs (Tc $- T_a$ vs. VPD). The upper baseline could be determined after cutting the plant and taking the canopy temperature the next day, as the plant no longer transpires. The theoretical procedure requires a measurement of the net flux density of the radiation (R_n) and an aerodynamic resistance factor, in addition to the air temperature and relative humidity of the air that is required by the empirical approach [36].

The theoretical development of the CWSI is based on the well-known equation of energy balance of the surface $(R_n - G = H + \lambda E)$, where R_n: net radiation, G: soil heat flux, H: sensible heat flux, and λE: latent heat flux of evaporation. All the terms are in Wm^{-2}. The terms H and λE are expressed in Equations (1) and (2), which are based on two assumptions (see [36,45–47]):

$$H = \rho C_p (T_c - T_a)/r_a \tag{1}$$

$$\lambda E = \left[\rho C_p (e_s - e_a) \right] / \left[\gamma (r_a + r_c) \right] \tag{2}$$

where ρ : density of air (kg m^{-3}), C_p: specific heat moist air (1.013 kJ kg^{-1} $^\circ$C^{-1}), T_c : canopy temperature ($^\circ$C), T_a: air temperature ($^\circ$C), e_s: saturated vapor pressure (kPa), e_a: actual vapor pressure (kPa), γ : psychrometric constant (kPa $^\circ$C^{-1}), r_a: aerodynamic resistance (sm^{-1}), and r_c: bulk surface resistance (s m^{-1}).

Clothier et al. [48] showed that the soil heat flux estimation can be calculated from R_n measurements with reasonable accuracy. The ratio G/R_n for full canopies (more than 45 cm) is nearly constant at 0.1. Thus, the soil heat flux does not contribute significantly to the energy balance above the canopy. Taking $G = 0.1R_{net}$ the equation becomes $I_c R_n = H + \lambda E$, where $I_c = 0.9$ is an interception coefficient.

$$T_c - T_a = \frac{r_a I_c R_n}{\rho C_p} \cdot \frac{\gamma^*}{\Delta + \gamma^*} - \frac{VPD}{\Delta + \gamma^*} \tag{3}$$

where γ^* (kPa $°C^{-1}$) is the modified psychrometric constant and VPD is the vapor pressure deficit.

Equation (4) represents the slope of the saturated vapor pressure relation:

$$\Delta = (e_{s_C} - e_s)/(T_c - T_a) \tag{4}$$

where e_{s_C} and e_s represent the saturated vapor pressure of the canopy temperature and the air temperature, respectively.

The upper limit $T_c - T_a$ is determined when the canopy resistance tends to infinity $(T_c - T_a \to \infty)$ and Equation (3) becomes

$$[T_c - T_a]_{ul} = \frac{r_a I_{cu} R_n}{\rho C_p} \tag{5}$$

The lower limit, determined by setting the bulk surface resistance $r_c = 0$, is

$$T_c - T_a = \frac{r_a I_{cl} R_n}{\rho C_p} \cdot \frac{\gamma}{\Delta + \gamma} - \frac{VPD}{\Delta + \gamma} \tag{6}$$

Thom and Oliver [49] proposed an effective aerodynamic resistance r_{ae} that represents a semi-empirical equation for r_a:

$$r_{ae} = 4.72l\{\ln[(z - d)/z_0]\}^2/(1 + 0.54u) \tag{7}$$

The crop water stress index can be estimated using the equation below:

$$CWSI = \frac{(T_c - T_a) - (T_c - T_a)_{ll}}{(T_c - T_a)_{ul} - (T_c - T_a)_{ll}} \tag{8}$$

2.2. Structure and Architecture of the GWD Project

The GreenWaterDrone system is functionally and physically divided into four subsystems, as follows (Figure 2):

- The ground measurements subsystem (MMS), which is applied only during the calibration phase, consists of micrometeorological stations and their integrated/peripheral sensors, which are required to collect microclimatic and soil measurements of the crop field. Data collected from a station is used to calibrate and approximate the CWSI of a specific crop under the local climatic regime for one growing season. Data is communicated to the system over available mobile WAN infrastructures (2/3/4 G, IoT).
- The aerial measurement subsystem (UAS) consists of two types of UAVs. A quadcopter platform UAS1 uses an autonomous microstation to collect raw spatial data from the crop foliage and environment (infrared temperature, air temperature, relative humidity, accurate coordinates, and elevations). A fixed-wing platform UAS2 is required to collect thermal, multispectral, and photogrammetry images over large crop areas. Field data collected by the UAS is communicated to the system over available mobile WAN infrastructures (2/3/4 G, IoT). Both scheduled (e.g., during calibration and normal operation) and emergency (e.g., extreme weather conditions) flights are managed by the GWD System Administrator via the FlightPR interface.
- The service support information system (BackEnd) implements the crop data management necessary for the storage, classification, management, and updating of field measurements, empirical irrigation data, spatial and crop quality data, field status

multimedia, end-user preferences, and interfaces with external services (satellite imagery, photogrammetry applications). In addition, it interconnects and supports all other subsystems and is responsible for providing the services of the system (alerting and multimedia content) to all types of supported end users.

- The service provision I/Fs (FrontEnd) includes appropriate web interfaces of the system to predefined types of GWD users, such as plain (farmers/agronomists), group (partnerships), and strategic (local/regional authorities) end users, with graded access to the three supported applications through different devices (PCs, smartphones, etc.) and relevant GUIs.

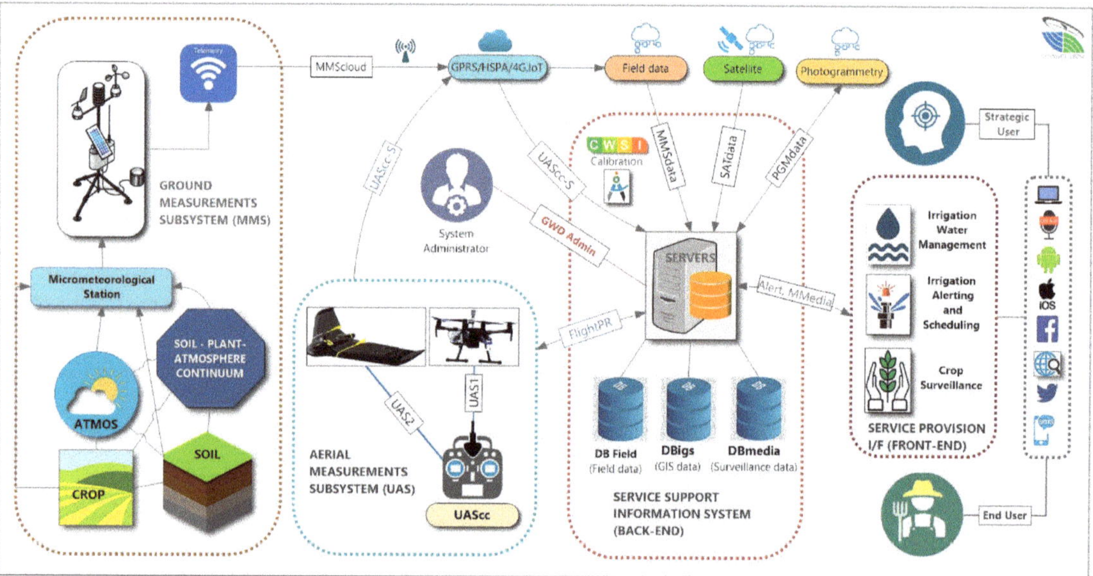

Figure 2. The concise physical architecture of the system with the basic physical entities, implementing the respective groups of functional entities and their placement in the above subsystems, as well as the relevant interfaces between entities and subsystems.

The GWD system supports the following applications to provide a comprehensive irrigation management and field surveillance framework to both the plain and strategic end users that are targeted:

1. Irrigation alerting and scheduling (IRRas): The plain end user (farmer/agronomist/ farmer partnership) receives alerts in near real time regarding the short-term need to irrigate (or not) a specific crop based on CWSI calculations and empirical irrigation scheduling.
2. Crop surveillance (CS): The plain end user (farmer/agronomist/farmer partnership) can view on-demand, multimedia content (e.g., photos/video relating to crop condition) of a field or receive alerts in near real time regarding the availability of synchronous video/photos of his crop in the case of a natural disaster or a security issue triggering an emergency drone flight.
3. Irrigation water management (IRRmgt): The strategic user (agricultural institute, local/regional authorities) may select zones (clusters) on a graphical interface with a map of the area covered by the GWD system (effectively calibrated crops in the area) and obtain irrigation requirements for specific crop patterns and periods, thus enabling the implementation of scenarios for future irrigation water policies.

3. Case Study—Materials and Methodology

3.1. Study Area

The study area was situated in the prefecture of Messinia (western Peloponnese), close to the city of Kyparissia in the Municipality of Trifilia (Figure 3). Based on long-term data obtained from the nearby station of Kalamata (37.07° N, 22.10° E, alt. 8 m a.s.l.) over the last century, the broader area has a humid climate according to Thornthwaite's climate classification system (UNEP 1992), which has, however, become drier nowadays compared to the past, as seen via the decreasing aridity index [50] values from 0.88–0.89 in the previous climatic periods (1900–1930 and 1930–1960) to 0.85 in the period 1960–1997 [51].

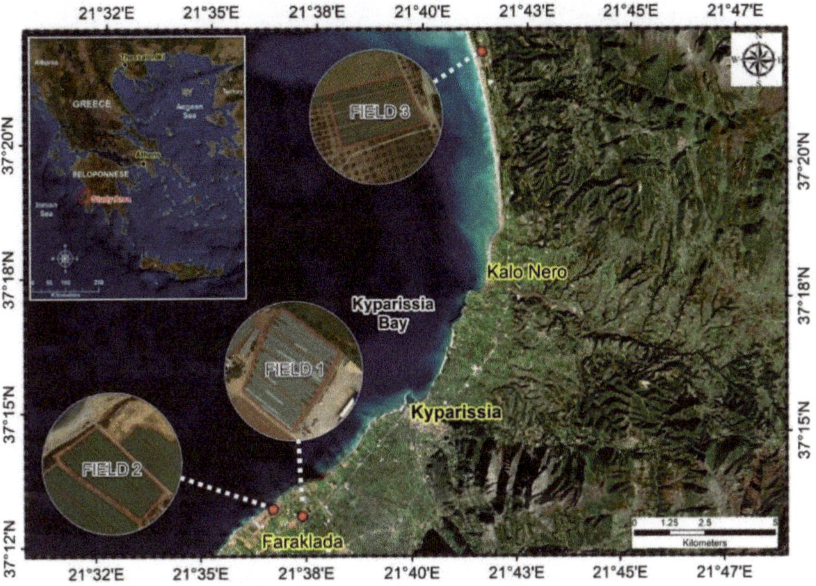

Figure 3. Location of the GWD project's experimental fields in Trifilia.

The experimentation was conducted in three fields with three different crops that are representative of the area, i.e., potato, watermelon, and tomato, which were utilized for the validation procedure of the GWD project methodology (Figure 3). Indicatively, in this work, potato field 2 (*Solanum tuberosum*; 1.25 hectares; 37°13'20.65'' N, 21°36'41.51'' E, alt. 1 m a.s.l.) was used throughout the next sections for demonstrating the preliminary results of the different imaging used in the project and GWD methodology validation during the 2019 growing season. The potato crop was selected by considering its relatively high water requirements, with its maximum in the hot and dry summer period, and the critical susceptibility of potato plants to water stress, which is attributed mainly to their relatively shallow root system (Figure 4a) [18]. The field had a small slope of 0–3% and a clay loam surface soil layer (0–20 cm depth), becoming sandy clay loam further down (20–60 cm depth; Figure 4b).

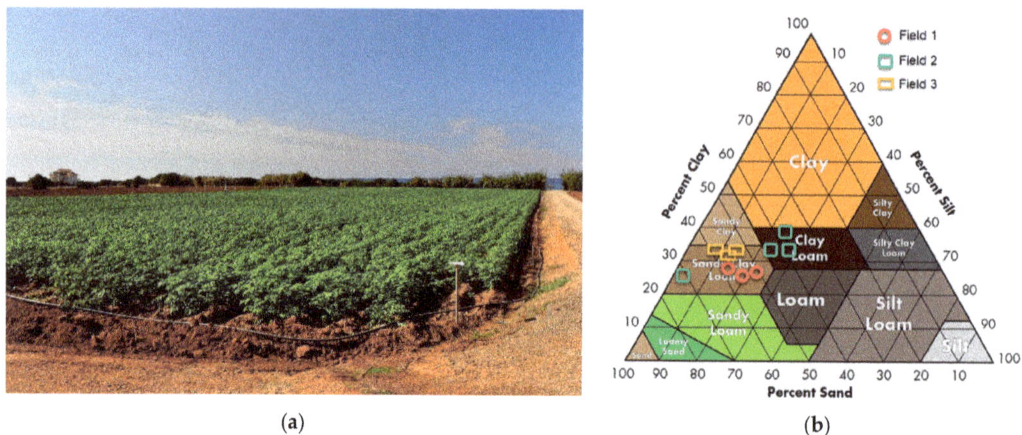

(a)   (b)

Figure 4. (**a**) Experimental field 2 with the potato cultivation; (**b**) categorization of the soils of the three experimental fields at depths of 0–20, 20–40, and 40–60 cm in the soil triangle. The soil texture classification according to the USDA (U.S. Department of Agriculture).

3.2. Ground Micrometeorological Station

Three automatic micrometeorological stations were installed in the middle of the fields (Figure 5) for observations of the upper atmospheric layer of the crops and measurements within the soil profile (at the root zone). This was an indispensable condition for the accurate estimations of CWSI and the determination of all the biophysical attributes of the cultivations during all stages of the growing season. This ground-based measuring system composed the core of the calibration for estimating the CWSI of each crop, which was the basis for determining the real-time irrigation needs according to the methodology proposed by the GWD project.

Figure 5. Three stations were installed in the middle of the experimental fields: (**a**) the first station in field 1, with early watermelon cultivation with a plastic cover for protection against the spring frost; (**b**) the second station in field 2 with the cultivation of spring potato in the middle growing stage; (**c**) the third station installed in field 3 with young tomato plants.

3.2.1. Measurements above Crops

The three ground micrometeorological stations (GMMS) were equipped with the following sensors:

i. A high-precision miniature air temperature (T) and relative humidity (RH) sensor model EE08 (Elektronik Ges.m.b.H, Austria) working over a wide range was installed at a 2 m level.

ii. The LPPYRA02 fitted on stations 2 and 3 is a spectrally first-class pyranometer that measures the shortwave (Rs) solar radiation flux density on a flat surface (in Wm^{-2}; global irradiance, sum of direct and diffuse radiation).

iii. A pyranometer following ISO 9060:2018 with the criteria of the WMO "Guide to Meteorological Instruments and Methods of Observation." An albedometer LP PYRA 05 (Delta OHM, Italy) was fitted on station 1 and measured the global radiation (in Wm^{-2}) and the albedo of any surface. It was manufactured incorporating two LPPYRA02 pyranometers into only one body.

iv. Net radiation pyranometer (RNET LP NET 07 Delta OHM, Italy) for the net flux density measurement (Rn) passing through a surface across the near-ultraviolet and the far-infrared spectral ranges.

v. Photosynthetically active radiation sensor PAR (QUANTUM SENSOR Model SQ-100-SS APOGEE, Italy) that measured the total radiation across the range of 400–700 nm expressed as photosynthetic photon flux density (in $\mu mol\ m^{-2}\ s$, equal to $\mu E\ m^{-2}\ s$).

vi. Two tipping bucket rain gauges, model HD2015 and model HD2015 (Delta OHM, Italy), for precipitation measurements in millimeters (Pre) were installed on stations 1 and 3. It should be noted the distance between meteostations 2 and 3 was 15 km along the north–south axis, while the distance between meteostations 1 and 2 was 1 km.

vii. Cup anemometers for horizontal wind speed (WS) measurement (in m/s; model 4.3519.00.167 Thies GmbH & Co, Germany) and wind direction (WS) sensor (in degrees; model 4.3140.51.010 Thies) were installed on the mast at a height of 3.4 m for the three meteostations.

Given the importance of measuring the temperature of the plant foliage (canopy) remotely, reliable infrared radiometer sensors were crucial for achieving the program's objectives. The model SI-111-SS (Apogee, Berkeley, CA, USA), which is an unamplified analog sensor with a standard field of view, was selected to measure the canopy temperature. The response time of the sensor is 0.6 s, with a measurement repeatability of less than 0.05 °C, and the calibration uncertainty is 0.2 °C. The spectral range of the measurement is between 8 and 14 μm, that is, the atmospheric window. The sensor has a 22° half-angle field of view. The placement height and the angle formed between the mast with the longitudinal axis of the infrared thermometer determine the elliptical area of measurement (in m^2) of the surface of the dense canopy.

Moreover, portable infrared thermometers (Model MI-210 Apogee, USA) were used for conducting spatial measurements in all fields' canopies during the flights of the UAV for calibration and comparative data analysis.

3.2.2. Measurements in the Root Zone

The soil moisture (SMois) ($v/v\%$) and soil temperature (Tsoil) were measured continuously at six soil layers (from 0–10 cm to 50–60 cm in 10 cm intervals) using the Drill and Drop probe (model 00620 Sentek Technologies, Australia). In this way, the actual changes in soil moisture were constantly obtained throughout the depth of the root zone at any time. Furthermore, a soil heat flux plate (HFP01 Hukseflux Thermal Sensor, Netherlands) was installed in the topsoil layer (at a 5 cm depth). Figure 6 graphically shows the soil moisture profile estimated using a geostatistical gridding method (Kriging), which was proven to be useful in many fields. The Kriging method estimates the surface at successive nodes in the grid using only a selection of the nearly closed data points and produces surfaces from irregularly spaced data.

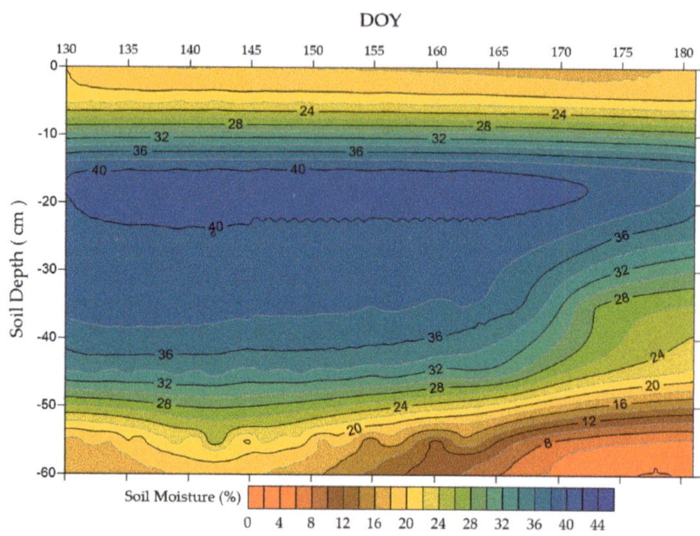

Figure 6. Soil water content isolines (%) in the irrigated spring potato crop (field 2) for 10 May–9 July 2019 (DOY: 130–181).

The method efficiently and naturally incorporates anisotropy and underlying trends from a data set by specifying the appropriate variogram model. To create this graph, 1080 daily soil moisture measurements were used from six depths that were monitored in the potato cultivation field for 51 days.

For the detailed fluctuations of the soil moisture of the potato field at depths of 0–10 and 10–20 cm (the two most variable layers), we used 17,284 average values of ten-minute continuous soil moisture measurements. These measurements are shown in Figure 6 and refer to the interval from 10 May to 9 July. The smooth, continuous, and detailed data-collection process, which took place throughout the growing season in the soil profiles, allowed us to accurately estimate the actual evapotranspiration (ETa) from the soil moisture profile in hourly time steps. Moreover, from the observations and characteristics of the plants in all stages (height, LAI, aerodynamic, and surface resistance), in combination with the micrometeorological observations just above the crop (Rs, Rn, albedo, soil heat flux, T, RH), we estimated the crop potential evapotranspiration (ET_c) with satisfactory accuracy. All the above estimates contributed to estimating the crop water stress index (CWSI), both practically and theoretically.

3.2.3. Data Acquisition System and Telemetry

The telemetry unit (model A753 addWAVE GPRS RTU, ADCON (Business Unit of OTT HydroMet GmbH)) provided the possibility of local storage in EPROM Memory for at least half a million measurements. The telemetry unit has an internal rechargeable battery that is charged using a solar panel, which is independent of the one used for the energy needs of the sensors. The duration of operation in normal operation is up to 14 days. According to the GSM standard, the frequency band is 850/900/1800/1900 MHz, and the maximum transmission distance is 36 km. To achieve the maximum accuracy (connecting sensors to a central station unit), especially in the connection of thermocouple sensors (pyranometers, soil heat flux plate) to a central station a unit HD978TR3 (Delta OHM), signal amplifiers were used.

The time step for storing the average data values for each sensor is ten minutes for all stations (144 average values per 24 h). From the ten-minute average values, we obtained and recorded hourly and daily average values for all sensors. All average values from each sensor were found using 4320 reads per 24 h. This means that almost all sensors were automatically excited to give a reading sample every 20 s (3 readings per minute). Of course, this programming considers the response time of each sensor, i.e., the ability of the sensor to respond to two different successive values in time.

3.3. Experimental Design—Unmanned Aerial Vehicle, Cameras, and IR-TH

In the present study, the quadcopter DJI Matrice 200 (DJI Technology Co., Ltd., Shenzhen, China) was used for the measurements of the multispectral and IR-TH data, and the fixed-wing Q100 Datahawk was employed for the photogrammetric mapping of the fields. The quadcopter was selected due to the compatibility of incorporating both multispectral and thermal cameras, and because the flights with infrared sensors require a low flight altitude, low-speed navigation, and high stability, which cannot be achieved with a fixed-wing platform [52]. The accuracy of the system is 10 cm in the vertical and 30 cm in the horizontal, which is considered suitable for PA applications, while at the same time, it offers the possibilities of low, slow, and stable flight. It has significant autonomy and a maximum take-off mass with a large payload (6.14 and 1.61 kg, respectively) and can withstand strong winds (up to 12 m/s), which ensures its smooth operation in demanding flight environments. In addition, the system incorporates an advanced obstacle recognition and avoidance subsystem, which consists of a combination of optical, infrared, and ultrasonic sensors, making the platform one of the safest for flight operations in existence, and incorporates inertial measurement units (IMU) sensors, supporting both automatic, semi-automatic, and manual flights.

The flight planning of the photogrammetric, multispectral, and thermal cameras was conducted with the ground control station software, GS PRO (for the IR), Pix4D (for the thermal camera), Field Agent (for the multispectral camera), and the SkyCircuits Plan and Flight (for the photogrammetric camera), which allow the user to generate a route over a grid path as a function of the field of view of the sensor, flight altitude, flight speed, direction, degree of overlap between images, and ground resolution.

3.3.1. UAV for IR Spatial Row Data Measurements

The DJI Matrice quadcopter was selected to install an autonomous aerial micrometeorological system (MicroStation) consisting of a high-performance data logger (Symetron type Stylitis-12), as well as four sensors: IR radiometer (SI-111-SS Apogee), miniature Thermo-Hygrometer EE08, and an accurate GPS (Figure 7). Through the "Opton 4" Symetron software, the data logger operation was managed and controlled. Furthermore, all the appropriate equations and precalibrated factors for any crop were incorporated into the CWSI estimation software.

The radiation shield plates (which protected the temperature and RH sensors), the watertight box (which protected the data logger), and all the sensors' mounts were made from lightweight synthetic materials. All these components were designed in a 3D CAD program and printed using a 3D printer in the lab. Particular attention was paid to the placement of the thermo-hygrometer so that it was not affected by the flow of the drone propellers (Figure 7). The vertical positioning of the radiation shield on top of the UAV ensured mild air vortexing and mixing, avoiding the violent airflow from the downside of the drone due to the propellers.

One of the most important parts of the IR instrument sampling workflow was the calculation and design of the most suitable flight planning in order to have the best point samples coverage and complementarity to ideally cover the whole cultivated area. The flight height depends on the sensor type and the characteristics of the crop. For a specific IRT sensor, the visible surface area is a function of height. For the sensor SI-111 that was placed on the UAV in such a way to be vertical to the foliage for a flight height of 4 m (h), the surface area was estimated to be 8.2 m^2 (E), and the diameter of the circular surface was about 3.2 m (Figure 8).

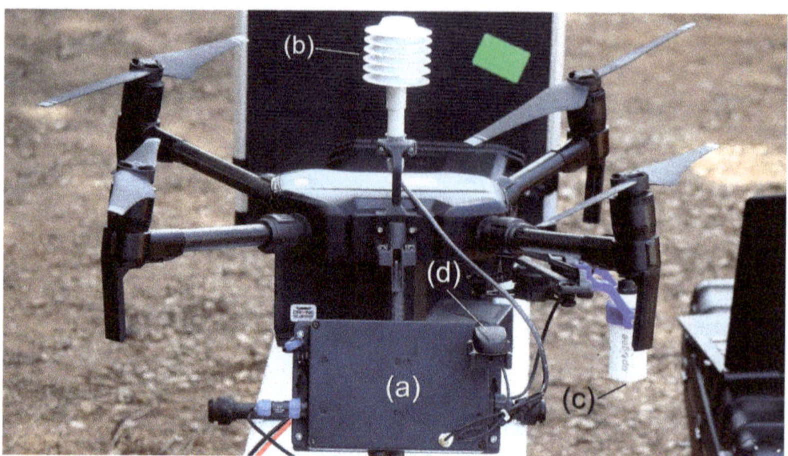

Figure 7. (**a**) Watertight box (data logger Stylitis-12 inside); (**b**) temperature and relative humidity (thermo-hygrometer) sensor in the radiation shield; (**c**) infrared sensor and the mounting base; (**d**) GPS sensor.

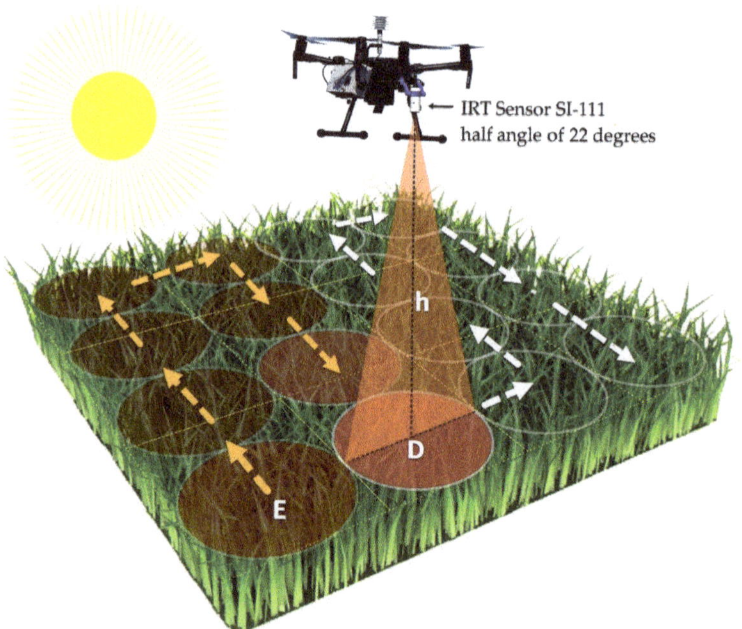

Figure 8. Graphic illustration of the measurement process with an infrared radiometer adapted to a UAV over dense cultivation foliage under a clear sky. The flight height depends on the sensor type and the characteristics of the crop. For a specific IRT sensor, the visible surface area is a function of height. For the sensor SI-111 that was placed vertically to the foliage with a height of flight h = 4 m, the surface area was E = 8.2 m^2, and the diameter of the circular surface was about 3.2 m.

3.3.2. UAVs for Photogrammetric, Multispectral, and Thermal Measurements

The photogrammetric camera that was used for the present study was the SONY rx 100, with a 20.2 megapixels CMOS lens that has a ground resolution of 3.2 cm/pixel. The camera was adjusted on a fixed-wing Q100 Datahawk UAV (Figure 9a). By processing the images that were produced, an accurate description of the canopy height and a digital surface model (DSM) were created. This information is significant in endeavoring to explore the surface slope inclination and, indirectly, the conditions of drainage and water flow in the soil, as well as the growth conditions of the crop (canopy height and density).

Figure 9. (**a**) The fixed-wing Q100 Datahawk; (**b**) the multispectral Sentera camera; (**c**) the thermal infrared camera Zenmuse XT2.

The DSM from UAV image stereo pairs used the structure from motion (SfM) algorithm, which is a photogrammetric technique that was initially developed for archaeological sites [53–55]. The algorithm assesses the unknown camera orientations through a comparison of multiple detected image feature points in multiple images. It subsequently produces the 3D structural model of a scene from the overlapping two-dimensional (2D) image sequences that are taken from various spots and orientations [56].

The multispectral camera that was used in the present study was the Sentera AGX710 (12.3 megapixels) multi-spectral sensor with five bands (Sentera Inc., Minneapolis, MN, USA), which cover the blue, green, red, red edge, and near-infrared parts of the electromagnetic spectrum (Figure 9b). Their central wavelengths are 446 nm, 548 nm, 650 nm, 720 nm, and 840 nm, respectively. A 15 cm × 15 cm white reference panel (MicaSense Inc., Seattle, DC, USA) with a 60% nominal reflectance was used for the radiometric correction. UAV imaging was conducted in sunny weather and the period between 11 a.m. and 12 p.m. was chosen for imaging to minimize sunshade. To certify correct image acquisition during the flight, the FieldAgent Mobile App from Sentera was used to plan the flight path and automate the operation of the UAV. A constant flight height was maintained at 30 m and the flight speed was set to 3 m/s. The ground sample distance of the imagery was approximately 1 cm/pixel and a 90% overlap between two images both for side-lap and front-lap was implemented. In total, about 820 images were taken for the experimental field and stacked into a single image for analysis.

Agisoft Metashape Professional (v 1.5.5) was preferred as the Structure for Motion Multi-View Stereo (SfM-MVS) processing software to generate the digital surface model (DSM) and the correspondent normalized difference red-edge index (NDRE) vegetation index.

NDRE is a narrowband greenness VI that was designed to provide a measure of the overall amount and quality of photosynthetic material in vegetation, which is essential for understanding the state of vegetation (photosynthetic capacity and canopy chlorophyll content, structure, and plant health status). NDRE (Equation (9)) was measured by using the NIR and red-edge band (RE; 705 nm), where RE replaces the red band in the widely used normalized difference vegetation index (NDVI) equation. Compared with the NDVI, the NDRE index as a widely used red-edge-based VI was shown to be more resistant to the saturation problem and is more sensitive than the NDVI to chlorophyll content in vegetation [57].

$$\text{NDRE} = \frac{\left(B_{NIR} - B_{RedEdge}\right)}{\left(B_{NIR} + B_{RedEdge}\right)} \tag{9}$$

The final image nominal spatial resolution, expressed as a ground sample distance (GSD), was about 1.2 cm [58].

The final NDRE image ideally demonstrates the plant health status and the existing gaps in the canopy cover and along the cultivation lines, which are important to distinguish and exclude since the reflection from the soil significantly affects the calculation results of the CWSI. Thus, the results of the NDRE image were used to create a binary mask of the soil. Sunlit soil parts were extracted from the image by utilizing a threshold value on a local spatial statistic that characterized the illumination levels [18,59]. This final mask was applied on the co-registered TIR image to produce the thermal product with pixels belonging only to sunlit leaves.

The thermal camera used in the present study was the DJI Zenmuse XT2 (12 MP), which incorporates a high-resolution forward-looking infrared FLIR Tau 2 thermal sensor (FLIR Systems, Wilsonville, OR, USA; resolution 640 × 512 pixels, lens 9 mm; thermal sensitivity of <50 mK) and a 4 K visual camera (1/1.7″ active-pixel CMOS) with a leading stabilization and machine intelligence technology (Figure 9c) [60]. In general, the Zenmuse XT2 camera used in the present study is a very high technological and sophisticated camera that performs radiometric calibration of the acquired thermal images through its advanced digital system, which records the absolute temperature. The calibration of the FLIR camera was performed annually according to the factory specifications, which is carried out by the official supplier by measuring targets with known temperatures and comparing the known and the measured temperatures (e.g., boiling water and melting ice) [16,61]. The thermal lens has a focal length of 19 mm, which avoids previous deformations and obtains rectilinear images. The second lens has a focal length of 8 mm and it also retrieves high-resolution RGB images [62]. The ground sample distance of the imagery was approximately 2 cm/pixel and an 85% overlap between two images for both side-lap and front-lap was implemented. In total, about 382 images were acquired for the experimental field and stacked into a single image for analysis.

UAV thermal imaging was performed in sunny weather from 12:00 to 1:00 p.m. to minimize sunshade and because it has general unique image-acquiring conditions since the maximum temperature differences occur between the soil and vegetation [63,64].

The thermal images acquired by UAV were mosaicked using Pix4D mapper (version 4.5 from PIX4D, Prilly, Switzerland). This software offers an automated SfM procedure, taking a set of images as input and routinely going through the steps of feature identification, matching, and bundle adjustment. The process aligned the images that were acquired by the thermal camera. A polygon mesh was calculated from the dense 3D point, and the pixel values of each image were then projected onto the mesh to create an orthomosaic. When combined with the GPS locations, this procedure allowed for the establishment of a high-resolution orthophoto and a digital elevation model (DEM) of the crop field [33,65].

The final steps considered the exclusion of the soil cover and the gaps among the canopy by using the very high resolution multispectral and true color orthomosaic. Then, the interpolation of the selected points for the selected area comprised the most homogeneous and appropriate parts of the canopy for the following measurements.

For the better calibration of the thermal image and IR data derived from the UAV, field samples and continuous measurements of the meteorological station using the same IR instrument were utilized and evaluated (Figure 10).

Figure 10. The field measurements that were taken using the IR sensors on (**a**) the meteorological station and (**b**) a portable device.

The information on GPS locations for each image was obtained from the UAV during collection; however, to achieve higher accuracy, eight rectangular aluminum plates were used as ground control points (GCPs), which were equally spread around the potato field and were surveyed using an RTK GNSS with a range of <1 cm in the horizontal plane and 1.7 cm in the vertical axis (Figure 11).

Figure 11. (**a**) The RTK GNSS that was used for the collection of the GCPs; (**b**) the rectangular aluminum plates that were used as ground control points on the field.

Finally, for the temperature calibration, a linear regression model for the precise acquisition height (30 m) of the UAV thermal data was applied since the temperature obtained using the IR sensors considerably decreased with increasing flight height [28,66,67]. Thus, a linear regression model using ground truth temperatures (measured with the handheld

IR) and those obtained from the UAV was formed following analogous results that were in the recent literature, which found that the linear regression model based on data acquired at a 1 m height had a slope of 1.0, while that based on data acquired at a 50 m height had a slope of 1.4 [16].

4. Preliminary Results and Discussion

The results listed below came from flights that took place over an autumn potato field (25 October 2019). The pilot flights were conducted around midday. The meteorological conditions prevailing during the day were characterized by a clear sky with maximum global solar and net radiation flux densities of 670 and 395 Wm^{-2}, respectively, at 13:00 (Figure 12a). The air temperature attributes were normal for the season with a daily Tavg = 19.1 °C, Tmax = 24.8 °C, and Tmin = 14.4 °C (Figure 12b). The VPD, ET_c, and wind speed hourly averages depicted in Figure 12c indicate that during the flights, the wind speed was about 1.5 m/s and the ET_c maximum rate was 0.43 mm/h. For the specific day, the total ET_c reached 2.7 mm/day. The relatively low crop evapotranspiration rate was in line with the small changes in the soil moisture profile presented in Figure 12d.

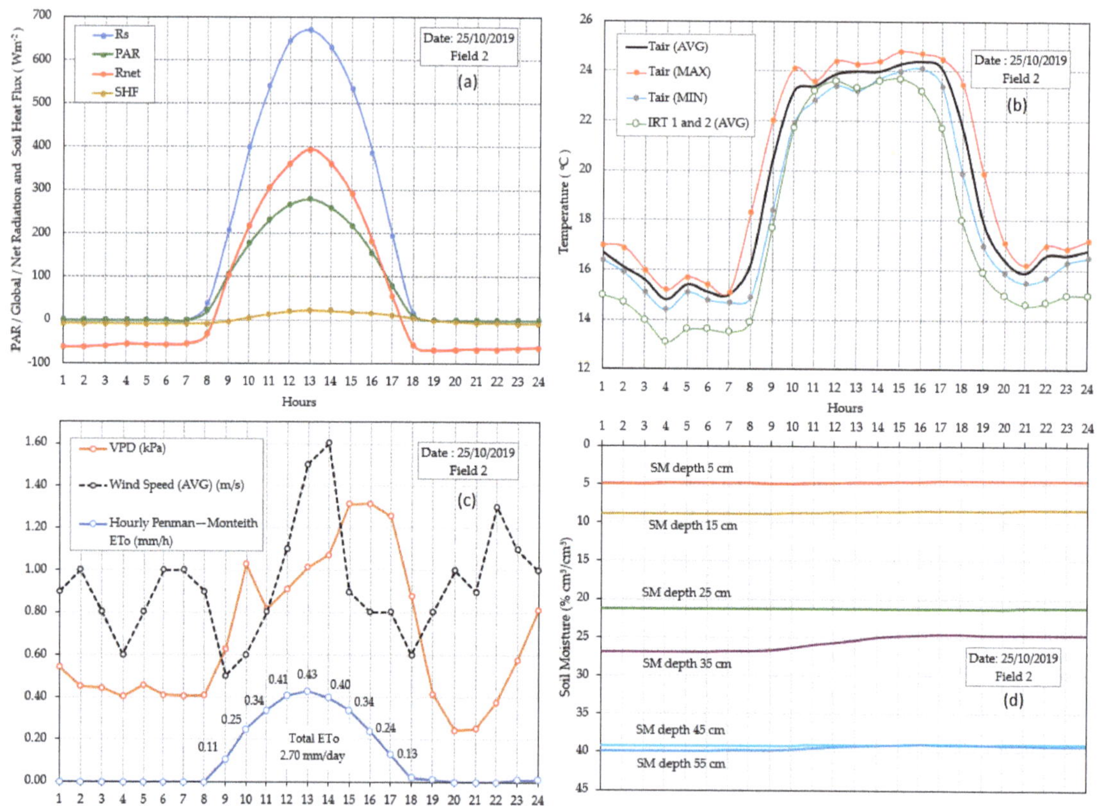

Figure 12. Hourly values for the date 25/10/2019 for the (**a**) global solar radiation Rs, net radiation Rnet, photosynthetically active radiation PAR, and soil heat flux SHF; (**b**) air and canopy temperatures; (**c**) vapor pressure deficit VPD, wind speed WS, crop evapotranspiration; (**d**) soil moisture profile for six depths measured in the potato field.

4.1. Photogrammetry—Multispectral and Thermal Imagery

The photogrammetric analysis provided very high-resolution information about the canopy conditions by creating the RGB true-color imagery of the field (Figure 13a), which revealed all the aspects of the canopy attributes and its surrounding area.

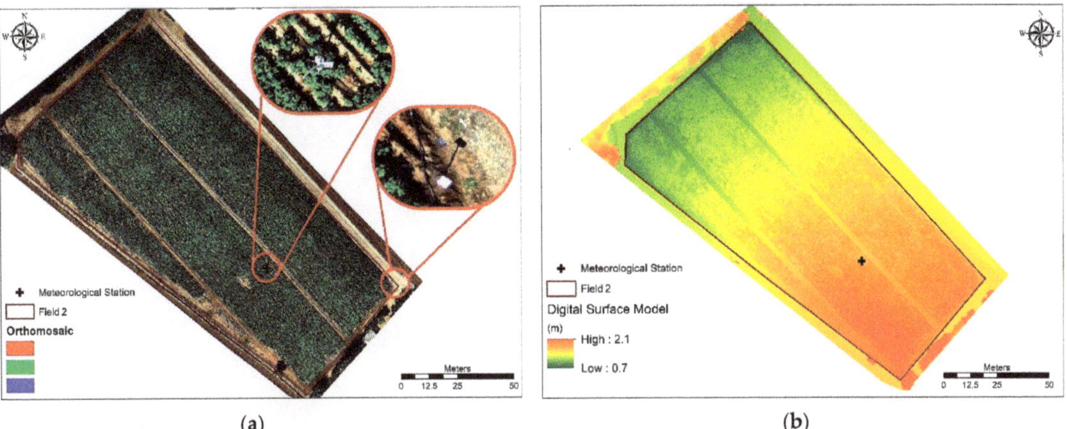

Figure 13. (**a**) The very high-resolution true color orthomosaic and (**b**) the very high-resolution digital surface model of field 2.

The DSM indicated a gentle inclination of the field, which was ideal for the appropriate drainage of the field and avoiding overlogging situations, where the elevation varied from 1.1 m at the south-southeastern part to 0.4 m at the north-northwestern part of the field (Figure 13b).

The result of the NDRE equation provided an image with continuum pixel values that ranged from -1 to 1, where negative values corresponded to non-vegetated surfaces, in this case, soil cover, while positive values corresponded to the vegetation reflectance (Figure 14). The values between 0.25 and 0.64 were related to healthy photosynthetic vegetation of potato canopy, while lower values, especially those <0.2, were related to stressed vegetation or bare soil. Values higher than 0.64 represented the dense tree cover in the northwestern and southeastern parts of the field. As it was mentioned before, the NDRE image was utilized to mask the soil cover from the co-registered TIR image (XT2) and the interpolation image derived from the IR sensor point values to produce the thermal products with pixels belonging only to sunlit leaves.

The thermal image (Figure 15) that was taken by the Zenmuse XT2 on the UAV was used to calculate the canopy temperature (T_c) of the experimental field. The image was classified (using custom classification), keeping only the pixels of the crop canopy area to separate and present the reliable foliage temperature values. The canopy temperatures (from the thermal image) ranged between approximately 21.2 and 24.3 °C (range of about 3 °C), which was very close to the station average IRT temperature (23.9 °C) and the average spatial IRT (21.8 °C), respectively. These measurements were taken during a day with clear sky conditions and relatively low temperatures. The crop was in the middle stage of growth (almost full canopy cover), where the soil and atmospheric conditions are presented in the following charts.

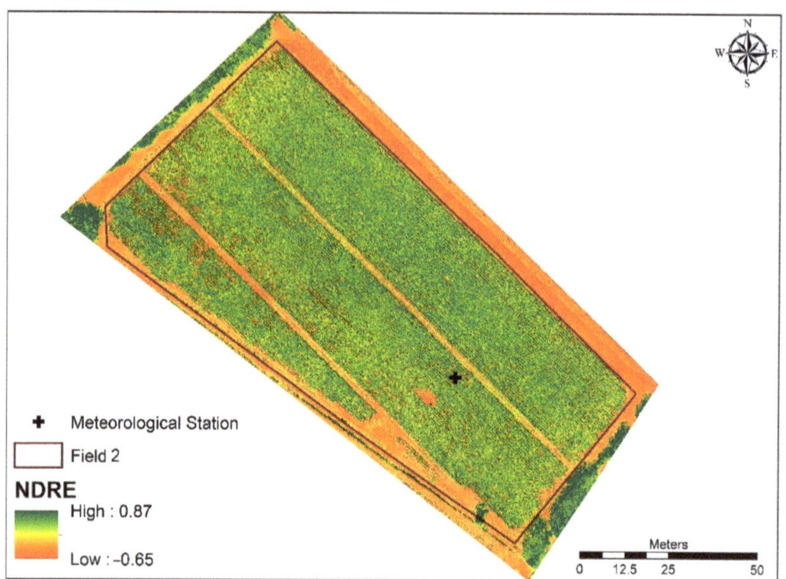

Figure 14. The red-edge normalized difference vegetation index of field 2, which was calculated using the multispectral camera data.

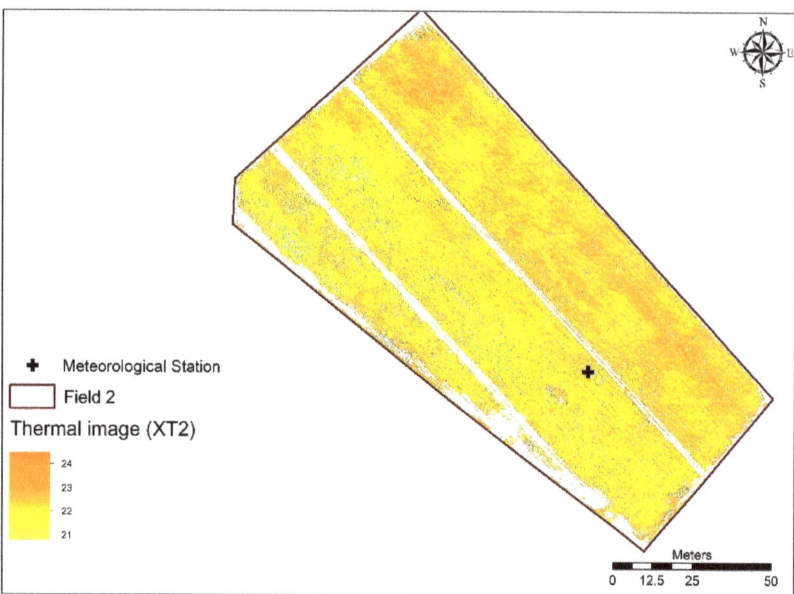

Figure 15. Thermal image acquired by the Zenmuse XT2 camera showing only the crop vegetation's temperature values by masking all the non-vegetation pixels.

Although thermal RS has various potential advantages in crop monitoring, there are several practical difficulties in its use, including atmospheric attenuation and absorption, calibration, climatic conditions, crop growth stages, and complex soil and plant interaction that have to be adequately addressed.

The high-resolution multispectral and thermal imagery acquired using the UAV platform ideally enables the extraction and identification of crop conditions (vegetation vigor, spatial distribution, canopy cover, and bare ground gaps) and the temperature of the canopy. Furthermore, they demonstrate great capability for the recognition and monitoring of drought stress and were used in several different crops, such as potato, cotton, soybean, cotton, maize, vineyards, and orchards, over the last two decades [16,18,26,27,68–71]. Additionally, several studies showed that cloud cover and other haze or dew conditions can influence the quality of the thermal data obtained (due to lower contrast and poor radiometric resolution) [72–74]. Hence, these variables, along with relative humidity, altitude, and viewing angle, should be thoroughly examined while acquiring thermal data [71,73]. Moreover, the acquisition time during midday was demonstrated by several scientists that were the optimal time for thermal image acquisition [75].

4.2. Direct Canopy IRT Measurements and Spatial CWSI Estimation

The first pilot flight with the UAV that brought the aMMS (aerial micrometeorological system) equipment took place on 25 October 2019 over field 2 during late potato cultivation. It was essentially the first comprehensive process for estimating the CWSI spatial mapping. In the same field, during the early cultivation of potatoes in May 2019, a previous estimation of CWSI calibration coefficients from specific days with ideal weather conditions was undertaken. We used all ground station data and parameters to determine the lower and upper baselines using theoretical and practical approaches. The day of the UAV flight was a clear sunny day with relatively low temperature levels, as the period was in the middle of autumn. The crop was in the middle stage of development. The UAV flight for the spatial measurements of the canopy temperature, air temperature, and relative humidity started at 12:14 p.m., which corresponded to the local solar time of the ground station (lat. 37.222194 N, long. 21.611806 E in degrees). The flight duration was 21 min, and 217 primary spatial measurements of infrared temperature from the crop canopy, air temperature, and relative humidity were collected from a height of 4 m above the ground (about 3.5 m from the top of the foliage). The flight was autonomous and scheduled following the standard flight plan described in Section 3.3.1 (Figure 8). The speed of the quadcopter was 1 m/s, and the excitation of the sensors for measurement was every 6 s (measurement every 6 m along the flight line). All microclimatic data obtained from the ground station (GMMS) above the crop, as well as the mean spatial values (N = 217) from the sensors of the aerial micrometeorological system, are shown in Table 1.

Table 1. Ten-minute average climatic data from the ground micrometeorological station (GMMS) from field 2 and the averages values from 217 measurements of the air temperature, canopy infrared temperature, and relative humidity from the aerial micrometeorological system (AMMS).

Solar Time	Tair	IRT	RH	Wind Direction	Wind Speed	Solar Radiation	Net Radiation
	($^\circ$C)		(%)	($^\circ$)	(ms^{-1})	(Wm^{-2})	
12:15	23.9	23.4	66.6	314	1.6	675	399
12:25	24.0	24.0	64.1	320	1.5	667	394
12:35	24.1	24.2	62.2	320	1.2	665	393
GMMS *	24.0	23.9	64.3	318	1.4	669	395
AMMS *	24.3	21.8	56.0				

* Mean values of measurements.

Moreover, Table 2 presents the moisture conditions in the six layers (from 0–60 cm with 10 cm intervals), and the soil heat flux density at a depth of 8 cm. As shown in Figure 16, the surface layer (0–10 cm) had lost 67% of its moisture from the total height of the available water between two irrigation events. Usually, the irrigation frequency for this period was every 3–4 days, as shown in Figure 16 for the last three irrigation events (16, 20, and 23 October 19).

Table 2. Soil water moisture (vol%) for the depth of 60 cm of the soil profile in the potato field.

Time	Volumetric Soil Water Content (%) of the Soil Layer (in cm)						Soil Heat Flux
	0–10	10–20	20–30	30–40	40–50	50–60	(Wm^{-2})
12:30 p.m.	4.7	8.6	25.4	39.2	39.1	21.3	23.7
12:40 p.m.	4.7	8.6	25.3	39.2	39.1	21.3	23.5
12:50 p.m.	4.7	8.5	25.2	39.2	39.1	21.3	23.2
Average	**4.7**	**8.6**	**25.3**	**39.2**	**39.1**	**21.3**	**23.5**

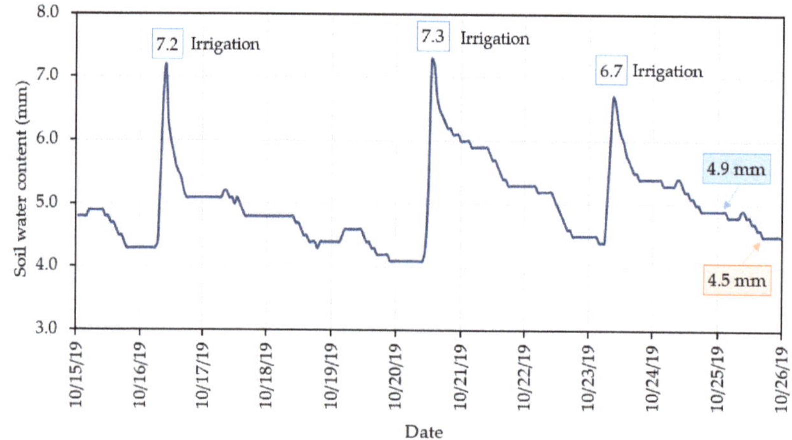

Figure 16. The fluctuation of the soil moisture in the surface layer of the soil (0–10 cm) after three consecutive irrigations.

Figure 17 presents the isothermal lines from a spatial-observation-derived analysis (N = 217) using the geostatistical gridding Kriging method, which produces maps from atypical, spaced data.

The method is very flexible and produces an accurate data grid by a smoothing interpolator depending on the user-specified parameters. The spacing nodes grid size used 86 rows × 100 columns (total nodes: 8600). Some univariate grid statistics for the canopy IR temperature for the grid data (in °C) were min: 21.603, max: 22.133, mean: 21.886, root mean square: 21.8869, standard deviation: 0.1486, average abs. deviation: 0.1252, and relative mean diff.: 0.0078.

The exact area (enclosed area by the boundary line) of field 2 was 9940.3 m^2. Each observation was taken from about 4 m above the crop and corresponded to the mean value of the infrared temperature of the circular surface of the cone base radius (target) of about 3 m (Figure 8).

The evaluation of the CWSI index during the clear sky and dry hours of the day offered a real and more direct method to determine the ideal time for irrigation, as it effectively found and normalized the short-term microclimatic changes of the environment in which the crop grew up and responded reliably to maintaining the optimal water content of the soil porosity into the rhizosphere zone. This significantly improved the irrigation efficiency and water-saving at all stages of plant growth. The spatial estimates of the CWSI are shown in Figure 18.

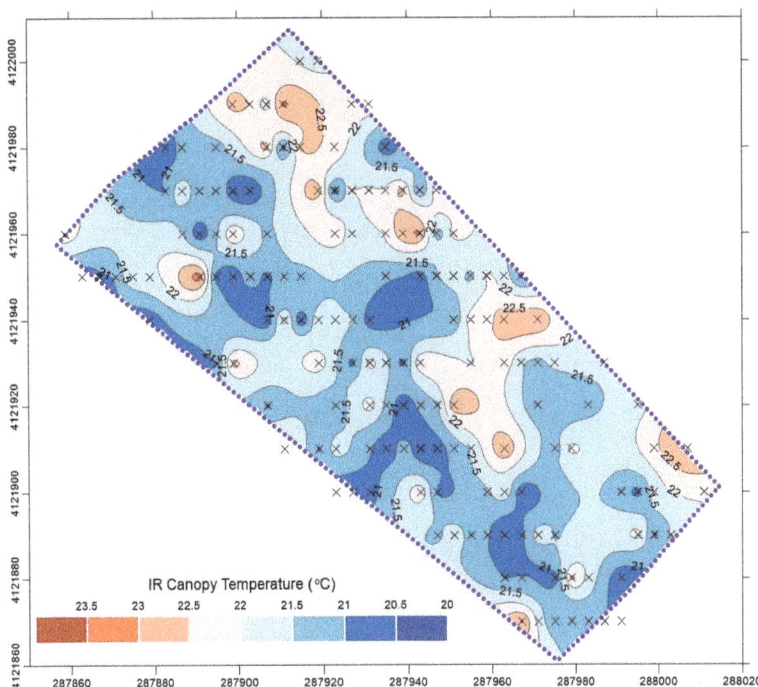

Figure 17. The temperature values that were derived from the IR sensor measurements.

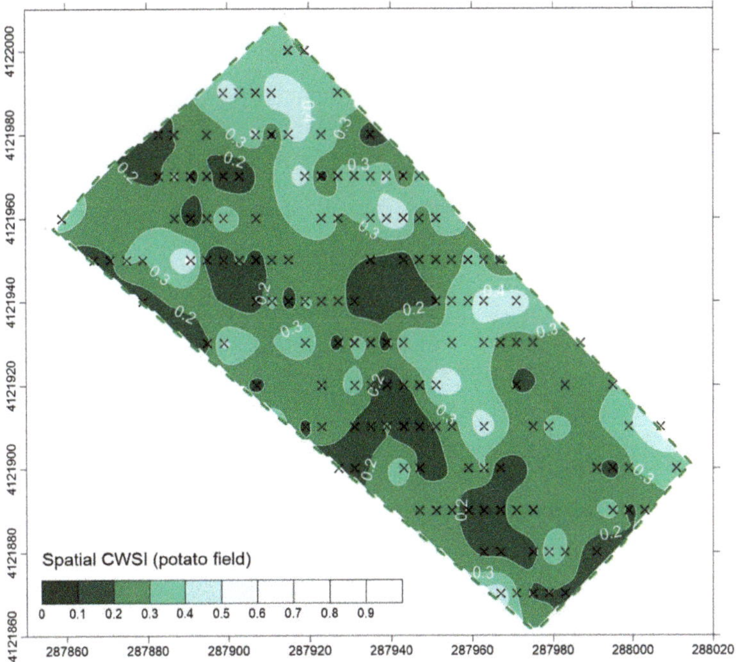

Figure 18. Spatial estimates (isolines of CWSI) from the first pilot flight with AMMS.

The calibration resulting from the spring potato cultivation (non-stressed lower base-line: $T_c - T_a = -1.74 \times VPD - 1.23$, stressed upper baseline $T_c - T_a = 2.32\ °C$) was used in the first pilot approach of the spatial CWSI in the autumn potato cultivation. The high relative humidity that prevailed during the calibration season, in combination with the low temperatures, gave a limited range of vapor pressure deficit of the atmosphere (VPD), leading to a high slope of the lower non-stressed baseline bottom line with a relatively high dispersion of the observed measurements ($r^2 = 0.613$). However, this calibration provided satisfactory values for the spatial CWSI estimates in conjunction with the actual water loss from the ground profile just before the next irrigation (Figure 16 and Table 2).

5. Conclusions

The integration of new, improved methods and tools in modern agriculture, combined with the limitations of the water availability for irrigation, pushes research into new and smart approaches. A major advantage in this context is that, nowadays, it is both feasible and affordable to use proven and reliable methods that required expensive instruments and high installation costs in the past. Furthermore, in agricultural production, the involvement of irrigation water is crucial in maintaining the quality and conservation of it as a natural resource.

Nowadays, the use of UAVs in agriculture is universal. The continuous evolution of UAV technology with lighter materials and the evolution of battery technology allows for long flight periods with heavier equipment loads. However, spatial data from crops are mostly acquired only through multispectral and thermal cameras to date.

Taking advantage of the UAV evolution, GreenWaterDrone (GWD) proposes an innovative system, integrating aerial measurements of canopy temperatures for the near-real-time calculation of crop irrigation needs (crop water stress detection) and dynamic crop surveillance, achieving measurements with high temporal and spatial resolution, while at the same time, minimizing the intervention in crop activities and system maintenance costs. The GWD solution is especially suited for relatively small fields with increased crop diversity that cannot be efficiently addressed by satellite remote sensing. An onboard aerial micrometeorological system (AMMS) allows for calculating the known CWSI (crop water stress index) in spatial estimation for many crops in the area. Using the thermal images in parallel with spatial measurements by the infrared radiometers is an advantage that allows us to compare and calibrate the methodology at an accurate level. Furthermore, it must be noted that the estimation of the CWSI at the implementation level is achieved by the spatial estimates of the infrared thermometers from UAVs (AMMS). Moreover, it must be mentioned that the multispectral and photogrammetric cameras significantly assist the results by providing information concerning the canopy cover and field conditions useful for crop surveillance.

In this work, we present the GWD methodology, along with the equipment specifications used. All processes at all steps of applying the methodology on a pilot potato field are described, and preliminary results from the application are presented. From the pilot study, the functionality of the proposed system (GWD) was certified (accuracy of the UAV path and flight altitude, reliability of the aerial data acquisition system, communication stability between UAVs and ground base). Our findings indicated that the canopy temperatures derived from the ground meteorological station (GMMS), the AMMS, and the portable IRT radiometers produced a suitable thermal image from the surface of the crop. However, it should be noted that the scheduled flight took place in the middle of the autumn season when the crop evapotranspiration rate was very low (ET_c was 2.7 mm/d and the soil moisture profile remained almost unchanged).

The project subsystems can be useful for supporting applications that are significant for irrigation water management and programming, such as irrigation alerting and scheduling, crop surveillance, and irrigation water management. However, more efforts are necessary to make these technologies more user-friendly and available for all end users, covering different advantages for a precise crop water stress evaluation. Still, the stakeholders, farmers, and industry responded positively to this effort's augmented interest. Furthermore, similar pilot campaigns need to be conducted in other regions on different crops to enhance the applicability of the GWD system. There is ongoing work by the authors to collect data for calibrating the methodology for other crops.

Author Contributions: Conceptualization, S.A., E.P., N.P., P.P., I.C. and G.K.; methodology, S.A., E.P., N.P., I.C. and P.P.; software, S.A., E.P., N.P. and I.C.; validation, S.A. and E.P.; formal analysis, S.A., E.P. and N.P.; resources, S.A., E.P., N.P. and I.C.; data curation, S.A., E.P., S.V., E.-M.P., P.P. and A.P.; writing—original draft preparation, S.A., E.P., N.P. P.P., I.C., G.K., S.V. and E.-M.P.; writing—review and editing, S.A., E.P., N.P. and P.P.; supervision, S.A. and E.P. All authors read and agreed to the published version of the manuscript.

Funding: This research was co-financed by the European Regional Development Fund of the European Union and Greek national funds through the Operational Program Competitiveness, Entrepreneurship, and Innovation under the call RESEARCH—CREATE—INNOVATE (project code: T1EDK-04021).

Institutional Review Board Statement: Not applicable.

Informed Consent Statement: Not applicable.

Data Availability Statement: Data sharing is not applicable for this article.

Acknowledgments: We want to thank all the Department of Agriculture, Economics, and Veterinary Medicine of Trifylia, and especially the director of the Department, Antonios Paraskevopoulos, for their tremendous support throughout the project GreenWaterDrone. Furthermore, we want to acknowledge the farmers Nikos Xirokostas, Charalabos Papadopoulos, and George Kalianis for their collaboration in our experiments in their field.

Conflicts of Interest: The authors declare no conflict of interest.

References

1. Bin Abdullah, K. Use of water and land for food security and environmental sustainability. *Irrig. Drain.* **2006**, *55*, 219–222. [CrossRef]
2. Mulla, D.; Miao, Y. *Precision Farming*; CRC Press: Boca Raton, FL, USA, 2015; pp. 161–178.
3. Yang, Y.; Shang, S.; Jiang, L. Remote sensing temporal and spatial patterns of evapotranspiration and the responses to water management in a large irrigation district of North China. *Agric. For. Meteorol.* **2012**, *164*, 112–122. [CrossRef]
4. Peña-Arancibia, J.L.; Mainuddin, M.; Kirby, J.M.; Chiew, F.H.S.; McVicar, T.R.; Vaze, J. Assessing irrigated agriculture's surface water and groundwater consumption by combining satellite remote sensing and hydrologic modelling. *Sci. Total Environ.* **2016**, *542*, 372–382. [CrossRef] [PubMed]
5. Vecchio, Y.; De Rosa, M.; Adinolfi, F.; Bartoli, L.; Masi, M. Adoption of precision farming tools: A context-related analysis. *Land Use Policy* **2020**, *94*, 104481. [CrossRef]
6. Tsouros, D.C.; Bibi, S.; Sarigiannidis, P.G. A review on UAV-based applications for precision agriculture. *Information* **2019**, *10*, 349. [CrossRef]
7. Alexandris, S.; Proutsos, N. How significant is the effect of the surface characteristics on the Reference Evapotranspiration estimates? *Agric. Water Manag.* **2020**, *237*, 106181. [CrossRef]
8. Egea, G.; Padilla-Díaz, C.M.; Martinez-Guanter, J.; Fernández, J.E.; Pérez-Ruiz, M. Assessing a crop water stress index derived from aerial thermal imaging and infrared thermometry in super-high density olive orchards. *Agric. Water Manag.* **2017**, *187*, 210–221. [CrossRef]
9. Xie, Y.; Sha, Z.; Yu, M. Remote sensing imagery in vegetation mapping: A review. *J. Plant. Ecol.* **2008**, *1*, 9–23. [CrossRef]
10. Kavvadias, A.; Psomiadis, E.; Chanioti, M.; Gala, E.; Michas, S. Precision agriculture—Comparison and evaluation of innovative very high resolution (UAV) and LandSat data. In Proceedings of the 7th International Conference on Information and Communication Technologies in Agriculture, Food and Environment (HAICTA 2015), Kavala, Greece, 17–20 September 2015; Volume 1498.
11. Kavvadias, A.; Psomiadis, E.; Chanioti, M.; Tsitouras, A.; Toulios, L.; Dercas, N. Unmanned Aerial Vehicle (UAV) data analysis for fertilization dose assessment. In Proceedings of the SPIE—Remote Sensing for Agriculture, Ecosystems, and Hydrology XIX, Warsaw, Poland, 11–14 September 2017; Volume 10421.
12. Gago, J.; Douthe, C.; Coopman, R.E.; Gallego, P.P.; Ribas-Carbo, M.; Flexas, J.; Escalona, J.; Medrano, H. UAVs challenge to assess water stress for sustainable agriculture. *Agric. Water Manag.* **2015**, *153*, 9–19. [CrossRef]
13. Jones, H.; Sirault, X. Scaling of Thermal Images at Different Spatial Resolution: The Mixed Pixel Problem. *Agronomy* **2014**, *4*, 380–396. [CrossRef]

14. Nahry, A.H.E.; Ali, R.R.; Baroudy, A.A.E. An approach for precision farming under pivot irrigation system using remote sensing and GIS techniques. *Agric. Water Manag.* **2011**, *98*, 517–531. [CrossRef]

15. Anderson, K.; Gaston, K.J. Lightweight unmanned aerial vehicles will revolutionize spatial ecology. *Front. Ecol. Environ.* **2013**, *11*, 138–146. [CrossRef]

16. Zhang, L.; Niu, Y.; Zhang, H.; Han, W.; Li, G.; Tang, J.; Peng, X. Maize Canopy Temperature Extracted From UAV Thermal and RGB Imagery and Its Application in Water Stress Monitoring. *Front. Plant. Sci.* **2019**, *10*, 1270. [CrossRef]

17. Zhang, C.; Kovacs, J.M. The application of small unmanned aerial systems for precision agriculture: A review. *Precis. Agric.* **2012**, *13*, 693–712. [CrossRef]

18. Rud, R.; Cohen, Y.; Alchanatis, V.; Levi, A.; Brikman, R.; Shenderey, C.; Heuer, B.; Markovitch, T.; Dar, Z.; Rosen, C.; et al. Crop water stress index derived from multi-year ground and aerial thermal images as an indicator of potato water status. *Precis. Agric.* **2014**, *15*, 273–289. [CrossRef]

19. Cucho-Padin, G.; Rinza, J.; Ninanya, J.; Loayza, H.; Quiroz, R.; Ramírez, D.A. Development of an Open-Source Thermal Image Processing Software for Improving Irrigation Management in Potato Crops (*Solanum tuberosum* L.). *Sensors* **2020**, *20*, 472. [CrossRef] [PubMed]

20. Veysi, S.; Naseri, A.A.; Hamzeh, S.; Bartholomeus, H. A satellite based crop water stress index for irrigation scheduling in sugarcane fields. *Agric. Water Manag.* **2017**, *189*, 70–86. [CrossRef]

21. Lebourgeois, V.; Labbé, S.; Jacob, F.; Bégué, A. Atmospheric corrections of low altitude thermal infrared airborne images acquired over a tropical cropped area. In Proceedings of the International Geoscience and Remote Sensing Symposium (IGARSS), Boston, MA, USA, 7–11 July 2008; Institute of Electrical and Electronics Engineers Inc.: Piscataway, NJ, USA, 2008; Volume 3, pp. 672–675.

22. da Silva, B.B.; Ferreira, J.A.; Rammana, T.V.; Rodríguez, V.P. Crop Water Stress Index and Water-Use Efficiency for Melon (Cucumis melo L.) on Different Irrigation Regimes. *Agric. J.* **2007**, *2*, 31–37.

23. Sepulcre-Cantó, G.; Zarco-Tejada, P.J.; Jiménez-Muñoz, J.C.; Sobrino, J.A.; De Miguel, E.; Villalobos, F.J. Detection of water stress in an olive orchard with thermal remote sensing imagery. *Agric. For. Meteorol.* **2006**, *136*, 31–44. [CrossRef]

24. Gutiérrez, S.; Diago, M.P.; Fernández-Novales, J.; Tardaguila, J. Vineyard water status assessment using on-the-go thermal imaging and machine learning. *PLoS ONE* **2018**, *13*, e0192037. [CrossRef]

25. Baluja, J.; Diago, M.P.; Balda, P.; Zorer, R.; Meggio, F.; Morales, F.; Tardaguila, J. Assessment of vineyard water status variability by thermal and multispectral imagery using an unmanned aerial vehicle (UAV). *Irrig. Sci.* **2012**, *30*, 511–522. [CrossRef]

26. Moller, M.; Alchanatis, V.; Cohen, Y.; Meron, M.; Tsipris, J.; Naor, A.; Ostrovsky, V.; Sprintsin, M.; Cohen, S. Use of thermal and visible imagery for estimating crop water status of irrigated grapevine. *J. Exp. Bot.* **2006**, *58*, 827–838. [CrossRef]

27. Zhang, L.; Zhang, H.; Niu, Y.; Han, W. Mapping Maize Water Stress Based on UAV Multispectral Remote Sensing. *Remote Sens.* **2019**, *11*, 605. [CrossRef]

28. Yang, W.; Li, C.; Yang, H.; Yang, G.; Feng, H.; Han, L.; Niu, Q.; Han, D. Monitoring of canopy temperature of maize based on UAV thermal infrared imagery and digital imagery. *Nongye Gongcheng Xuebao/Trans. Chin. Soc. Agric. Eng.* **2018**, *34*, 68–75. [CrossRef]

29. Gonzalez-Dugo, V.; Zarco-Tejada, P.; Nicolás, E.; Nortes, P.A.; Alarcón, J.J.; Intrigliolo, D.S.; Fereres, E. Using high resolution UAV thermal imagery to assess the variability in the water status of five fruit tree species within a commercial orchard. *Precis. Agric.* **2013**, *14*, 660–678. [CrossRef]

30. Bellvert, J.; Zarco-Tejada, P.J.; Girona, J.; Fereres, E. Mapping crop water stress index in a "Pinot-noir" vineyard: Comparing ground measurements with thermal remote sensing imagery from an unmanned aerial vehicle. *Precis. Agric.* **2014**, *15*, 361–376. [CrossRef]

31. Berni, J.A.J.; Zarco-Tejada, P.J.; Suárez, L.; Fereres, E. Thermal and narrowband multispectral remote sensing for vegetation monitoring from an unmanned aerial vehicle. *IEEE Trans. Geosci. Remote Sens.* **2009**, *47*, 722–738. [CrossRef]

32. Matese, A.; Baraldi, R.; Berton, A.; Cesaraccio, C.; Di Gennaro, S.; Duce, P.; Facini, O.; Mameli, M.; Piga, A.; Zaldei, A. Estimation of Water Stress in Grapevines Using Proximal and Remote Sensing Methods. *Remote Sens.* **2018**, *10*, 114. [CrossRef]

33. Santesteban, L.G.; Di Gennaro, S.F.; Herrero-Langreo, A.; Miranda, C.; Royo, J.B.; Matese, A. High-resolution UAV-based thermal imaging to estimate the instantaneous and seasonal variability of plant water status within a vineyard. *Agric. Water Manag.* **2017**, *183*, 49–59. [CrossRef]

34. Jackson, R.D.; Idso, S.B.; Reginato, R.J.; Pinter, P.J. Canopy temperature as a crop water stress indicator. *Water Resour. Res.* **1981**, *17*, 1133–1138. [CrossRef]

35. Idso, S.B.; Jackson, R.D.; Pinter, P.J.; Reginato, R.J.; Hatfield, J.L. Normalizing the stress-degree-day parameter for environmental variability. *Agric. Meteorol.* **1981**, *24*, 45–55. [CrossRef]

36. Jackson, R.D.; Kustas, W.P.; Choudhury, B.J. A reexamination of the crop water stress index. *Irrig. Sci.* **1988**, *9*, 309–317. [CrossRef]

37. Awais, M.; Li, W.; Cheema, M.J.M.; Hussain, S.; AlGarni, T.S.; Liu, C.; Ali, A. Remotely sensed identification of canopy characteristics using UAV-based imagery under unstable environmental conditions. *Environ. Technol. Innov.* **2021**, *22*, 101465. [CrossRef]

38. Mwinuka, P.R.; Mbilinyi, B.P.; Mbungu, W.B.; Mourice, S.K.; Mahoo, H.F.; Schmitter, P. The feasibility of hand-held thermal and UAV-based multispectral imaging for canopy water status assessment and yield prediction of irrigated African eggplant (*Solanum aethopicum* L.). *Agric. Water Manag.* **2021**, *245*, 106584. [CrossRef]

39. Shao, G.; Han, W.; Zhang, H.; Liu, S.; Wang, Y.; Zhang, L.; Cui, X. Mapping maize crop coefficient Kc using random forest algorithm based on leaf area index and UAV-based multispectral vegetation indices. *Agric. Water Manag.* **2021**, *252*, 106906. [CrossRef]
40. Zhou, Z.; Majeed, Y.; Diverres Naranjo, G.; Gambacorta, E.M.T. Assessment for crop water stress with infrared thermal imagery in precision agriculture: A review and future prospects for deep learning applications. *Comput. Electron. Agric.* **2021**, *182*, 106019. [CrossRef]
41. de Castro, A.I.; Shi, Y.; Maja, J.M.; Peña, J.M. UAVs for Vegetation Monitoring: Overview and Recent Scientific Contributions. *Remote Sens.* **2021**, *13*, 2139. [CrossRef]
42. Han, Y.; Tarakey, B.A.; Hong, S.-J.; Kim, S.-Y.; Kim, E.; Lee, C.-H.; Kim, G. Calibration and Image Processing of Aerial Thermal Image for UAV Application in Crop Water Stress Estimation. *J. Sens.* **2021**, *2021*, 1–14. [CrossRef]
43. Zhang, L.; Han, W.; Niu, Y.; Chávez, J.L.; Shao, G.; Zhang, H. Evaluating the sensitivity of water stressed maize chlorophyll and structure based on UAV derived vegetation indices. *Comput. Electron. Agric.* **2021**, *185*, 106174. [CrossRef]
44. Irmak, S.; Haman, D.Z.; Bastug, R. Determination of Crop Water Stress Index for Irrigation Timing and Yield Estimation of Corn. *Agron. J.* **2000**, *92*, 1221–1227. [CrossRef]
45. Thom, A.S. Momentum, mass and heat exchange of vegetation. *Q. J. R. Meteorol. Soc.* **1972**, *98*, 124–134. [CrossRef]
46. Monteith, J.; Unsworth, M. *Principles of Environmental Physics: Plants, Animals, and the Atmosphere*; Academic Press: Cambridge, MA, USA, 2013.
47. Brutsaert, W. *Evaporation into the Atmosphere: Theory, History and Applications*; Springer: Berlin/Heidelberg, Germany, 2013.
48. Clothier, B.E.; Clawson, K.L.; Pinter, P.J.; Moran, M.S.; Reginato, R.J.; Jackson, R.D. Estimation of soil heat flux from net radiation during the growth of alfalfa. *Agric. For. Meteorol.* **1986**, *37*, 319–329. [CrossRef]
49. Thom, A.S.; Oliver, H.R. On Penman's equation for estimating regional evaporation. *Q. J. R. Meteorol. Soc.* **1977**, *103*, 345–357. [CrossRef]
50. Thornthwaite, C.W. An Approach toward a Rational Classification of Climate. *Geogr. Rev.* **1948**, *38*, 55. [CrossRef]
51. Tsiros, I.X.; Nastos, P.; Proutsos, N.D.; Tsaousidis, A. Variability of the aridity index and related drought parameters in Greece using climatological data over the last century (1900–1997). *Atmos. Res.* **2020**, *240*, 104914. [CrossRef]
52. McEvoy, J.F.; Hall, G.P.; Mcdonald, P.G. Evaluation of unmanned aerial vehicle shape, flight path and camera type for waterfowl surveys: Disturbance effects and species recognition. *PeerJ* **2016**, *4*, e1831. [CrossRef] [PubMed]
53. Mouazen, A.M.; Alexandridis, T.; Buddenbaum, H.; Cohen, Y.; Moshou, D.; Mulla, D.; Nawar, S.; Sudduth, K.A. Monitoring. In *Agricultural Internet of Things and Decision Support for Precision Smart Farming*; Elsevier Inc.: Amsterdam, The Netherlands, 2020; pp. 35–138. ISBN 9780128183731.
54. Verhoeven, G.; Doneus, M.; Briese, C.; Vermeulen, F. Mapping by matching: A computer vision-based approach to fast and accurate georeferencing of archaeological aerial photographs. *J. Archaeol. Sci.* **2012**, *39*, 2060–2070. [CrossRef]
55. Alexiou, S.; Deligiannakis, G.; Pallikarakis, A.; Papanikolaou, I.; Psomiadis, E.; Reicherter, K. Comparing High Accuracy t-LiDAR and UAV-SfM Derived Point Clouds for Geomorphological Change Detection. *ISPRS Int. J. Geo-Inf.* **2021**, *10*, 367. [CrossRef]
56. Kung, O.; Strecha, C.; Beyeler, A.; Zufferey, J.-C.; Floreano, D.; Fua, P.; Gervaix, F. The accuracy of automatic photogrammetric techniques on ultra-light UAV imagery. In Proceedings of the UAV-g 2011—Unmanned Aerial Vehicle in Geomatics, Zürich, Switzerland, 14–16 September 2011.
57. Darra, N.; Psomiadis, E.; Kasimati, A.; Anastasiou, A.; Anastasiou, E.; Fountas, S. Remote and Proximal Sensing-Derived Spectral Indices and Biophysical Variables for Spatial Variation Determination in Vineyards. *Agronomy* **2021**, *11*, 741. [CrossRef]
58. Andriolo, U.; Gonçalves, G.; Bessa, F.; Sobral, P. Mapping marine litter on coastal dunes with unmanned aerial systems: A showcase on the Atlantic Coast. *Sci. Total Environ.* **2020**, *736*, 139632. [CrossRef] [PubMed]
59. Rud, R.; Shoshany, M.; Alchanatis, V. Spatial-spectral processing strategies for detection of salinity effects in cauliflower, aubergine and kohlrabi. *Biosyst. Eng.* **2013**, *114*, 384–396. [CrossRef]
60. Melis, M.T.; Da Pelo, S.; Erbì, I.; Loche, M.; Deiana, G.; Demurtas, V.; Meloni, M.A.; Dessì, F.; Funedda, A.; Scaioni, M.; et al. Thermal remote sensing from UAVs: A review on methods in coastal cliffs prone to landslides. *Remote Sens.* **2020**, *12*, 1971. [CrossRef]
61. Processing Thermal Images—Support. Available online: https://support.pix4d.com/hc/en-us/articles/360000173463 -Processing-thermal-images (accessed on 15 June 2021).
62. López, A.; Jurado, J.M.; Ogayar, C.J.; Feito, F.R. A framework for registering UAV-based imagery for crop-tracking in Precision Agriculture. *Int. J. Appl. Earth Obs. Geoinf.* **2021**, *97*, 102274. [CrossRef]
63. Bellvert, J.; Zarco-Tejada, P.J.; Marsal, J.; Girona, J.; González-Dugo, V.; Fereres, E. Vineyard irrigation scheduling based on airborne thermal imagery and water potential thresholds. *Aust. J. Grape Wine Res.* **2016**, *22*, 307–315. [CrossRef]
64. García-Tejero, I.F.; Costa, J.M.; Egipto, R.; Durán-Zuazo, V.H.; Lima, R.S.N.; Lopes, C.M.; Chaves, M.M. Thermal data to monitor crop-water status in irrigated Mediterranean viticulture. *Agric. Water Manag.* **2016**, *176*, 80–90. [CrossRef]
65. Tang, Z.; Parajuli, A.; Chen, C.J.; Hu, Y.; Revolinski, S.; Medina, C.A.; Lin, S.; Zhang, Z.; Yu, L.X. Validation of UAV-based alfalfa biomass predictability using photogrammetry with fully automatic plot segmentation. *Sci. Rep.* **2021**, *11*, 3336. [CrossRef] [PubMed]
66. Bian, J.; Zhang, Z.; Chen, J.; Chen, H.; Cui, C.; Li, X.; Chen, S.; Fu, Q. Simplified Evaluation of Cotton Water Stress Using High Resolution Unmanned Aerial Vehicle Thermal Imagery. *Remote Sens.* **2019**, *11*, 267. [CrossRef]

67. Sagan, V.; Maimaitijiang, M.; Sidike, P.; Eblimit, K.; Peterson, K.; Hartling, S.; Esposito, F.; Khanal, K.; Newcomb, M.; Pauli, D.; et al. UAV-Based High Resolution Thermal Imaging for Vegetation Monitoring, and Plant Phenotyping Using ICI 8640 P, FLIR Vue Pro R 640, and thermoMap Cameras. *Remote Sens.* **2019**, *11*, 330. [CrossRef]
68. Vamvakoulas, C.; Argyrokastritis, I.; Papastylianou, P.; Papatheohari, Y.; Alexandris, S. Crop water stress index relationship with soybean seed, protein and oil yield under varying irrigation regimes in a Mediterranean environment. *Isr. J. Plant. Sci.* **2020**, *67*, 1–13. [CrossRef]
69. Cohen, Y.; Alchanatis, V.; Sela, E.; Saranga, Y.; Cohen, S.; Meron, M.; Bosak, A.; Tsipris, J.; Ostrovsky, V.; Orolov, V.; et al. Crop water status estimation using thermography: Multi-year model development using ground-based thermal images. *Precis. Agric.* **2015**, *16*, 311–329. [CrossRef]
70. Cohen, Y.; Alchanatis, V.; Saranga, Y.; Rosenberg, O.; Sela, E.; Bosak, A. Mapping water status based on aerial thermal imagery: Comparison of methodologies for upscaling from a single leaf to commercial fields. *Precis. Agric.* **2017**, *18*, 801–822. [CrossRef]
71. Berni, J.A.J.; Zarco-Tejada, P.J.; Sepulcre-Cantó, G.; Fereres, E.; Villalobos, F. Mapping canopy conductance and CWSI in olive orchards using high resolution thermal remote sensing imagery. *Remote Sens. Environ.* **2009**, *113*, 2380–2388. [CrossRef]
72. Thomson, S.J.; Ouellet-Plamondon, C.M.; DeFauw, S.L.; Huang, Y.; Fisher, D.K.; English, P.J. Potential and Challenges in Use of Thermal Imaging for Humid Region Irrigation System Management. *J. Agric. Sci.* **2012**, *4*, p103. [CrossRef]
73. Khanal, S.; Fulton, J.; Shearer, S. An overview of current and potential applications of thermal remote sensing in precision agriculture. *Comput. Electron. Agric.* **2017**, *139*, 22–32. [CrossRef]
74. Fuchs, M.; Kanemasu, E.T.; Kerr, J.P.; Tanner, C.B. Effect of Viewing Angle on Canopy Temperature Measurements with Infrared Thermometers. *Agron. J.* **1967**, *59*, 494–496. [CrossRef]
75. Alchanatis, V.; Cohen, Y.; Cohen, S.; Moller, M.; Sprinstin, M.; Meron, M.; Tsipris, J.; Saranga, Y.; Sela, E. Evaluation of different approaches for estimating and mapping crop water status in cotton with thermal imaging. *Precis. Agric.* **2010**, *11*, 27–41. [CrossRef]

 *hydrology*

Article

Estimation of Daily Potential Evapotranspiration in Real-Time from GK2A/AMI Data Using Artificial Neural Network for the Korean Peninsula

Jae-Cheol Jang *, Eun-Ha Sohn, Ki-Hong Park and Soobong Lee

National Meteorological Satellite Center, Korea Meteorological Administration, Jincheon 27803, Korea; ehsohn@korea.kr (E.-H.S.); parkkihong@korea.kr (K.-H.P.); sblee88@korea.kr (S.L.)
* Correspondence: jaecheol00@korea.kr; Tel.: +82-07-7850-5915

Abstract: Evapotranspiration (ET) is a fundamental factor in energy and hydrologic cycles. Although highly precise in-situ ET monitoring is possible, such data are not always available due to the high spatiotemporal variability in ET. This study estimates daily potential ET (PET) in real-time for the Korean Peninsula, via an artificial neural network (ANN), using data from the GEO-KOMPSAT 2A satellite, which is equipped with an Advanced Meteorological Imager (GK2A/AMI). We also used passive microwave data, numerical weather prediction (NWP) model data, and static data. The ANN-based PET model was trained using data for the period 25 July 2019 to 24 July 2020, and was tested by comparing with in-situ PET for the period 25 July 2020 to 31 July 2021. In terms of accuracy, the PET model performed well, with root-mean-square error (RMSE), bias, and Pearson's correlation coefficient (R) of 0.649 mm day^{-1}, -0.134 mm day^{-1}, and 0.954, respectively. To examine the efficiency of the GK2A/AMI-derived PET data, we compared it with in-situ ET measured at flux towers and with MODIS PET data. The accuracy of the GK2A/AMI-derived PET, in comparison with the flux tower-measured ET, showed RMSE, bias, and Pearson's R of 1.730 mm day^{-1}, 1.212 mm day^{-1}, and 0.809, respectively. In comparison with the in-situ PET, the ANN model produced more accurate estimates than the MODIS data, indicating that it is more locally optimized for the Korean Peninsula than MODIS. This study advances the field by applying an ANN approach using GK2A/AMI data and could play an important role in examining hydrologic energy for air-land interactions.

Keywords: evapotranspiration; GK2A/AMI; artificial neural network; Korean Peninsula

Citation: Jang, J.-C.; Sohn, E.-H.; Park, K.-H.; Lee, S. Estimation of Daily Potential Evapotranspiration in Real-Time from GK2A/AMI Data Using Artificial Neural Network for the Korean Peninsula. *Hydrology* **2021**, *8*, 129. https://doi.org/10.3390/hydrology8030129

Academic Editors: Aristoteles Tegos and Nikolaos Malamos

Received: 14 July 2021
Accepted: 23 August 2021
Published: 27 August 2021

Publisher's Note: MDPI stays neutral with regard to jurisdictional claims in published maps and institutional affiliations.

1. Introduction

Evapotranspiration (ET) reflects fundamental components of hydrologic and energy cycles of the Earth and is a key element in hydrological resource management [1]. As climate change has progressed, trends in drought and flood have shown different spatial variability, and the importance of hydrological system monitoring has been emphasized [2]. Accordingly, it is fundamental to quantify and monitor ET. However, since water resources are directly affected by regional hydrologic systems and meteorology, ET shows high spatial and temporal variability [3].

A major application of ET is drought monitoring. Climate change has altered drought trends, increasing the intensity, frequency, and extent of droughts [4]. Thereafter, numerous indices for drought monitoring have been developed, with several, such as the standardized precipitation evapotranspiration index [5], precipitation evapotranspiration difference condition index [6], reconnaissance drought index [7], and combined terrestrial evapotranspiration index [8], directly associated with ET. Based on these drought indices, many studies were conducted to investigate the long-term variability of water budget under specific climate change conditions [9], effects of climate elasticity of ET on water balance [10], spatiotemporal variability of drought characteristics [11], and impacts of drought events on agricultural production [12]. In addition to various applications of ET, the methods to estimate ET with higher accuracy and spatiotemporal resolution have also been studied.

ET can be classified depending on the soil moisture condition. Potential ET (PET) is defined as the water vapor transpired and evaporated from vegetation and soil in unlimited soil moisture conditions [13]. Actual ET (AET) represents the water vapor transferred from a surface under limited soil moisture conditions. Weighing lysimeters are the most accurate AET measuring instruments [14]. Although they measure AET directly, the available AET data are substantially limited for end-users [15]. To ensure the versatility of ET data, various ET estimating models have been developed that can be broadly classified into three types [16]: (1) fully physical combination models that deal with mass and energy transfer principles [17,18]; (2) semi-physical models that account for mass or energy transfer principles and are based on temperature and radiation [19,20]; and (3) black-box models that are based on empirical relationship, artificial neural networks (ANN), fuzzy, and genetic algorithm. Although there are various ET estimating models, the most widely used method is the Penman–Monteith (PM) method [21,22]. The PM method is a fully physical model developed by Penman [17] and later modified by Monteith [18]. This model is recommended as the global reference model for ET monitoring by the Food and Agriculture Organization of the United Nations (FAO).

Although in-situ ET measurements are highly precise, the spatial variability of ET is high, and the availability of in-situ ET measurements is limited [23]. Remotely sensed data have been used to address this problem. Satellite data have broad spatial coverage with high temporal resolution and produce reliable products [24]. MODerate resolution Imaging Spectroradiometer (MODIS) derives the operative ET products with 500 m spatial and 8-days temporal resolution [25]. Several studies have estimated the spatial distribution of ET using low Earth orbit (LEO) satellites with optical-infrared and microwave sensors [26,27]. When calculating ET using LEO satellites, external input data, such as meteorological data, are generally necessary [28]. In particular, because LEO satellites observe the Earth's surface at specific local times, it is difficult for the instantaneous observation to monitor the environmental conditions all day and all weather [29]. Therefore, due to the high temporal variability of ET, LEO satellite-derived ET has inevitable limitations for routine monitoring of daily ET and surface energy fluxes [11,30]. In addition, since LEO satellites apply the physical-based model or energy conservation-based model for estimating ET, there exist uncertainties of external input data for applying the model [25–27]. Using geostationary orbit (GEO) satellites data can compensate for the limitations associated with the temporal resolution of LEO satellite data. However, it is difficult to resolve the uncertainties of external input data and the data contaminated by weather conditions, including clouds and aerosols [29].

The Korean Peninsula is located on the margin of Northeast Asia, bordering the northwest Pacific Ocean (Figure 1a). Since it is located in a monsoon region, where meteorological droughts occur during the summer monsoon, the droughts tend to propagate into agricultural or hydrological droughts [31]. In particular, the Korean Peninsula land cover type showed complex spatial distribution comprising of diverse vegetation cover types (Figure 1b). Furthermore, in the Korean Peninsula, the drought frequency has increased, and drought trends and characteristics vary regionally [32]. The Korean Peninsula has various land cover types and specific terrain properties; these factors make it particularly difficult to monitor daily ET even employing both in-situ measurement and remotely sensed data. Due to frequent cloud cover and rainfall, it is challenging to observe the land surface using optical-infrared satellites in the summer monsoon season [33]. Therefore, to overcome this limitation, numerical model data and ancillary data have been used to retrieve ET [34,35].

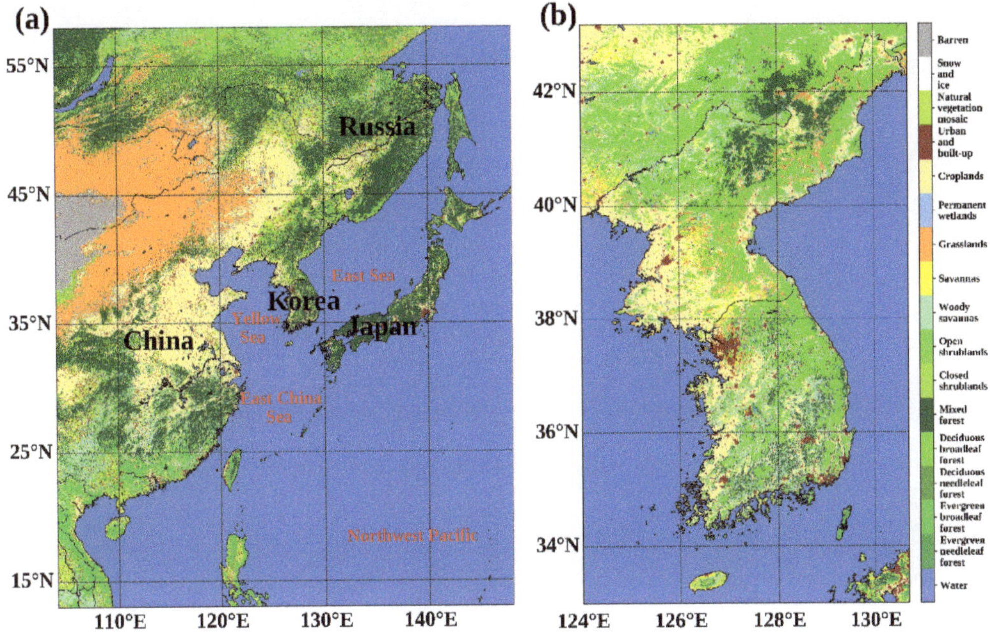

Figure 1. MODIS land cover from the Annual International Geosphere-Biosphere Programme over (**a**) Northeast Asia and (**b**) the Korean Peninsula in 2019.

In order to manage hydrological resources over the Korean Peninsula, Korea Meteorological Administration (KMA) monitors the ET in real-time using in-situ measurements and numerical model data. In-situ measurements exhibit good performance with high temporal resolution every hour; however, its availability is limited due to the point observation. For complementing the limitation of in-situ measurements, KMA calculates the spatial distribution of ET using numerical model data based on geophysical models. Numerical model data-derived ET is suitable for analyzing droughts with a large time scale. In contrast, the accuracy of the ET changes depending on the numerical model data, and it is difficult to calculate the ET that reflects various topographical characteristics of the Korean Peninsula due to the sparse spatial resolution of the numerical model data.

Although physical-based models show good performance, due to numerous associated meteorological parameters, it is difficult to estimate accurate ET, especially in remote sensing applications. Then, over the last few decades, many researchers have identified that machine learning (ML) approaches were an effective method to overcome the complexity of ET estimation [29]. Because ML techniques solve the non-linear relationship between input and output variables, a lot of ML techniques have been proposed to estimate ET for hydrological applications [36], such as k-nearest neighbors [37], support vector machine [38], random forest [37], and artificial neural network (ANN) [39]. Previously, most studies applied ML approaches to in-situ measurements; however, many recent studies have also applied ML approaches to remote sensing data [40–42].

In this study, considering the spatiotemporal variability in ET, we developed a model that estimates daily PET based on ANN using the GEOstationary Korea Multi-Purpose SATellite 2A (GEO-KOMPSAT 2A, GK2A). The objective was to retrieve real-time daily ET with a spatial resolution of 1 km for hydrological resource monitoring on the Korean Peninsula. To reflect the complex relationships and nonlinearity between the GK2A-derived data and ET, we used precipitation data and the digital elevation data as input data for the ANN. Daily PET from KMA was used as reference data for ANN model training.

The accuracy of the model was verified by comparing modeled data with ET from in-situ measurements of the KMA and National Institute of Forest Science (NIFoS) for the period excluding the period of training data.

2. Data and Methods

2.1. Remote Sensing Data

2.1.1. GEO-KOMPSAT 2A (GK2A)

GK2A, launched on 4 December 2018 and operated by the KMA National Meteorological Satellite Center (NMSC), is equipped with the Advanced Meteorological Imager (AMI). AMI is the optical-infrared sensor with 16 channels and its spatial resolution ranges from 0.5 to 2.0 km depending on wavelength (Table 1). Since GK2A/AMI observes the Earth with a high spatiotemporal resolution, it is more capable of monitoring the Earth's hydrological system than previous GEO satellite (Communication, Ocean and Meteorological Satellite, COMS) operated by KMA NMSC and other LEO satellites [43]. We used seven GK2A/AMI operational products: Reflected Shortwave Radiation (RSR), Downward Shortwave Radiation (DSR), Absorbed Shortwave Radiation (ASR), Outgoing Longwave Radiation (OLR), Downward Longwave Radiation (DLR), Upward Longwave Radiation (ULR), and Normalized Difference Vegetation Index (NDVI).

Table 1. Specifications of the GEO-KOMPSAT 2A Advanced Meteorological Imager (GK2A/AMI) spectral channels.

Channel No.	Channel Name	Wavelength Range (μm)	Resolution (km)
1	VIS004	0.431–0.479	1.0×1.0
2	VIS005	0.5025–0.5175	1.0×1.0
3	VIS006	0.625–0.66	0.5×0.5
4	VIS008	0.8495–0.8705	1.0×1.0
5	NR013	1.373–1.383	2.0×2.0
6	NR016	1.601–1.619	2.0×2.0
7	SW038	3.74–3.96	2.0×2.0
8	WV063	6.061–6.425	2.0×2.0
9	WV069	6.89–7.01	2.0×2.0
10	WV073	7.258–7.433	2.0×2.0
11	IR087	8.44–8.76	2.0×2.0
12	IR096	9.543–9.717	2.0×2.0
13	IR105	10.25–10.61	2.0×2.0
14	IR112	11.08–11.32	2.0×2.0
15	IR123	12.15–12.45	2.0×2.0
16	IR133	13.21–13.39	2.0×2.0

2.1.2. Precipitation Data

Integrated Multi-satellitE Retrievals for Global Precipitation Measurement (IMERG) version 6 were used to calculate ET, even for areas for which precipitation data were not available. The IMERG precipitation products are derived from the global precipitation measurement constellation comprising the various passive microwave sensors including the meteorological operational satellite series, polar operational environmental satellite series, and global change observation mission 1st-water satellite [44]. The data from various passive microwave satellites are merged into $0.1° \times 0.1°$ resolution every 30 min. We used the standardized precipitation index for six months (SPI6), derived from the precipitation product of IMERG, rather than daily precipitation data.

2.2. Numerical Model and Elevation Data

Since 2010, the KMA has used numerical weather prediction (NWP) systems from the Unified Model (UM). NWP model data from UM systems, operated by KMA in real-time, could be classified depending on spatial coverage and boundary conditions, and we used Local Data Assimilation and Prediction System (LDAPS) over the Korean Peninsula in this study. LDAPS is based on boundary conditions derived by three-dimensional variational data assimilation and its spatial resolution of 1.5 km [45]. LDAPS has 70 vertical layers and provides 36-h predictions (at every 00, 06, 12, and 18 UTC), and additional 3-h predictions (at every 03, 09, 15, and 21 UTC). We used four meteorological parameters—air temperature (Ta), surface temperature (Ts), relative humidity (RH), and wind speed (WS)—from LDAPS version 10.1. Furthermore, Shuttle Radar Topography Mission (SRTM) digital elevation model (DEM) data were used to reflect the effect of elevation on ET, and its spatial resolution was an arc-second, approximately 30 m [46].

2.3. In-Situ Measurements

PM equation calculates the PET using micrometeorological data, and the eddy covariance (EC) systems estimate the AET based on energy flux observations [47]. PET derived from the PM method was used for model training and validation. On the other hand, since the AET derived from EC systems was different from PET, we only used the AET data for testing the availability of the PET model.

Since the Korean Peninsula has specific geographic characteristics, each region shows different weather conditions and climate properties. KMA operates 81 Automated Surface Observing System (ASOS) stations in real-time. In this study, we used 42 of these that monitor ET_0 based on the PM equation (we hereafter refer to ET obtained using the PM equation as PM-ET) (Figure 2a). ASOS stations observe the following meteorological parameters every hour: Ta, Ts, RH, WS, soil temperature, precipitation, surface pressure, and net solar radiation (https://data.kma.go.kr/cmmn/main.do, accessed on 13 July 2021).

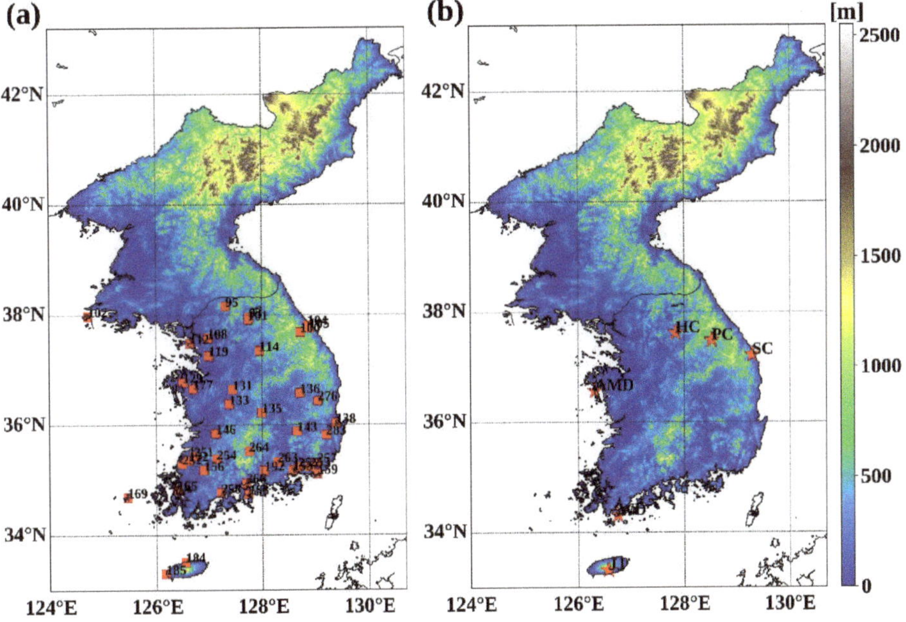

Figure 2. Distribution of the digital elevation model (DEM), where the red squares and stars indicate the (**a**) Automated Surface Observing System (ASOS) stations and (**b**) flux towers, respectively.

To evaluate the ANN model-derived ET, we used ET calculated using the EC method. The NIFoS operates six flux towers to monitor ET on the Korean Peninsula (Figure 2b). These flux towers observe meteorological parameters every 30 min (http://know.nifos. go.kr/know/service/flux/fluxIntro.do, accessed on 13 July 2021). Using these direct observations of vertical flux and meteorological data, it is possible to calculate ET via the

EC method. From the ASOS stations and flux towers, we selected only those variables observed for full 24-h periods.

2.4. Processing

Figure 3 illustrates the process used here to estimate and evaluate daily ET using GK2A/AMI data. We preprocessed the input data; the preprocessed data were then subsampled (at 1 km resolution) around the Korean Peninsula. We constructed matchups between the subsampled data and PM-ET, and classified them into two datasets (training and testing) depending on the acquisition date. For the ANN model training, we used five-fold cross-validation; 80% of the data were used to optimize the weights and biases of the model, and 20% were used to verify the accuracy and monitor the loss function of the model, to minimize overfitting. To enable the ANN model to reflect seasonal variation, we set the training period for the training data to 1 year (25 July 2019 to 24 July 2020). ANN model performance was assessed using PM-ET and EC-ET data for the period 25 July 2019 to 31 July 2021.

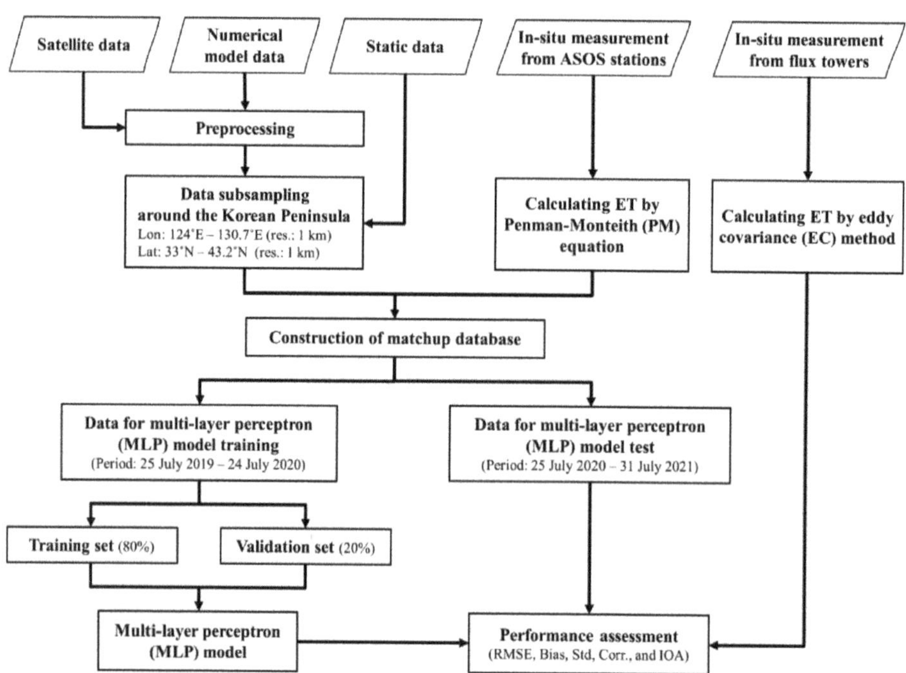

Figure 3. Flowchart illustrating the construction and assessment of the evapotranspiration (ET) retrieval artificial neural network (ANN) model.

To estimate daily ET via GK2A/AMI data, we used 22 parameters as input variables of the ANN model (Table 2). The GK2A/AMI operational products include the preprocessed daily means of six radiation variables (RSR, DSR, ASR, OLR, DLR, and ULR) and the 16 days maximum NDVI. The GPM IMERG precipitation product was preprocessed to generate SPI6. We used four UM LDAPS variables (Ta, Ts, RH, and WS) affecting ET. To take into account diurnal variation in ET, we preprocessed NWP variables to daily mean, daily minimum, and daily maximum. As static data, we used extraterrestrial solar radiation (ESR) and a DEM to account for seasonal variation and the terrain effect, respectively.

Table 2. Spatial and temporal resolution, processing method, and input data source for the artificial neural network (ANN) model, where Mean, Max, Min, and Sum indicate average, maximum, minimum, and cumulative values, respectively.

Data	Variables	Spatial Resolution (Temporal Resolution)	Processing	Source
GK2A/AMI	RSR DSR ASR OLR DLR ULR	2 km × 2 km (10 min)	$Mean_{1day}$	KMA NMSC
	NDVI	2 km × 2 km (1 day)	Max_{16days}	
GPM IMERG	SPI6	10 km × 10 km (1 day)	$Sum_{6months}$	NASA
UM LDAPS	Ta Ts RH WS	1.5 km × 1.5 km (3 h)	$Mean_{1day}$, Max_{1day}, Min_{1day}	Met Office
Static data	ESR	–	–	–
	DEM	30 m × 30 m	–	NASA

2.4.1. Extraterrestrial Solar Radiation (ESR)

ESR indicates solar radiation incident outside the Earth from the Sun. ESR is a key parameter for estimating ET, and can be calculated using the latitude and the day of the year as follows [21,48]:

$$R_a = \frac{24 \times 60}{\pi} G_{SC} d_r (\omega_S \sin \varphi \sin \delta + \cos \varphi \cos \delta \sin \omega_S),\tag{1}$$

where R_a refers to ESR; G_{SC} denotes the solar constant; d_r represents the inverse of the relative distance between the Earth and the Sun; ω_S indicates the Sun and sunset hour angle; φ and δ refer to latitude and solar declination, respectively.

2.4.2. Penman–Monteith Evapotranspiration (PM-ET)

We calculated hourly PM-ET from in-situ KMA ASOS station measurements. To account for diurnal variability in ET, we also derived daily PM-ET from hourly PM-ET. It is possible to estimate hourly PM-ET as follows [21]:

$$ET_o = \frac{0.408\Delta(R_n - G) + \gamma \frac{37}{T_{hr}+273} u_2 (e^o(T_{hr}) - e_a)}{\Delta + \gamma(1 + 0.34u_2)},\tag{2}$$

where ET_o indicates hourly ET; T_{hr} and u_2 represents hourly mean air temperature and hourly mean wind speed, respectively; Δ denotes the saturation slope vapor pressure at T_{hr}; γ and R_n denote the psychrometric constant and the net radiation at the surface, respectively; G and e^o refer to the soil heat flux density and the saturation vapor pressure at T_{hr}, respectively; e_a indicates hourly mean actual vapor pressure. KMA ASOS station calculated hourly PM-ET every hour, and the cumulative PM-ET over 24 h was used as the daily PM-ET.

2.4.3. Standardization of Input Variables

In an ANN model, when the input variables are linearly related, it is not necessary to standardize or normalize them. However, when the input variables show a non-linear relationship in the ANN model, before using input variables, it is important to standardize or normalize them [49]. When using the variables without standardization or normalization, large values of the input variables would cause very small weighting factors, and small values of the input variables would result in very large weighting factors, which could cause some problems during training and optimizing process [50]; Using extremely small weights would cause the uncertainties of floating-point calculations on computer; not using extremely small initial weights would make the improvement of the model by the backpropagation algorithm insignificantly small [51]. There are no fixed methods of standardization that should be used in specific applications; in this study, standardization was applied to input variables as follows:

$$V' = \frac{(V - V_{mean})}{V_{std}},$$ (3)

where V' and V indicate the standardized input variable and unstandardized input variable, respectively; V_{mean} represents the mean of input variable; V_{std} denotes the standard deviation of the input variable.

2.5. ANN Model

2.5.1. Model Structure

We used a multilayer perceptron (MLP), ANN, to estimate daily ET. MLP involves feedforward backpropagation networks with a simple structure and high performance; they have therefore been used for diverse applications using satellite data [52,53]. These neurons are interconnected, with weights and biases that enable repetitive learning. Each hidden layer has an activation function computing the neuronal weights and biases. An optimizer algorithm trains the network and minimizes the error, by correcting the weights and biases via a backpropagation process [54]. We developed a five-layer MLP model with hidden layers of 200 neurons. In MLP model training, input values of neurons in the previous layer transfer to a neuron in the current layer, and a neuron combines the input values with weights and biases as follows [51]:

$$n_j = \sum x_i w_{ij} - b_j,$$ (4)

where n_j represent the net of the weighted input for the jth neuron; x_i indicate the input transferred from the ith neuron; w_{ij} refers to the weight connected from the ith neuron to the jth neuron; b_j means the bias of the jth neuron. In n_j, for being a final output for passing to the next layer, it should be activated by the activation function [49]. The activation function can be a diverse discrete or continuous function; we used the exponential linear unit (ELU), showing fine performance with a fast learning rate and significantly better generalization as follows [55]:

$$f(x) = \begin{cases} x & if \ x > 0 \\ \alpha(\exp(x) - 1) & if \ x \le 0 \end{cases}$$ (5)

where α represents the hyperparmeter controlling the value where an ELU saturates for negative n_j; x denotes the input value and indicates the n_j.

For improving and accelerating the convergence, we used the batch normalization (BN) layer between each hidden layer [56]. The normalization is calculated based on the dimension of the batch and BN ensures that the input of each hidden layer is distributed in the same way. Their performance dramatically depends on the batch size, and setting a larger batch size generally yields better performance [57]. We used a method for stochastic optimization (ADAM) as the optimizer algorithm [58]. The parameters and hyperparameters of the MLP model are summarized in Table 3. To train and run the MLP model, we used Keras with the TensorFlow back-end in Python.

Table 3. Parameters and hyperparameters of the multilayer perceptron (MLP) model.

Parameter		Hyperparameter	
Activation	ELU	Alpha	1
Optimizer	ADAM	Learning rate	10^{-4}
		Beta1	0.9
		Beta2	0.999
		Epsilon	10^{-7}
Loss function	RMSE		
Epochs	100		
Batch size	500		

2.5.2. Mean Decrease Accuracy (MDA)

In a black-box model such as an ANN model, it is difficult to analyze the information and structure of the model in detail. However, it is possible to rank the importance that each input variable occupies in the model. In this study, in order to analyze the trained MLP model, we conducted a permutation test of each input variable. This test randomly permutes the list of a variable and measures the decrease of model accuracy; this process was conducted repeatedly with each variable; finally, the Mean Decrease Accuracy (MDA; also known as the permutation importance) was calculated with each variable [59]. A variable with a larger MDA is interpreted as an important variable in the model because the accuracy of the variable greatly affects the accuracy of the model. We used the MDA in terms of the increase in RMSE when each variable was randomly permutated.

2.6. Statistical Analysis

Daily ET, estimated via MLP, was compared with PM-ET and EC-ET. To quantitatively evaluate the MLP-derived daily ET, we used the bias [60], root-mean-square error (RMSE) [36], mean absolute error (MAE) [36], standard deviation (STD) [60], normalized RMSE (nRMSE) [61], Pearson's correlation coefficient (R) [36], and the Index of Agreement (IOA) [62]. The detailed equations are as follows:

$$\text{Bias} = \frac{1}{N} \sum_{i=1}^{N} (E_i - O_i),$$ (6)

$$\text{RMSE} = \sqrt{\frac{\sum_{i=1}^{N} (E_i - O_i)^2}{N}},$$ (7)

$$\text{MAE} = \frac{1}{N} \sum_{i=1}^{N} |E_i - O_i|,$$ (8)

$$\text{STD} = \sqrt{\frac{\sum_{i=1}^{N} (E_i - O_i - Bias)^2}{N}},$$ (9)

$$\text{nRMSE} = \frac{\sqrt{\frac{\sum_{i=1}^{N} (E_i - O_i)^2}{N}}}{\frac{\sum_{i=1}^{N} O_i}{N}},$$ (10)

$$R = \frac{\sum_{i=1}^{N} (E_i - \overline{E})(O_i - \overline{O})}{\sqrt{\sum_{i=1}^{N} (E_i - \overline{E})^2} \sqrt{\sum_{i=1}^{N} (O_i - \overline{O})^2}},$$ (11)

$$\text{IOA} = 1 - \frac{\sum_{i=1}^{N} (E_i - O_i)^2}{\sum_{i=1}^{N} (|E_i - \overline{O}| + |O_i - \overline{O}|)^2},$$ (12)

where E_i and O_i represent the estimated ET and observed ET, respectively; the subscript i denotes the ith data point; N refers to the number of data; $\overline{E}$ and $\overline{O}$ represent the mean of the estimated ET and observed ET, respectively.

3. Results

3.1. Input Data Correlation

Figure 4 describes the correlations between the input variables used in estimating daily ET, and ET from the KMA ASOS stations, for the Korean Peninsula. Fifteen of the variables (ESR, DSR, ASR, DLR, OLR, ULR, NDVI, Ta_{mean}, Ta_{min}, Ta_{max}, Ts_{mean}, Ts_{min}, Ts_{max}, WS_{mean}, and WS_{max}) were positively correlated with daily ET from the KMA ASOS stations. As the radiation incident on the surface and the temperature increases, evaporation increases, because sufficient energy to convert water into water vapor is provided, and transpiration increases because vegetation activity accelerates [63]. Seven variables (i.e., RSR, SPI6, RH_{mean}, RH_{min}, RH_{max}, WS_{min}, and DEM) were negatively correlated with daily ET from the KMA ASOS stations. As higher RH is associated with less water vapor transported from the water surface, RH was negatively correlated with ET. Since precipitation increases surface water content and inhibits evaporation, SPI6 was negatively correlated to ET. As RSR increases, the radiation incident on the surface decreases, reducing both evaporation and transpiration. The mean, maximum, and minimum WS showed different correlations with ET; this could be because the complex topography of the Korean Peninsula, in terms of spatiotemporal variability in WS, causes uncertainty of the LDAPS model WS estimates. Overall, the positive correlations were stronger than the negative correlations. Relative to ET, DSR had the strongest positive correlation (0.86), and RH_{mean} had the largest negative correlation (-0.45).

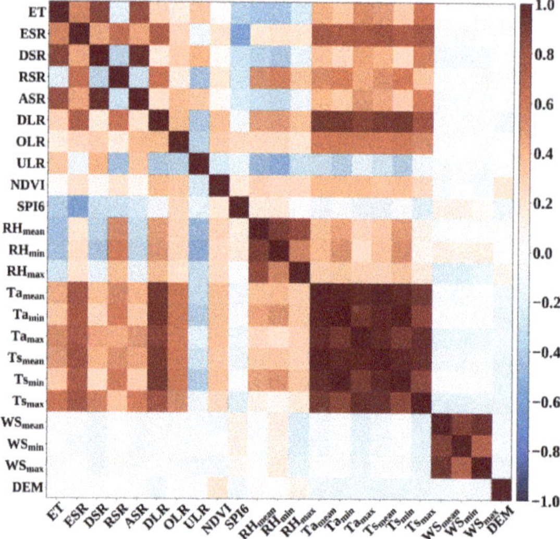

Figure 4. Correlation coefficient between the input variables in the matchups from 25 July 2019 to 24 July 2020.

3.2. MLP Model

Figure 5 describes the MLP model training history. Training the MLP model involves minimizing RMSE (the loss function) by optimizing neuronal bias and weight. Up to training epoch 50, RMSE and MAE decreased rapidly, but after epoch 70, the accuracy slightly improved. By training epoch 100, the change in RMSE and MAE of both the training and validation datasets were almost negligible.

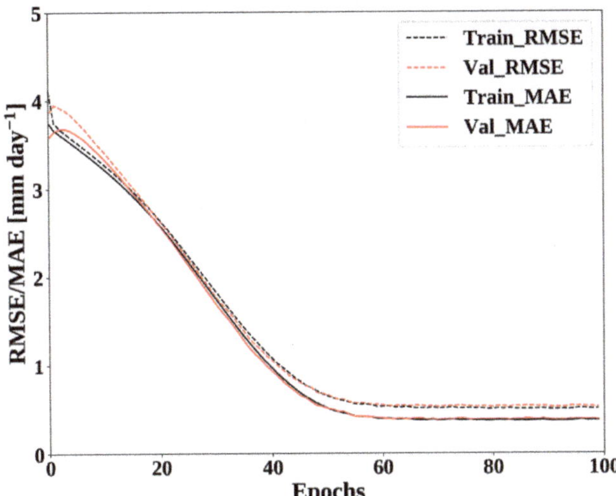

Figure 5. Changes in RMSE and MAE with respect to the training epochs of the ANN model training epochs, where Red and black lines indicate validation and training data sets, respectively; Dotted and solid lines represent RMSE and MAE, respectively.

Figure 6 shows the MDA of 22 input variables in the ANN model. ESR and RSR showed high MDA (>1.5 mm day^{-1}), which means that ET is predominantly affected by radiation energy. ESR, which is used directly in the PM equation, showed an MDA of 1.63 mm day^{-1}. RSR, which measures the shortwave radiation that emits outside the Earth, is principally controlled by clouds and surface albedo. These land and meteorological conditions directly affect the parameters in the PM equation, which explains the high MDA values of RSR and DSR, at 1.56 and 0.72 mm day^{-1}, respectively.

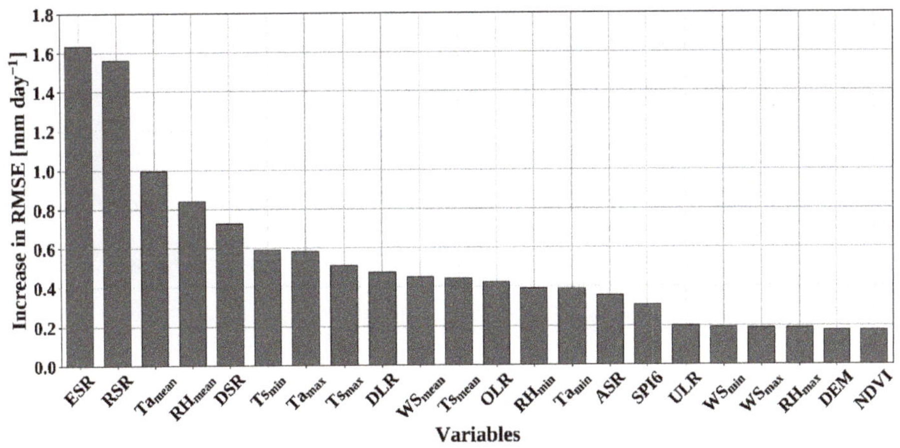

Figure 6. Mean Decrease Accuracy (MDA), expressed as the increase in RMSE, of ET derived from the ANN model.

Ta, Ts, and RH are directly related to ET estimation, via the PM equation. However, since they are derived from numerical model data with uncertainty, they showed relatively low MDAs, from 0.99 to 0.19 mm day^{-1}. The variables describing WS, which are used directly in the PM-ET estimation, showed lower MDAs (<0.45 mm day^{-1}) than the other meteorological variables. This reflects the fact that it is difficult to simulate transient changes in wind caused by sudden gusts or topography using numerical model-based wind data. PET reflects the rate of ET when sufficient soil moisture is available; hence it does not account for vegetation and terrain characteristics. As a result, NDVI and DEM showed lower MDA values (<0.18 mm day^{-1}).

3.3. Evaluation against KMA Stations

We compared GK2A-derived daily ET for the Korean Peninsula with PM-ET derived from KMA ASOS stations for the period 25 July 2020 to 31 July 2021 (Figure 7). KMA ASOS-derived PM-ET (mm day^{-1}) ranged from 0.28 to 14.41, and GK2A/AMI-derived PET (mm day^{-1}) ranged from 0.00 to 11.10. In comparison with PM-ET derived from KMA ASOS stations, the total number of matchup data was 15,414, and GK2A/AMI-derived PET showed accuracy (mm day^{-1}) of 0.649 (RMSE), 0.488 (MAE), 0.636 (STD), and −0.134 (bias) with nRMSE of 0.168, indicating the MLP model tended to underestimate relative to the in-situ PM-ET overall. In particular, at PET values less than 2.0 mm day^{-1}, the tendency of underestimation of the MLP model was remarkable. Although the MLP model shows the tendency to underestimate, its underestimation was slight overall and it shows good performance estimating PM-ET from the KMA ASOS stations; Pearson's *R* was 0.954, and IOA was 0.975.

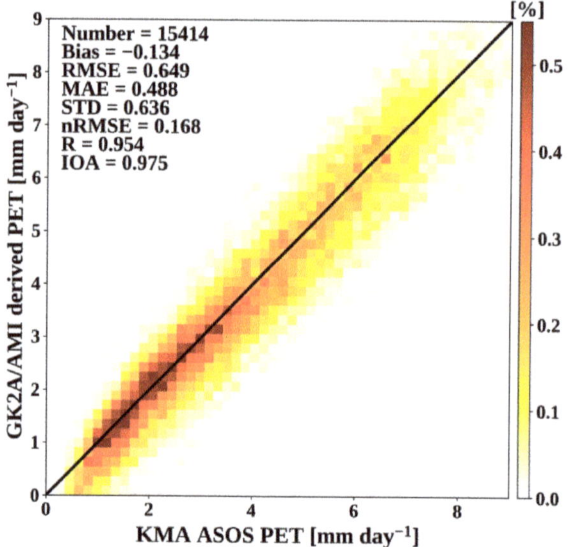

Figure 7. Comparison between GK2A/AMI satellite-derived PET estimates and PM-ET from ASOS stations operated by KMA, for the period 25 July 2020 to 31 July 2021. The color represents the proportion of the data relative to the total number of matchups.

From 25 July 2020 to 31 July 2021, we verified the accuracy of PET derived from GK2A/AMI by comparing them with the PM-ET from the KMA ASOS stations (Figure 8). RMSE (mm day^{-1}) ranged from 0.449 (at station 136) to 0.871 (at station 185), nRMSE ranged from 0.117 (at station 159) to 0.237 (at station 169), and STD (mm day^{-1}) ranged from 0.449 (at station 136) to 0.861 (at station 185) (Figure 8a–c). Bias (mm day^{-1}) ranged from −0.568 (at station 172) to 0.215 (at station 108) (Figure 8d). Pearson's *R* ranged from 0.891 (at station 181) to 0.979 (at station 136), and IOA ranged from 0.939 (at station 185) to 0.988 (at station 136). Overall, the PET estimated from GK2A/AMI using the MLP model were accurate relative to the PM-ET from KMA ASSOS stations (Figure 8e,f).

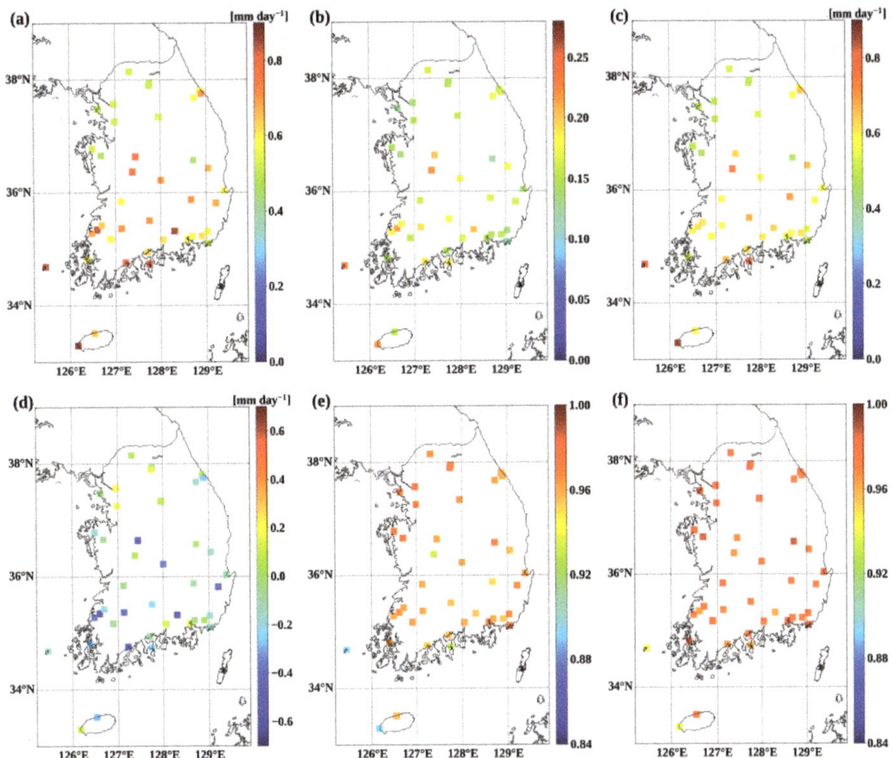

Figure 8. Spatial representation of the comparison between GK2A/AMI satellite-derived PET, and PET from ASOS stations operated by KMA, for the period 25 July 2020 to 31 July 2021. Accuracy is represented by (**a**) RMSE, (**b**) nRMSE, (**c**) STD, (**d**) bias, (**e**) Pearson's *R*, and (**f**) IOA.

We examined the seasonal characteristics of GK2A/AMI-derived PET. We simply classified the seasons into two classes; we hereafter referred to the period when monthly mean value of observed PET was less than 3 mm day^{-1} as cold seasons (November to February), and the period when monthly mean value of observed PET was more than 3 mm day^{-1} as warm seasons (March to October). In the cold seasons, KMA ASOS-derived PM-ET and GK2A/AMI-estimated PET both had lower values than in the warm seasons (Table 4). In cold seasons, RMSE (mm day^{-1}) ranged from 0.399 to 0.671, Pearson's *R* ranged from 0.881 to 0.908, and nRMSE ranged from 0.193 to 0.244 (Table 5). On the other hand, in warm seasons, RMSE (mm day^{-1}) ranged from 0.585 to 0.804, Pearson's *R* ranged from 0.901 to 0.960, and nRMSE ranged from 0.116 to 0.207. Regardless of seasons, the model was found to show low RMSE less than 0.81 mm day^{-1} and high Pearson's *R* more than 0.88, indicating that the model simulates the in-situ PET with high accuracy.

When compared to the warm seasons, the cold seasons show good performance in terms of RMSE, MAE, and STD, but poor performance in terms of nRMSE, Pearson's *R*, and IOA. These seasonal differences are caused by the seasonal variation of PET. As shown in Table 4, the lower the temperature, the lower the water vapor evaporated from soil and transpired by vegetation; the variation of PET in the warm seasons is higher than in the cold seasons [64,65]. Therefore, the low variation of PET in the cold seasons causes low RMSE, MAE, and STD; however, due to the small magnitude of PET in cold seasons, even a small error substantially affects the ratio-dependent accuracy score such as nRMSE, Pearson's *R*, and IOA.

Table 4. Comparison of observed PET and GK2A/AMI satellite-derived PET estimates.

Month	Observed PET (mm day^{-1})			Estimated PET (mm day^{-1})		
	Minimum	Maximum	Mean	Minimum	Maximum	Mean
August 2020	0.52	9.74	4.29	0.18	8.32	4.09
September 2020	0.69	8.66	3.78	0.00	8.25	3.27
October 2020	0.79	9.44	3.90	0.00	7.88	3.37
November 2020	0.39	6.96	2.76	0.00	5.98	2.52
December 2020	0.34	5.06	2.06	0.06	4.47	2.09
January 2021	0.30	6.52	1.87	0.00	4.76	1.84
February 2021	0.50	10.16	2.97	0.00	7.61	2.83
March 2021	0.28	9.04	3.67	0.08	8.72	3.59
April 2021	0.66	12.28	5.44	0.42	10.59	5.48
May 2021	0.54	14.41	5.14	0.65	11.10	5.27
June 2021	0.40	11.30	5.25	0.94	10.32	5.32
July 2021	0.52	10.22	5.50	0.92	9.47	5.31

Table 5. Accuracy (in terms of bias, RMSE, MAE, STD, nRMSE, R, and IOA), of the GK2A/AMI satellite-derived estimated PET, with respect to the month.

Month	No.	Bias (mm day^{-1})	RMSE (mm day^{-1})	MAE (mm day^{-1})	STD (mm day^{-1})	nRMSE	R	IOA
August 2020	1260	−0.208	0.671	0.510	0.638	0.156	0.949	0.968
September 2020	1234	−0.506	0.782	0.645	0.597	0.207	0.931	0.940
October 2020	1286	−0.529	0.804	0.651	0.605	0.206	0.901	0.913
November 2020	1241	−0.237	0.575	0.446	0.524	0.208	0.908	0.941
December 2020	1289	0.027	0.399	0.304	0.398	0.193	0.881	0.937
January 2021	1291	−0.028	0.456	0.353	0.455	0.244	0.883	0.932
February 2021	1170	−0.142	0.671	0.466	0.625	0.216	0.885	0.936
March 2021	1294	−0.073	0.585	0.448	0.581	0.160	0.954	0.974
April 2021	1250	0.035	0.763	0.582	0.762	0.140	0.928	0.963
May 2021	1293	0.131	0.704	0.512	0.692	0.137	0.960	0.979
June 2021	1249	0.067	0.609	0.457	0.605	0.116	0.955	0.977
July 2021	1275	−0.186	0.710	0.530	0.685	0.129	0.940	0.965

4. Discussions

4.1. NIFoS Flux Towers

Because the ANN-based daily ET model was trained using the PM-ET data from the KMA ASOS stations, we examined the availability of the GK2A/AMI-derived PET by comparing it with EC-ET data. We compared daily PET derived from GK2A/AMI for the Korean Peninsula with EC-ET derived from NIFoS flux tower, for the period 25 July 2020 to 31 July 2021 (Figure 9). NIFoS flux tower-derived EC-ET (mm day^{-1}) ranged from 0.02 to 9.82, and GK2A/AMI-derived PET (mm day^{-1}) ranged from 0.00 to 10.06. In comparison with EC-ET derived from NIFoS flux tower, the total number of matchup data was 654, and GK2A/AMI-derived PET showed the accuracy (mm day^{-1}) of 1.730 (RMSE), 1.409 (MAE), 1.235 (STD), and 1.212 (bias) with nRMSE of 0.525, indicating the PET derived from GK2A/AMI using the MLP model tended to overestimate relative to the EC-ET derived from NIFoS flux tower overall. The model performed in following the trend in the EC-ET data; Pearson's R was 0.809, and IOA was 0.822.

In theoretical conditions, the PET derived from the PM method was not expected to match with the AET derived from the EC method. Although the differences depend on the environmental conditions and PET retrieval methods, the PM method generally overestimated ET compared with EC-ET in both hourly and daily time scales [47]. However, the comparison result shows a high correlation with both variables and between the input parameters for both variables, which indicates that PM-ET and EC-ET are affected by the same factors [66,67]. Because the PM method quantifies water vapor loss in sufficient soil moisture conditions, it overestimates ET relative to EC-ET under the dry conditions [67].

However, in sufficient soil moisture conditions on rainy days, the PM method nonetheless overestimated ET relative to EC-ET [49,68]. Furthermore, the differences between PM-ET and EC-ET depend on the environmental conditions, the tendency to overestimate ET was strong with intense net radiation and water vapor deficit [67,69]. Another possible reason for the overestimation is that PM-ET does not consider the complicated structure of the forest. The comparison result between PM-ET and EC-ET depended on the reference level, and the accuracy of PM-ET increased with the reference level of measurement [47]. The PM method assumes that the vegetation is a single big leaf, and ET occurs on a surface with zero plane displacement. However, vegetation conditions vary depending on the spatiotemporal environment, and ET occurs in the forest floor to the top of vegetation. On the other hand, during the vegetation growing season with low leaf area index, surface and underground ET take a substantial part of the water vapor cycle. Because of that, PM-ET could underestimate ET at a small leaf area index, compared with EC-ET [47]. Another possible reason for overestimation is that the PM method cannot accurately include the resistance due to the surface canopy or soil conditions [69]. Since PM-ET data depend highly on surface conductance; its overestimation could cause the overestimation of ET [47,70]. Although the PM model overestimated ET, it showed a high correlation with the EC-ET data. Since the model accounts for radiative and aerodynamic conditions, it might produce more reliable estimates of AET than other PET models [71].

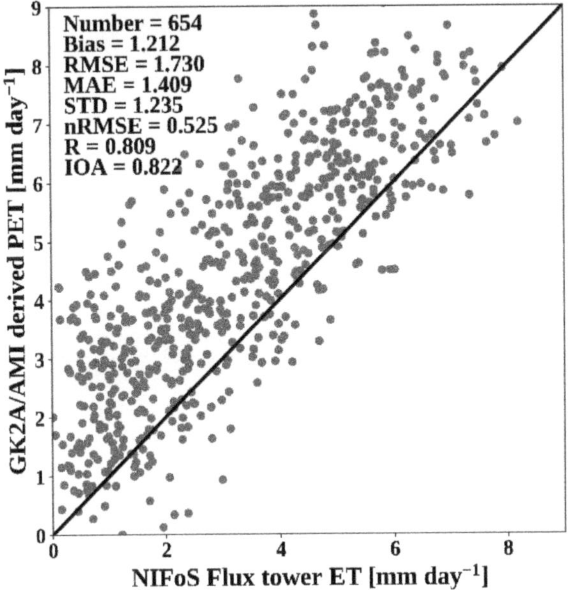

Figure 9. Comparison between GK2A/AMI satellite-derived and NIFoS flux tower-derived ET from 25 July 2020 to 31 July 2021.

4.2. Comparison with MODIS

To validate the GK2A/AMI-derived daily PET data, we compared it with the Terra/MODIS PET product. Because Terra/MODIS produces an 8-days PET composite, we produced 8-days aggregates of daily PET data from the GK2A/AMI satellite and from the KMA ASOS stations. In the KMA ASOS stations, when the number of daily PET data for 8-days was less than 8, it was excluded from the validation data. We then compared the Terra/MODIS PET data with the KMA ASOS station and GK2A/AMI satellite PET data, for the period 27 July 2020 to 27 July 2021 (Figure 10).

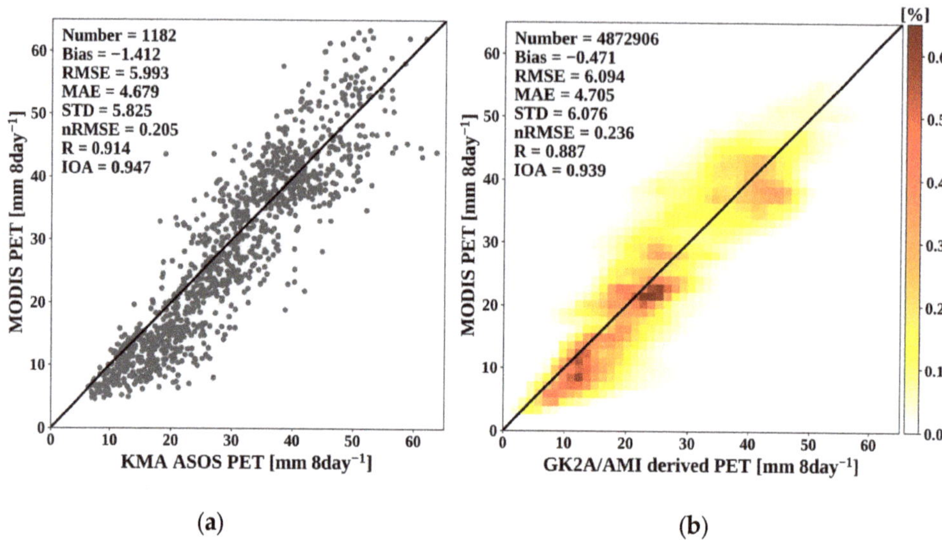

Figure 10. Validation of Terra/MODIS PET data for the period 27 July 2020 to 27 July 2021, relative to PET from (**a**) ASOS stations operated by KMA, and (**b**) the GK2A/AMI satellite.

In comparison with the KMA ASOS station PET data, the Terra/MODIS PET data showed accuracy (mm 8 day^{-1}) of 5.993 (RMSE), 4.679 (MAE), 5.825 (STD), and −1.412 (bias) with an nRMSE of 0.205; Pearson's R was 0.914 and IOA was 0.947, indicating the Terra/MODIS PET data tended to underestimate PET relative to KMA ASOS (Figure 10a). The underestimation of the Terra/MODIS PET data was remarkably shown in the PET of less than 20 mm 8 day^{-1}. In previous studies, the MODIS-based PET product was converted to daily PET and compared with PM-ET. The assessment of MODIS-based PET product varied on the land cover and showed Pearson's R of 0.71 to 0.94 [72,73]. Although the previous studies and this study used the verification with a daily and 8-day product, respectively, the high Pearson's R means that MODIS-based PET product is useful for ET monitoring on the Korean Peninsula. In comparison with the GK2A/AMI-derived PET data, the Terra/MODIS PET data showed accuracy (mm 8 day^{-1}) of 6.094 (RMSE), 4.705 (MAE), 6.076 (STD), and −0.471 (bias) with an nRMSE of 0.236; Pearson's R was 0.887 and IOA was 0.939, indicating the Terra/MODIS PET data tended to underestimate PET relative to GK2A/AMI (Figure 10b). The underestimation of the Terra/MODIS PET data was remarkably shown in the PET of less than 20 mm 8 day^{-1}, indicating the comparing result of GK2A was consistent with that of KMA ASOS.

For the assessment of the spatial distribution of GK2A/AMI-derived PET, we verified the accuracy of Terra/MODIS PET relative to the PET data for each KMA ASOS station and GK2A/AMI coordinate for the period 27 July 2020 to 27 July 2021 (Figure 11). In comparison with the KMA ASOS station data, RMSE (mm 8 day^{-1}) ranged from 3.056 (at station 119) to 10.061 (at station 105); bias (mm 8 day^{-1}) ranged from −5.692 (at station 105) to 1.075 (at station 177); and Pearson's R ranged from 0.748 (at station 185) to 0.981 (at station 119) (Figure 11a–c). Relative to the GK2A/AMI-derived PET, RMSE (mm 8 day^{-1}) ranged from 1.445 to 17.039, bias (mm 8 day^{-1}) from −14.549 to 13.627, and Pearson's R from 0.305 to 0.991 (Figure 11d–f). In Terra/MODIS PET, the result compared with KMA ASOS PET (Figure 11a–c) was consistent with that of GK2A/AMI-PET (Figure 11d–f). In particular, in the eastern region of the Korean Peninsula, it showed high RMSE, negative bias, and low Pearson's R compared with the other area in the Korean Peninsula.

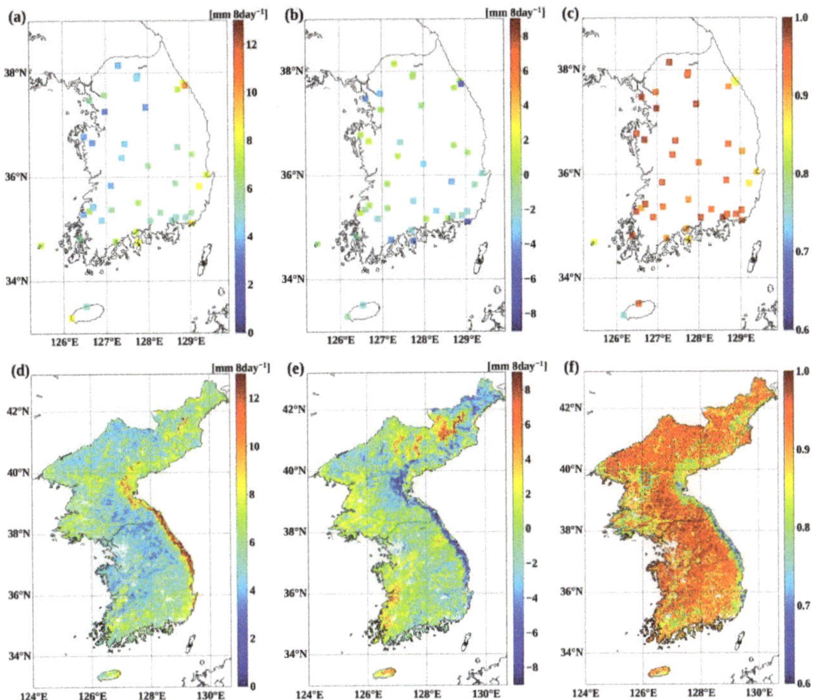

Figure 11. Accuracy of Terra/MODIS PET for the period 27 July 2020 to 27 July 2021 at (**a–c**) the ASOS stations operated by KMA, and (**d–f**) the coordinates of the GK2A/AMI satellite. The accuracy is represented by (**a,d**) RMSE, (**b,e**) bias, and (**c,f**) Pearson's *R*.

The KMA ASOS station-derived PM-ET data showed a Pearson correlation of 0.914 with Terra/MODIS PET (Figure 10a), and 0.954 with GK2A/AMI-derived PET (Figure 7). While the Terra/MODIS PET algorithm is optimized for global coverage, our MLP model was locally optimized for the Korean Peninsula. Furthermore, since our MLP model used daily remotely sensed and numerical model product not related to cloud, the GK2A-derived PET shows fine temporal resolution and has no masked value due to cloud relative to Terra/MODIS product. Therefore, the GK2A/AMI-derived PET performed better than Terra/MODIS for estimating PET on the Korean Peninsula. Relative to the GK2A/AMI-derived PET and in-situ PM-ET data, the consistency of the Terra/MODIS PET data decreased remarkably for the eastern region of the Korean Peninsula (Figure 11). In the eastern coastal area of the Korean Peninsula, elevation decreases dramatically (Figure 2). In contrast to the lack of consistency with the Terra/MODIS PET data, the GK2A/AMI-derived PET and in-situ PM-ET were highly correlated (Pearson's *R* > 0.879), regardless of the topography (Figure 8). This result indicates that Terra/MODIS did not reflect the local terrain characteristics of the Korean Peninsula, due to its global optimization. Thus, for ET monitoring with high spatiotemporal variability on the Korean Peninsula, the real-time daily GK2A/AMI-derived PET was more suitable (due to local optimization) than the 8-days Terra/MODIS PET product.

4.3. Previous Studies on the Korean Peninsula

The Korean Peninsula comprises various vegetation cover types and shows specific agrometeorological characteristics, and it is able to perform agrometeorological analysis using ET data. When investigating the ensemble model of virtual water content based on ET, it was found that the ensemble virtual water content and production of rice and maize decreased in future projections, which affected future water consumption on the Korean Peninsula [74]. Birhanu et al. [75], when constructing hydrological models, investigated the effect of model complexity and ET calculation methods on model performance based on the in-situ measurement. Um et al. [76] estimated the spatial distribution of ET based on in-situ measurements using the hybrid Kriging method and revealed various ET characteristics depending on the distance from the coast and elevation level above the ground surface. Jung et al. [77] developed the physiological modules to simulate the canopy photosynthesis and ET process and established the relationship of photosynthesis and ET with crop production based on satellite data and in-situ measurements. Similar to this study, Kim et al. [41] developed the ML model estimating daily PET for the Korean Peninsula using satellite data and NWP data. MODIS-based monthly vegetation index data, multi-microwave satellite-derived precipitation data, and LDAPS data were used as input data of the random forest model. The model showed accuracy (mm day^{-1}) of 1.038 (RMSE), 0.790 (MAE), and 0.007 (bias) with Pearson's R of 0.870. The model developed in this study not only has better accuracy but also has the advantage of retrieval in real-time.

5. Conclusions

This paper presents an ANN model that retrieves daily PET in real-time for the Korean Peninsula, using GK2A/AMI-derived data, microwave composite data, and NWP data. We used the data from 25 July 2019 to 24 July 2020 for model training, and 25 July 2020, to 31 July 2021 for model testing. In comparison with the KMA ASOS station-derived PM-ET, the ANN-based GK2A-derived PET showed high accuracy (mm day^{-1}) of 0.649 (RMSE) and -0.134 (bias); Pearson's R of 0.954; and IOA of 0.975. In validating the spatial distribution, the ANN model-estimated daily PET showed high accuracy at all KMA ASOS stations. To assess the efficiency of the GK2A/AMI-derived PET, we verified it using NIFoS flux tower-derived EC-ET, which showed that GK2A/AMI-derived PET overestimated ET. Furthermore, we assessed the performance of our ANN model by comparing it with operational Terra/MODIS PET products with 8-days temporal resolution. Because it was locally optimized, our ANN model outperformed Terra/MODIS PET over the Korean Peninsula. GK2A/AMI-derived PET performed particularly better than the Terra/MODIS PET product for the eastern coastal region of the Korean Peninsula, where elevation changes dramatically.

Although GK2A/AMI-derived PET showed high accuracy, it is necessary to extend its spatial coverage for overcoming its local optimization. When applying the additional in-situ measurements on other areas to the model, it is possible to improve the model in terms of spatial coverage. Furthermore, in order to develop the model estimating ET, we used and optimized the MLP model, but it is able to apply diverse ANN methods such as recurrent neural network, convolutional neural network, and long short-term memory. When applying and validating various ANN methods, it is possible to improve the accuracy of the model estimating ET.

ET is a key indicator to investigate the effects of the meteorological drought on vegetation activities. GK2A/AMI-derived 2-dimensional ET is thought to be a useful tool in examining the drought affecting the Korean Peninsula. In further studies, we will attempt to investigate drought on the Korean Peninsula by examining the relationship of GK2A/AMI-derived ET and precipitation data with vegetation information. This study contributes to understanding air-land interactions, and the development of ANN approaches using satellite and NWP data.

Author Contributions: Conceptualization, J.-C.J.; methodology, validation, and writing—original draft preparation, J.-C.J.; writing—review and editing, E.-H.S., K.-H.P. and S.L. All authors have read and agreed to the published version of the manuscript.

Funding: This work was funded by the Korea Meteorological Administration's Research and Development Program "Technical Development on Weather Forecast Support and Convergence Service using Meteorological Satellites" under Grant (KMA2020-00120).

Institutional Review Board Statement: Not applicable.

Informed Consent Statement: Not applicable.

Data Availability Statement: GK-2A/AMI data used in this study are available on http://datasvc.nmsc.kma.go.kr/datasvc/html/main/main.do?lang=en (accessed on 13 July 2021). In-situ data from KMA ASOS stations and NIFoS flux towers used in this study are available on https://data.kma.go.kr/cmmn/main.do and http://know.nifos.go.kr/know/service/flux/fluxIntro.do, respectively (accessed on 13 July 2021).

Conflicts of Interest: The authors declare no conflict of interest.

Abbreviations

ADAM	A Method for Stochastic Optimization
AET	Actual Evapotranspiration
AMI	Advanced Meteorological Imager
ANN	Artificial Neural Networks
ASR	Absorbed Shortwave Radiation
BN	Batch Normalization
COMS	Communication, Ocean and Meteorological Satellite
DEM	Digital Elevation Model
DLR	Downward Longwave Radiation
DSR	Downward Shortwave Radiation
EC	Eddy Covariance
ELU	Exponential Linear Unit
ESR	Extraterrestrial Solar Radiation
ET	Evapotranspiration
FAO	Food and Agriculture Organization of the United Nations
GEO	Geostationary Orbit
GK2A	GEOstationary Korea Multi-Purpose SATellite 2A
IMERG	Integrated Multi-satellitE Retrievals for Global Precipitation Measurement
KMA	Korea Meteorological Administration
LDAPS	Local Data Assimilation and Prediction System
LEO	Low Earth Orbit
MDA	Mean Decrease Accuracy
ML	Machine Learning
MLP	Multilayer Perceptron
MODIS	Moderate Resolution Imaging Spectroradiometer
NDVI	Normalized Difference Vegetation Index
NIFoS	National Institute of Forest Science
NMSC	National Meteorological Satellite Center
NWP	Numerical Weather Prediction
OLR	Outgoing Longwave Radiation
PET	Potential Evapotranspiration
PM	Penman-Monteith
RSR	Reflected Shortwave Radiation
SPI6	Standardized Precipitation Index for Six Months
SRTM	Shuttle Radar Topography Mission
ULR	Upward Longwave Radiation
UM	Unified Model

References

1. Zhao, L.L.; Xia, J.; Xu, C.Y.; Wang, Z.G.; Sobkowiak, L.; Long, C. Evapotranspiration estimated methods in hydrological simulation. *J. Geogr. Sci.* **2013**, *23*, 359–369. [CrossRef]
2. Fan, Z.; Thomas, A. Decadal changes of reference crop evapotranspiration attribution: Spatial and temporal variability over China 1960–2011. *J. Hydrol.* **2018**, *560*, 461–470. [CrossRef]
3. Pilgrim, D.H.; Chapman, R.G.; Doran, D.G. Problems of rainfall-runoff modelling in arid and semiarid regions. *Hydrol. Sci. J.* **1988**, *33*, 379–400. [CrossRef]

4. Iglesias, A.; Garrote, L. Adaptation strategies for agricultural water management under climate change in Europe. *Agric. Water Manag.* **2015**, *155*, 113–124. [CrossRef]

5. Vicente-Serrano, S.M.; Beguería, S.; López-Moreno, J.I. A multiscalar drought index sensitive to global warming: The standardized precipitation evapotranspiration index. *J. Clim.* **2010**, *23*, 1696–1718. [CrossRef]

6. Tian, L.; Leasor, Z.T.; Quiring, S.M. Developing a hybrid drought index: Precipitation Evapotranspiration Difference Condition Index. *Clim. Risk Manag.* **2020**, *29*, 100238. [CrossRef]

7. Tsakiris, G.; Vangelis, H. Establishing a drought index incorporating evapotranspiration. *Eur. Water* **2005**, *9*, 3–11.

8. Elbeltagi, A.; Kumari, N.; Dharpure, J.K.; Mokhtar, A.; Alsafadi, K.; Kumar, M.; Mehdinejadiani, B.; Ramezani Etedali, H.; Brouziyne, Y.; Towfiqul Islam, A.R.M.; et al. Prediction of Combined Terrestrial Evapotranspiration Index (CTEI) over Large River Basin Based on Machine Learning Approaches. *Water* **2021**, *13*, 547. [CrossRef]

9. Tadese, M.; Kumar, L.; Koech, R. Long-term variability in potential evapotranspiration, water availability and drought under climate change scenarios in the Awash River Basin, Ethiopia. *Atmosphere* **2020**, *11*, 883. [CrossRef]

10. Avanzi, F.; Rungee, J.; Maurer, T.; Bales, R.; Ma, Q.; Glaser, S.; Conklin, M. Climate elasticity of evapotranspiration shifts the water balance of Mediterranean climates during multi-year droughts. *Hydrol. Earth Syst. Sci.* **2020**, *24*, 4317–4337. [CrossRef]

11. Liu, C.; Yang, C.; Yang, Q.; Wang, J. Spatiotemporal drought analysis by the standardized precipitation index (SPI) and standardized precipitation evapotranspiration index (SPEI) in Sichuan Province, China. *Sci. Rep.* **2021**, *11*, 1280. [CrossRef]

12. Abdelmalek, M.B.; Nouiri, I. Study of trends and mapping of drought events in Tunisia and their impacts on agricultural production. *Sci. Total Environ.* **2020**, *734*, 139311. [CrossRef]

13. Yao, N.; Li, Y.; Dong, Q.G.; Li, L.; Peng, L.; Feng, H. Influence of the accuracy of reference crop evapotranspiration on drought monitoring using standardized precipitation evapotranspiration index in mainland China. *Land Degrad. Dev.* **2020**, *31*, 266–282. [CrossRef]

14. Allen, R.G.; Fisher, D.K. Low-Cost Electronic Weighing Lysimeters. *Trans. ASAE* **1990**, *33*, 1823–1833. [CrossRef]

15. Xu, C.Y.; Chen, D. Comparison of seven models for estimation of evapotranspiration and groundwater recharge using lysimeter measurement data in Germany. *Hydrol. Process.* **2005**, *19*, 3717–3734. [CrossRef]

16. Srivastava, A.; Sahoo, B.; Raghuwanshi, N.S.; Singh, R. Evaluation of Variable-Infiltration Capacity Model and MODIS-Terra Satellite-Derived Grid-Scale Evapotranspiration Estimates in a River Basin with Tropical Monsoon-Type Climatology. *J. Irrig. Drain. Eng.* **2017**, *143*, 04017028. [CrossRef]

17. Penman, H.L. Natural evaporation from open water, bare soil and grass. *Proc. R. Soc. Lond. A* **1948**, *194*, 120–145.

18. Monteith, J.L. Evaporation and environment. *Symp. Soc. Exp. Biol.* **1965**, *19*, 205–224. [PubMed]

19. Priestley, C.H.B.; Taylor, R.J. On the assessment of surface heat flux and evaporation using large scale parameters. *Mon. Weather Rev.* **1972**, *100*, 81–92. [CrossRef]

20. Hargreaves, G.H.; Samani, Z.A. Reference crop evapotranspiration from temperature. *Appl. Eng. Agric.* **1985**, *1*, 96–99. [CrossRef]

21. Allen, R.G.; Pereira, L.S.; Raes, D.; Smith, M. *Crop Evapotranspiration: Guidelines for Computing Crop Water Requirements*; Food and Agriculture Organization of the United Nations: Rome, Italy, 1998.

22. Allen, R.G.; Smith, M.; Perrier, A.; Pereira, L.S. An update for the definition of reference evapotranspiration. *ICID Bull.* **1994**, *43*, 1–34.

23. Wilby, R.L.; Yu, D. Rainfall and temperature estimation for a data sparse region. *Hydrol. Earth Syst. Sci.* **2013**, *17*, 3937–3955. [CrossRef]

24. Courault, D.; Seguin, B.; Olioso, A. Review on estimation of evapotranspiration from remote sensing data: From empirical to numerical modeling approaches. *Irrig. Drain. Syst.* **2005**, *19*, 223–249. [CrossRef]

25. Mu, Q.; Heinsch, F.A.; Zhao, M.; Running, S.W. Development of a global evapotranspiration algorithm based on MODIS and global meteorology data. *Remote Sens. Environ.* **2007**, *111*, 519–536. [CrossRef]

26. McCabe, M.F.; Wood, E.F. Scale influences on the remote estimation of evapotranspiration using multiple satellite sensors. *Remote Sens. Environ.* **2006**, *105*, 271–285. [CrossRef]

27. Long, D.; Longuevergne, L.; Scanlon, B.R. Uncertainty in evapotranspiration from land surface modeling, remote sensing, and GRACE satellites. *Water Resour. Res.* **2014**, *50*, 1131–1151. [CrossRef]

28. Gomis-Cebolla, J.; Jimenez, J.C.; Sobrino, J.A.; Corbari, C.; Mancini, M. Intercomparison of remote-sensing based evapotranspiration algorithms over amazonian forests. *Int. J. Appl. Earth Obs. Geoinf.* **2019**, *80*, 280–294. [CrossRef]

29. Chia, M.Y.; Huang, Y.F.; Koo, C.H.; Fung, K.F. Recent advances in evapotranspiration estimation using artificial intelligence approaches with a focus on hybridization techniques—A Review. *Agron. Basel* **2020**, *10*, 101. [CrossRef]

30. Zhang, L.; Lemeur, R. Evaluation of daily evapotranspiration estimates from instantaneous measurements. *Agric. For. Meteorol.* **1995**, *74*, 139–154. [CrossRef]

31. Bae, H.; Ji, H.; Lim, Y.-J.; Ryu, Y.; Kim, M.-H.; Kim, B.-J. Characteristics of drought propagation in South Korea: Relationship between meteorological, agricultural, and hydrological droughts. *Nat. Hazards* **2019**, *99*, 1–16. [CrossRef]

32. Azam, M.; Maeng, S.; Kim, H.; Lee, S.; Lee, J. Spatial and Temporal Trend Analysis of Precipitation and Drought in South Korea. *Water* **2018**, *10*, 765. [CrossRef]

33. Jang, K.; Kang, S.; Lim, Y.; Jeong, S.; Kim, J.; Kimball, J.S.; Hong, S.Y. Monitoring daily evapotranspiration in Northeast Asia using MODIS and a regional Land Data Assimilation System. *J. Geophys. Res. Atmos.* **2013**, *118*, 927–940. [CrossRef]

34. Yang, F.; White, M.A.; Michaelis, A.R.; Ichii, K.; Hashimoto, H.; Votava, P.; Zhu, A.-X.; Nemani, R.R. Prediction of continental-scale evapotranspiration by combining MODIS and AmeriFlux data through support vector machine. *IEEE Trans. Geosci. Remote Sens.* **2006**, *44*, 3452–3461. [CrossRef]

35. Yuan, W.; Liu, S.; Yu, G.; Bonnefond, J.-M.; Chen, J.; Davis, K.; Desai, A.R.; Goldstein, A.H.; Gianelle, D.; Rossi, F.; et al. Global estimates of evapotranspiration and gross primary production based on MODIS and global meteorology data. *Remote Sens. Environ.* **2010**, *114*, 1416–1431. [CrossRef]

36. Fan, J.; Yue, W.; Wu, L.; Zhang, F.; Cai, H.; Wang, X.; Lu, X.; Xiang, Y. Evaluation of SVM, ELM and four tree-based ensemble models for predicting daily reference evapotranspiration using limited meteorological data in different climates of China. *Agric. For. Meteorol.* **2018**, *263*, 225–241. [CrossRef]

37. Granata, F.; Gargano, R.; de Marinis, G. Artificial intelligence-based approaches to evaluate actual evapotranspiration in wetlands. *Sci. Total Environ.* **2020**, *703*, 135653. [CrossRef]

38. Rashid Niaghi, A.; Hassanijalilian, O.; Shiri, J. Estimation of reference evapotranspiration using spatial and temporal machine learning approaches. *Hydrology* **2021**, *8*, 25. [CrossRef]

39. Torres, A.F.; Walker, W.R.; McKee, M. Forecasting daily potential evapotranspiration using machine learning and limited climatic data. *Agric. Water Manag.* **2011**, *98*, 553–562. [CrossRef]

40. Käfer, P.S.; da Rocha, N.S.; Diaz, L.R.; Kaiser, E.A.; Santos, D.C.; Veeck, G.P.; Robérti, D.R.; Rolim, S.B.A.; Oliveira, G.G. Artificial neural networks model based on remote sensing to retrieve evapotranspiration over the Brazilian Pampa. *J. Appl. Remote Sens.* **2020**, *14*, 038504. [CrossRef]

41. Kim, N.; Kim, K.; Lee, S.; Cho, J.; Lee, Y. Retrieval of daily reference evapotranspiration for croplands in South Korea using machine learning with satellite images and numerical weather prediction data. *Remote Sens.* **2020**, *12*, 3642. [CrossRef]

42. Cui, Y.; Song, L.; Fan, W. Generation of spatio-temporally continuous evapotranspiration and its components by coupling a two-source energy balance model and a deep neural network over the Heihe River Basin. *J. Hydrol.* **2021**, *597*, 126176. [CrossRef]

43. Jang, J.C.; Lee, S.; Sohn, E.H.; Noh, Y.J.; Miller, S.D. Combined dust detection algorithm for Asian Dust events over East Asia using GK2A/AMI: A case study in October 2019. *Asia Pac. J. Atmos. Sci.* **2021**, 1–20. [CrossRef]

44. Huffman, G.; Bolvin, D.; Braithwaite, D.; Hsu, K.; Joyce, R.; Xie, P. *NASA Global Precipitation Measurement (GPM) Integrated Multi-Satellite Retrievals for GPM (IMERG)*; Algorithm Theoretical Basis Document, Version 4.4; NASA: Greenbelt, MD, USA, 2014; p. 30.

45. Song, H.J.; Lim, B.; Joo, S. Evaluation of rainfall forecasts with heavy rain types in the high-resolution unified model over South Korea. *Weather Forecast.* **2019**, *34*, 1277–1293. [CrossRef]

46. Rabus, B.; Eineder, M.; Roth, A.; Bamler, R. The shuttle radar topography mission—A new class of digital elevation models acquired by spaceborne radar. *ISPRS J. Photogramm. Eng. Remote Sens.* **2003**, *57*, 241–262. [CrossRef]

47. Shi, T.T.; Guan, D.X.; Wu, J.B.; Wang, A.Z.; Jin, C.J.; Han, S.J. Comparison of methods for estimating evapotranspiration rate of dry forest canopy: Eddy covariance, Bowen ratio energy balance, and Penman-Monteith equation. *J. Geophys. Res.* **2008**, *113*, D19116. [CrossRef]

48. Duffie, J.A.; Beckman, W.A. *Solar Engineering of Thermal Process.*; John Wiley and Sons: New York, NY, USA, 1991.

49. Tayfur, G. Artificial neural networks for sheet sediment transport. *Hydrol. Sci. J.* **2002**, *47*, 879–892. [CrossRef]

50. Tayfur, G.; Singh, V.P. ANN and fuzzy logic models for simulating event-based Rainfall-Runoff. *J. Hydraul. Eng.* **2006**, *132*, 1321–1330. [CrossRef]

51. Dawson, C.; Wilby, R. An artificial neural network approach to rainfall-runoff modelling. *Int. Assoc. Sci. Hydrol. Bull.* **1998**, *43*, 47–66. [CrossRef]

52. Tamouridou, A.A.; Alexandridis, T.K.; Pantazi, X.E. Application of multilayer perceptron with automatic relevance determination on weed mapping using UAV multispectral imagery. *Sensors* **2017**, *17*, 2307. [CrossRef]

53. Zhang, B.; Zhang, M.; Kang, J.; Hong, D.; Xu, J.; Zhu, X. Estimation of PMx concentrations from Landsat 8 OLI images based on a multilayer perceptron neural network. *Remote Sens.* **2019**, *11*, 646. [CrossRef]

54. Jiang, W.; He, G.; Long, T.; Ni, Y.; Liu, H.; Peng, Y.; Lv, K.; Wang, G. Multilayer Perceptron Neural Network for Surface Water Extraction in Landsat 8 OLI Satellite Images. *Remote Sens.* **2018**, *10*, 755. [CrossRef]

55. Clevert, D.-A.; Unterthiner, T.; Hochreiter, S. Fast and Accurate Deep Network Learning by Exponential Linear Units (ELUs). *arXiv* **2015**, arXiv:151107289.

56. Ioffe, S.; Szegedy, C. Batch normalization: Accelerating deep network training by reducing internal covariate shift. In Proceedings of the International Conference on Machine Learning (ICML), Lille, France, 6–11 July 2015.

57. Chen, C.; Gong, W.; Chen, Y.; Li, W. Object detection in remote sensing images based on a scene-contextual feature pyramid network. *Remote Sens.* **2019**, *11*, 339. [CrossRef]

58. Kingma, D.P.; Ba, J. Adam: A method for stochastic optimization. *arXiv* **2014**, arXiv:1412.6980.

59. McCaffrey, D.R.; Hopkinson, C. Modeling Watershed-Scale Historic Change in the Alpine Treeline Ecotone Using Random Forest. *Can. J. Remote Sens.* **2020**, *46*, 715–732. [CrossRef]

60. Hupet, F.; Vanclooster, M. Effect of the sampling frequency of meteorological variables on the estimation of the reference evapotranspiration. *J. Hydrol.* **2001**, *243*, 192–204. [CrossRef]

61. Lang, D.; Zheng, J.; Shi, J.; Liao, F.; Ma, X.; Wang, W.; Chen, X.; Zhang, M. A comparative study of potential evapotranspiration estimation by eight methods with FAO Penman–Monteith method in southwestern China. *Water* **2017**, *9*, 734. [CrossRef]

62. Willmott, C.J.; Robeson, S.M.; Matsuura, K. A refined index of model performance. *Int. J. Climatol.* **2012**, *32*, 2088–2094. [CrossRef]
63. Mo, X.; Liu, S.; Lin, Z.; Wang, S.; Hu, S. Trends in land surface evapotranspiration across China with remotely sensed NDVI and climatological data for 1981–2010. *Hydrol. Sci. J.* **2015**, *60*, 2163–2177. [CrossRef]
64. Li, X.; Wang, L.; Chen, D.; Yang, K.; Wang, A. Seasonal evapotranspiration changes (1983–2006) of four large basins on the Tibetan Plateau. *J. Geophys. Res.* **2014**, *119*, 13079–13095. [CrossRef]
65. Li, M.; Chu, R.; Shen, S.; Islam, A.R.M.T. Quantifying climatic impact on reference evapotranspiration trends in the Huai River Basin of eastern China. *Water* **2018**, *10*, 144. [CrossRef]
66. Gharsallah, O.; Facchi, A.; Gandolfi, C. Comparison of six evapotranspiration models for a surface irrigated maize agro-ecosystem in Northern Italy. *Agric. Water Manag.* **2013**, *130*, 119–130. [CrossRef]
67. Fleischer, E.; Bölter, J.; Klemm, O. Summer evapotranspiration in western Siberia: A comparison between eddy covariance and Penman method formulations. *Hydrol. Process.* **2015**, *29*, 4498–4513. [CrossRef]
68. Anderson, R.G.; Wang, D.; Tirado-Corbalá, R.; Zhang, H.; Ayars, J.E. Divergence of reference evapotranspiration observations with windy tropical conditions. *Hydro. Earth Syst. Sci. Discuss.* **2014**, *11*, 6473–6518.
69. Gao, G.; Zhang, X.; Yu, T.; Liu, B. Comparison of three evapotranspiration models with eddy covariance measurements for a Populus euphratica Oliv. forest in an arid region of northwestern China. *J. Arid Land* **2016**, *8*, 146–156. [CrossRef]
70. Hughes, C.E.; Kalma, J.D.; Binning, P.; Willgoose, G.R.; Vertzonis, M. Estimating evapotranspiration for a temperate salt marsh Newcastle, Australia. *Hydrol. Process.* **2001**, *15*, 957–975. [CrossRef]
71. Li, S.; Kang, S.Z.; Zhang, L.; Zhang, J.H.; Du, T.S.; Tong, L.; Ding, R.S. Evaluation of six potential evapotranspiration models for estimating crop potential and actual evapotranspiration in arid regions. *J. Hydrol.* **2016**, *543*, 450–461. [CrossRef]
72. Sun, Z.; Wang, Q.; Ouyang, Z.; Watanabe, M.; Matsushita, B.; Fukushima, T. Evaluation of MOD16 algorithm using MODIS and ground observational data in winter wheat field in North China Plain. *Hydrol. Process.* **2007**, *21*, 1196–1206. [CrossRef]
73. Kim, J.Y.; Hogue, T.S. Evaluation of a MODIS-based potential evapotranspiration product at the point scale. *J. Hydrometeor.* **2008**, *9*, 444–460. [CrossRef]
74. Lim, C.-H.; Kim, S.H.; Choi, Y.; Kafatos, M.C.; Lee, W.-K. Estimation of the virtual water content of main crops on the Korean Peninsula using multiple regional climate models and evapotranspiration methods. *Sustainability* **2017**, *9*, 1172. [CrossRef]
75. Birhanu, D.; Kim, H.; Jang, C.; Park, S. Does the complexity of evapotranspiration and hydrological models enhance robustness? *Sustainability* **2018**, *10*, 2837. [CrossRef]
76. Um, M.J.; Kim, Y.; Park, D. Spatial and temporal variations in reference crop evapotranspiration in a Mountainous Island, Jeju, in South Korea. *Water* **2017**, *9*, 261. [CrossRef]
77. Jeong, S.; Ko, J.; Kang, M.; Yeom, J.; Ng, C.T.; Lee, S.H.; Lee, Y.G.; Kim, H.Y. Geographical variations in gross primary production and evapotranspiration of paddy rice in the Korean Peninsula. *Sci. Total Environ.* **2020**, *714*, 136632. [CrossRef]

hydrology

Article

Simplified Interception/Evaporation Model

Giorgio Baiamonte 

Department of Agricultural, Food and Forest Sciences (SAAF), University of Palermo, Viale delle Scienze, 90128 Palermo, Italy; giorgio.baiamonte@unipa.it

Abstract: It is known that at the event scale, evaporation losses of rainfall intercepted by canopy are a few millimeters, which is often not much in comparison to other stocks in the water balance. Nevertheless, at yearly scale, the number of times that the canopy is filled by rainfall and then depleted can be so large that the interception flux may become an important fraction of rainfall. Many accurate interception models and models that describe evaporation by wet canopy have been proposed. However, they often require parameters that are difficult to obtain, especially for large-scale applications. In this paper, a simplified interception/evaporation model is proposed, which considers a modified Merrian model to compute interception during wet spells, and a simple power-law equation to model evaporation by wet canopy during dry spells. Thus, the model can be applied for continuous simulation, according to the sub hourly rainfall data that is appropriate to study both processes. It is shown that the Merrian model can be derived according to a simple linear storage model, also accounting for the antecedent intercepted stored volume, which is useful to consider for the suggested simplified approach. For faba bean cover crop, an application of the suggested procedure, providing reasonable results, is performed and discussed.

Keywords: interception; linear storage model; evaporation; cover crop; water balance; faba bean

Citation: Baiamonte, G. Simplified Interception/Evaporation Model. *Hydrology* **2021**, *8*, 99. https://doi.org/10.3390/hydrology8030099

Academic Editors: Aristoteles Tegos and Nikolaos Malamos

Received: 6 June 2021
Accepted: 30 June 2021
Published: 2 July 2021

1. Introduction

According to Brutsaert [1], the interception process is determined by the rainfall fraction that moistens vegetation and that is temporarily stored on it, before evaporating. When the vegetation cover is fully saturated, the interception storage capacity is achieved. In practice, the interception storage capacity is denoted as rainfall left on the canopy at the end of the rainfall after all drip has ceased [2]. The water stored on the canopy may evaporate soon after, thus short-circuiting the hydrologic cycle.

Although most surfaces can store only a few millimeters of rainfall, which is often not much in comparison to other stocks in the water balance, interception is generally a significant process and its impact becomes evident at longer time scales [3]. Thus, interception storage is generally small, but the number of times that the storage is filled and depleted can be so large that the evaporation losses by wet canopy may become of the same order of magnitude as the transpiration flux [4].

Evaporation flux by canopy exerts a negative effect on plant water consumption by preventing water from reaching the soil surface, thus the plant roots [2,3]. In contrast, the remaining rainfall (i.e., the net rainfall) reaches the soil surface either as throughfall or by flowing down branches and stems as stemflow. Throughfall is the fraction of water that reaches the soil surface directly through the canopy gaps without hitting the canopy surfaces, or indirectly through dripping from the leaves and branches [5].

The interception may also exert important effects on surface runoff [6], providing a certain delay compared with the time of the beginning of the rain. For the *Dunnian* mechanism of runoff generation, Baiamonte [7] showed the effect induced by the interception process on the delay time, and emphasized that the effect is more frequent for well-drained soils [8,9] in humid regions, for low rainfall intensities and high groundwater table, when infiltration capacity exceeds the rainfall intensity. Indeed, the latter conditions also occurs for high density of cover crops or forest soils that are reach in organic carbon and are very structured [10].

The proportion of the precipitation that does not reach the ground, i.e., the interception loss, depends on the type of vegetation (forest, tree, or grassland), its age, density of planting and the season of the year. The interception loss also depends on rainfall regime, thus on climate. For example, in tall dense forest vegetation at temperate latitudes interception loss as large as 30–40% of the gross precipitation has been observed [11], whereas in tropical forest with high rainfall intensity was of the order 10–15% [12] and in heather and shrub also 10–15% [13]. In arid and semi-arid areas, where there is little vegetation, the interception loss is negligible.

Since interception is an important component of the water balance, a comprehensive evaluation of interception loss by prediction tools has been considered of great value in the study of hydrological processes, and different formulations, at hourly and event scale, have been introduced in the hydrologic literature. Muzylo et al. [14] wrote an interesting review paper, where the principal models proposed in the literature are described, and their characteristics reported (input temporal scale, output variable, number of parameters, layers, spatial scale).

Linsley et al. [15] modified the very simple interception model first introduced by Horton [16], which did not account for the amount of gross rainfall, since it assumed that the rainfall in each storm completely filled the interception storage. Linsley et al. [15] assumed that the interception loss approached exponentially to the interception capacity as the amount of rainfall increased [17]. Then, this simple sketch was applied and tested by Merrian [18], who studied the effect of fog intensity and leaf shape on water storage on leaves, by using a simple fog wind tunnel and leaves of aluminum and plastic. Merrian [18] found that drip measurements were reasonably close to values predicted, by using an exponential equation based on fog flow and leaf storage capacity.

Rutter et al. [19,20] were the first to model forest rainfall interception recognizing that the process was primarily driven by evaporation from the wetted canopy. The conceptual model developed by Rutter et al. [19,20] describes the interception loss in terms of both the structure of the forest and the climate in which it is growing. The model is physically based, thus it has potential for application in all areas, where there are suitable data. In Rutter et al. [20], the model's definitive version was developed by adding a stemflow module, in which a fraction of the rainfall input is diverted to a compartment comprising the trunks. Early applications of Rutter-type models were made by Calder [21] and Gash and Morton [22].

The rate of evaporation increases with solar radiation and temperature. The process also depends on the air humidity and the wind speed. The greater the humidity, the less the evaporation. Wind carries moist air away from the ground surface, so wind decreases the local humidity and allows more water to evaporate. Therefore, in the Rutter model, the evaporation flux was calculated from the form of the Monteith–Penman equation [23]. Later, Gash [24] provided a simplification of the data-demanding Rutter model. Although some of the assumptions of the Gash model may not be suitable, the model has been shown to work well under a variety of forest types, including different species and sites [11,25–27].

For agricultural crops and for grassland, where interception loss is of the order 13–19% [28], Von Hoyningen-Hüne [29] and Braden [30], proposed a general formula, which was used in the SWAP model [31] that however can be applied only at daily scale.

By the experimental point of view, the interception evaporation process requires monitoring intercepted mass and interception loss with high accuracy and time resolution, to provide accurate estimates. Net precipitation techniques, in which interception evaporation is determined from the difference between gross precipitation and throughfall, fulfill many of the requirements but usually have too-low accuracy and time resolution for process studies. Furthermore, for grassland, these techniques are unsuitable.

In this paper, we explored the rainfall partitioning in net rainfall and evaporation losses by canopy, by using a very simplified sketch of the interception process, which combines a modified exponential equation applied and tested by Merrian [18], accounting for the antecedent volume stored on the canopy, and a simple power-law equation to compute evaporation by wet canopy. We are aware that the considered approach is far from the sophisticated physically based developments that were performed to quantify interception and evaporations losses. However, the latter may require many parameters that are not easy to determine. It is shown that the simplified parsimonious approach may lead to a reasonable quantification of this important component of the hydrologic cycle, which can be useful when a rough estimate is required, in absence of a detailed characterization of the canopy and of the climate conditions. It is also shown that the Merrian model can be derived by considering a simple linear storage model. For faba bean cover crop, an application of the suggested procedure is performed and discussed.

2. Rainfall Data Set, Wet and Dry Spells

Rainfall time series were analyzed according to Agnese et al. [32], who applied the discrete three-parameter Lerch distribution to model the frequency distribution of interarrival times, *IT*, derived from daily precipitation time-series, for the Sicily region, in Italy. Agnese et al. [32] showed a good fitting of the Lerch distribution, thus evidencing the wide applicability of this kind of distribution [33], also allowing us to jointly model dry spells, *DS*, and wet spells, WS.

Since this work aimed at modelling by continuous rainfall data series the interception losses, during the *WS* and the evaporation losses during *DS*, only the frequency distribution of the two latter were considered, which are defined in the following, according to Agnese et al. [32].

For any rainfall data series, the ten minutes temporal scale, τ, which is appropriate to model both interception and evaporation processes, could be considered in order to describe clustering and intermittency characters of continuous rainfall data series.

Let $H = \{h_1, h_2, \ldots, h_n\}$, a time-series of rainfall data of size n, spaced at uniform time-scale τ. The sub-series of H can be defined as the event series, $E\{t_1, t_2, \ldots, t_{nE}\}$, where n_E ($0 < n_E < n$) is the size of E, which is an integer multiple of time-scale τ. The succession constituted by the times elapsed between each element of the E series, with exception of the first one and the immediately preceding one, is defined as the inter-arrival time-series, $IT\{T_1, T_2, \ldots, T_{nE}\}$, with size n_{E-1} (Figure 1). It can be observed the sequence of dry spells, *DS*, can be derived from *IT* dataset by using the relationship $\{DS_k\} = \{T_k\} - 1$ for any $T_k > 1$.

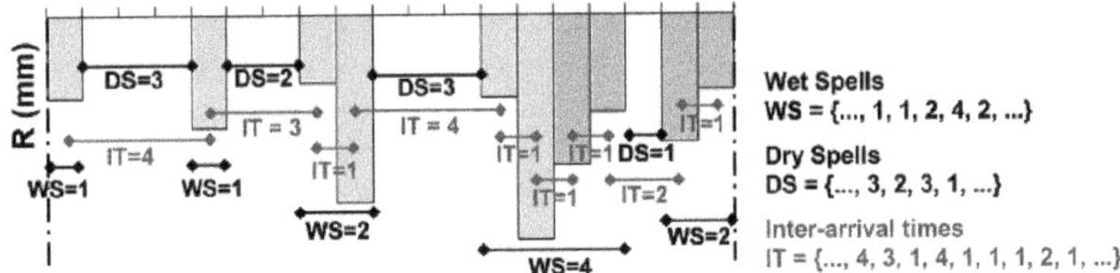

Figure 1. Sketch of inter-arrival times (*IT*), dry spells (*DS*) and wet spells (*WS*). Reprinted with permission from ref. [32], Copyright 2014, Elsevier (Amsterdam, The Netherlands).

Figure 1 shows an example of a sequence of wet and dry spells. In the context of this work, which aims at modelling the interception process during *WS* and the evaporation process from the canopy, as previously observed, only the frequency distribution of wet spells (*WS*) and dry spells (*DS*) were derived.

For the 2009 rainfall data series of Fontanasalsa station (Trapani, 37°56′37″ N, 12°33′12″ E, western Sicily, Italy), which will be considered for an example application, Figure 2 shows the complete characteristics of the rainfall regime. In particular, Figure 2a describes the *DS* distribution of frequency, *F*, whereas Figure 2b plots the *WS* distribution associated with the cumulated rainfall depth collected with 10 min time resolution. Both figures also illustrate the frequency of non-exceedance, corresponding the selected time resolution ($\tau = 1 = 10$ min) that equals 0.189 for *DS* and much higher for *WS* (0.596), indicating that a high fraction of rainfall with *WS* = 1 occurs, which could play an important role in the interception process. Figure 2b also plots the rainfall depth distribution corresponding to *WS*, where the frequency is calculated with respect to the yearly rainfall depth, h_{year}, which equals 885.2 mm. Therefore, in Figure 2b, to *WS* = 1 (F_1 = 0.115) corresponds a rainfall depth equals to 101.6 mm that, if associated with subsequent large enough *DS*, may potentially evaporate from the canopy.

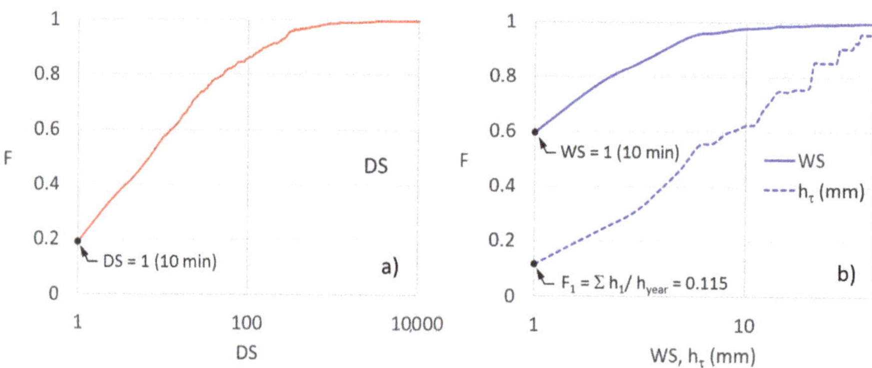

Figure 2. For the investigated year (2009), distributions of frequency, *F*, (**a**) of dry spells (*DS*), and (**b**) of wet spells (*WS*) and of the rainfall depth (h_τ) corresponding to *WS*.

3. Interception Rate and Stored Volumes during Wet Spells

The interception Merrian [18]'s model states that:

$$ICS = S\left[1 - exp\left(-\frac{R}{S}\right)\right]$$

(1)

where *S* (mm) is the interception capacity of the canopy, and *ICS* (mm) is the interception storage volume corresponding to the cumulated rainfall volume *R* (mm). Equation (1) can be applied for dry initial condition, i.e., when the no water volume is stored on the canopy at the beginning of the rainfall.

However, it is demonstrated here that the Merrian model can be derived by considering a simple linear stored model, also used in other hydrological contexts [34], which made it possible to also account for the antecedent interception volume. The latter is useful for the applications, when in between two consecutive *WSs*, *DS* is not long enough to result in full evaporation of the rainfall volume intercepted by the canopy starting from the end of a *WS*. Thus, a residual water volume, ICS_0, is still stored on the canopy at the beginning of the subsequent *WS*, and it needs to be considered as initial condition. Therefore, for the purpose of this study, which aims at estimating interception losses during *WS* and *DS* sequences of different amount and duration, Equation (1) needs to be extended to different antecedent water interception storage volume, ICS_0 (mm), before a new *WS* takes place, after a *DS* when evaporation process ceases.

Consider a simple linear reservoir, miming the interception storage volume (Figure 3). A stationary rainfall of intensity *r* (mm h^{-1}) uniformly distributed over the canopy, is applied to the reservoir. In the time *dt*, the interception storage, *ICS* (mm), stored in the canopy equals *dICS*. The interception capacity, i.e., the maximum rainfall volume that can be stored on the canopy, is denoted as *S* (mm). The duration of rainfall that needs to achieve the saturation of the canopy, without dripping out from the reservoir, t_s (h), can be expressed as:

$$t_s = \frac{S}{r}$$

(2)

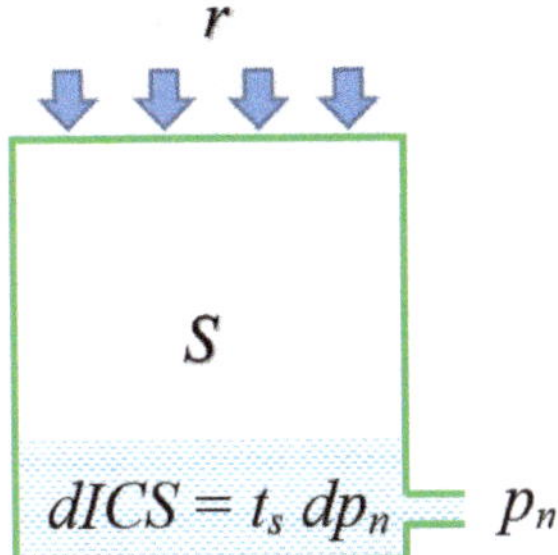

Figure 3. Sketch of the linear water tank describing the interception process, for a maximum water volume stored on the canopy, S (mm), corresponding to the interception capacity.

From a physical point of view, t_s is not very meaningful since it describes the duration of rainfall that is required for the canopy saturation, for dry antecedent conditions ($ICS_0 = 0$), and with no losses from the reservoir, e.g., no dripping and no net rainfall intensity p_n (mm h^{-1}), which actually occurs (Figure 3). However, t_s represents a characteristic time of the interception process that it is useful to introduce for the following derivations.

By assuming that the water volume stored in the reservoir linearly varies with the output net rainfall intensity, p_n, according to the time t_s (Equation (2)), the balance of the water volumes can be expressed as:

$$r \, dt - p_n \, dt = dICS = t_s \, dp_n \tag{3}$$

Separation of variables may be used to solve the ordinary differential equation:

$$dt = \frac{t_s}{r - p_n} dp_n \tag{4}$$

By assuming as initial conditions:

$$t = t_0 \qquad p_n = p_{n0} \tag{5}$$

Integration of Equation (4) provides:

$$\int_{t_0}^{t} dt = \int_{p_{n0}}^{p_n} \frac{t_s}{r - p_n} dp_n \tag{6}$$

Solving Equation (6) yields:

$$t = t_0 + t_s \, log \frac{r - p_{n0}}{r - p_n} \tag{7}$$

Equation (7) can be made explicit to derive the net rainfall intensity, p_n:

$$p_n = r - (r - p_{n0}) \, exp\left(-\frac{t - t_0}{t_s}\right) \tag{8}$$

Of course, knowledge of Equation (8) makes it possible to determine the interception intensity, ics (mm/h):

$$ics = r - p_n = (r - p_{n0}) \, exp\left(-\frac{t - t_0}{t_s}\right) \tag{9}$$

Due to the assumption of a linear storage model, the antecedent net rainfall intensity, p_{n0}, in Equations (8) and (9), is linked to the antecedent interception volume, ICS_0, by:

$$p_{n0} = \frac{ICS_0}{t_s} = r\frac{ICS_0}{S} \tag{10}$$

For a compacted graphical illustration, a constant rainfall intensity, r, and a zero time antecedent condition $t_0 = 0$ can be assumed. Thus, Equations (8) and (9) normalized with respect to r can be expressed as:

$$\frac{p_n}{r} = 1 - \left(1 - \frac{ICS_0}{S}\right) exp\left(-\frac{R}{S}\right) \tag{11}$$

$$\frac{ics}{r} = \left(1 - \frac{ICS_0}{S}\right) exp\left(-\frac{R}{S}\right) \tag{12}$$

where in Equations (8) and (9) t_s was substituted by Equation (2), and r by $R\,t$, with R (mm) the cumulated rainfall depth. Equations (11) and (12) are graphed in Figure 4a,b, respectively. For any fixed S, and for a fixed antecedent interception storage volume, ICS_0, Figure 4a shows that, at increasing R, the net rainfall intensity, p_n, attains the gross rainfall intensity r ($p_n/r \sim 1$), and that, at increasing ICS_0, the r achievement is slower and slower, as it could be expected. Of course, the normalized interception intensity ics/r plotted in Figure 4b provides complementary curves.

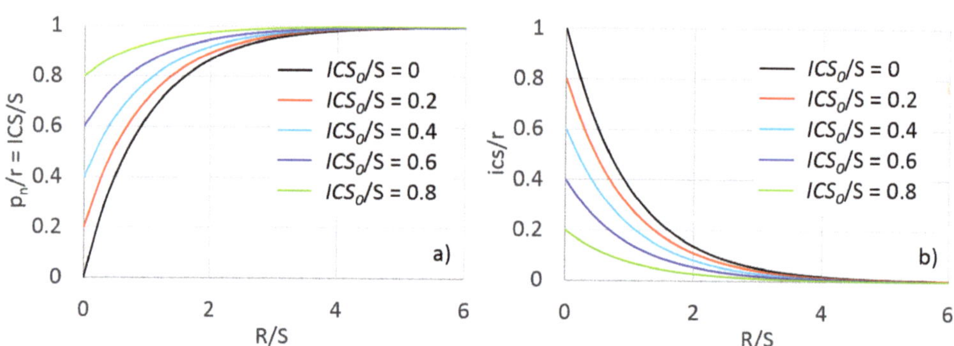

Figure 4. For different antecedent normalized water volume stored on the canopy, ICS_0/S, (**a**) normalized interception volume, ICS/S, corresponding the normalized net rainfall intensity, p_n/r, and (**b**) normalized interception intensity, ics/r, versus the normalized rainfall volume, R/S.

In order to determine the interception volume, ICS (mm), as a function of its antecedent amount, ICS_0 (mm), because of the assumption of a linear storage model (Equation (10)), it can be easily observed that Equation (11) also matches the interception storage volume, ICS, normalized with respect to the infiltration capacity, S:

$$\frac{ICS}{S} = 1 - \left(1 - \frac{ICS_0}{S}\right) exp\left(-\frac{R}{S}\right) \tag{13}$$

as displayed in the vertical axis label of Figure 4a. Therefore, Equation (13) is formally equivalent to Equation (11), but with a different meaning of the output parameters.

It is noteworthy to observe that, for dry antecedent condition ($ICS_0 = 0$), Equation (13) matches the interception storage model (Equation (1)) proposed by Merrian [18], which was developed only for dry initial conditions, as previously observed.

Equation (13) could also be derived analytically, with a lesser extend from a physical point of view, than that provided by the simple linear stored model previously described.

Indeed, denoting R_0 the antecedent rainfall volume corresponding to ICS_0, the Merrian's model (Equation (1)) yields:

$$ICS = S\left[1 - exp\left(-\frac{R}{S} - \frac{R_0}{S}\right)\right] \tag{14}$$

The antecedent rainfall volume, R_0, can be expressed as a function of ICS_0 by using Equation (1):

$$R_0 = -S\,log\left(1 - \frac{ICS_0}{S}\right) \tag{15}$$

Substituting Equation (15) into Equation (14), and normalizing ICS with respect to S, provides Equation (13), which was to be demonstrated.

As an example, for S = 0.5 mm and ICS_0 = 0, and for a constant rainfall intensity r = 1.75 mm/h, Figure 5 graphs ICS vs. the cumulated rainfall R (solid black line). For R_0 = 0.7 mm, corresponding to ICS_0 = 0.377 mm, Equation (14) is plotted in Figure 5 (red line), which of course is not physical meaningful for $R < 0$ (dashed line).

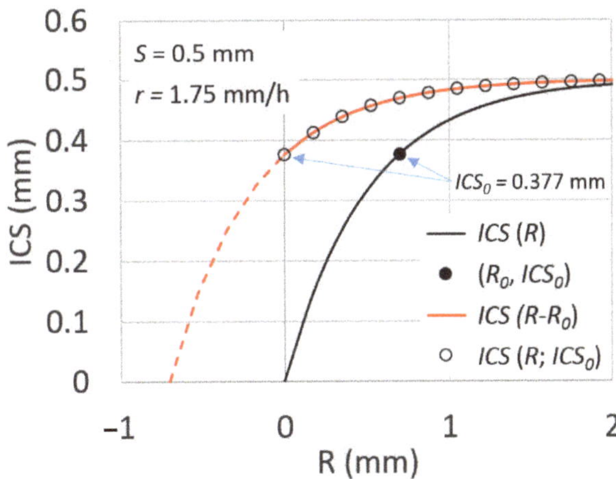

Figure 5. Relationship between the interception storage, ICS (mm), and the rainfall depth, R (mm), for dry antecedent condition, ICS_0 = 0, according to Merrian [18], and for a wet antecedent condition (ICS_0 = 0.377 mm).

For ICS_0 = 0.377 mm, Equation (13) is also plotted in Figure 5 (round circles) showing that Equation (13) matches Equation (14). Equation (13) will be used in the following to calculate the interception storage volume starting from any ICS_0 value.

4. Evaporation by Wet Canopy during Dry Spells

During a dry spell, DS, starting from an antecedent storage volume, $ICS_0 > 0$, the evaporation process from the wet canopy takes place. When the canopy is wetted by rain, evaporation of intercepted rainfall is largely a physical process that does not depend on the functioning of stomata. According to a realistic description of evaporation from canopies the Penman–Monteith equation [23,35] could be used, by imposing zero the surface (or canopy) resistance.

However, a different approach is used here, which is based on the physical circumstance highlighted well by Babu et al. [36] that evaporation by wet canopy comprises two stages. First, the drying process involves removal of unbound (free) moisture from the surface, and second it involves removal bound moisture from the interior of the leaf till a defined limit, corresponding to a critical moisture content. Apart from the second stage, which refers to water consumption by the canopy, and thus it is beyond the purpose of this study, the first stage comprises (i) a "preheat period", where the drying speed quickly increases, and then (ii) a "constant rate period", where evaporation takes place at the outside surface for the removal of unbound moisture (free water) from the surface of the leaf [36].

The evaporation mechanism by wet canopy of the first stage could be also described by the physical equations requiring the knowledge of climatic parameters and structure parameters of the canopy. However, in agreement with the simple sketch also considered for the interception process, in this simplified study the first stage evaporation mechanism is described by a simple power-law, according to two parameters. A similar power-law equation was also considered by Black et al. [37] to model the cumulative evaporation of an initially wet, deep soil. For the same power-law equation, Ritchie [38] reports the experimental parameters obtained by other researchers for different soils.

One limited experimental campaign, described in the next section for the faba bean, supported this choice and revealed that for fixed outdoor air temperature, T_{ex} (°C) the cumulated evaporation volume, per unit leaf surface area, E (mm), could be actually described by the power-law equation:

$$E = m\, t^{n} \text{ for a fixed } T_{ex} \text{ (°C)} \tag{16}$$

where t (h) is the time spent after the canopy interception capacity, S, is achieved, m is a scale parameter and n a shape parameter to be determined by experimental measurements.

In order to upscale Equation (16) to any values of air temperature, a regional equation developed for the Sicily region was considered [39]:

$$E_m = 0.38\, T_m^{1.93} \tag{17}$$

where E_m (mm) is the monthly evaporation depth and T_m (°C) is the monthly temperature. By scaling Equation (16) with Equation (17), provides:

$$E = LAI\, m\, t^{n} \left(\frac{T_m}{T_{ex}}\right)^{1.93} \tag{18}$$

where the leaf area index, LAI, was introduced, accounting for the actual leaf surface from which the water evaporates. Of course, Equation (18) gives the same experimental evaporation amount derived by Equation (17), when the experimental air temperature, T_{ex}, is equal to T_m.

It should be noted that Equation (18) does not account for wind speed, thus evaporation losses are linked to the wind experimental conditions for which m and n parameters are determined, otherwise losses are underestimated or overestimated for wind speeds lower and higher than the experimental ones, respectively. Moreover, application of this procedure in regions different from Sicily would require modifying Equations (17) and (18).

For $LAI = 2$ and $T_{ex} = 18$ °C, and for fixed experimental values of m and n parameters, Figure 6 shows evaporation losses, E, during the time, with the air temperature, T_m, as a parameter.

In order to apply the suggested procedure, according to the discrete nature of rainfall, the cumulated evaporation volume (Equation (18)) needs to be expressed in discrete terms by accounting, as per the interception model, for the antecedent conditions, at the aim to determine the E fraction, ΔE, which occurs during the dry spells, DS, starting from t_0:

$$\Delta E = LAI\, m \left(\frac{T_m}{T_{ex}}\right)^{1.93} \left((DS + t_0)^{n} - t_0^{n}\right) \tag{19}$$

where the antecedent initial condition, t_0, refers to the end of the wet spell, WS, when the evaporation process starts. Thus, t_0 needs to be calculated by Equation (19), by assuming that the evaporation initial condition, E_0, equals the interception capacity minus the water stored in the canopy as interception, $S - ICS$:

$$t_0 = \left(\left(\frac{T_{ex}}{T_m}\right)^{1.93} \frac{S - ICS}{LAI\, m}\right)^{1/n} \tag{20}$$

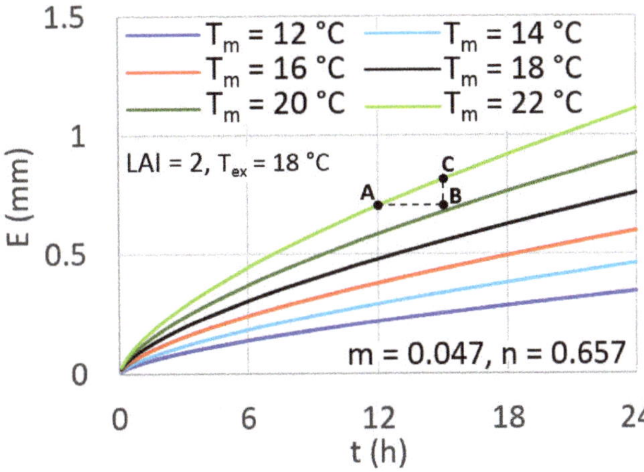

Figure 6. For $LAI = 2$ and an experimental temperature, $T_{ex} = 18\ °C$, evaporation losses by canopy per unit surface leaf, E (mm), during the time t (h), starting from the interception capacity ($E = 0$, $t = 0$), with air temperature, T_m (°C), as a parameter. For $T_m = 22\ °C$ and $DS = 3$ h, the figure also illustrates the evaporation loss, ΔE (mm), corresponding to the segment B–C, starting from an initial condition A = (t_0, E_0) drier than saturation.

By using Equations (18) and (20), an example of ΔE calculation (Equation (19)), which is useful for the applications that will be shown, is performed here. Let assume $m = 0.047$, $n = 0.657$, $T_{ex} = 22\ °C$ (Figure 6), and an evaporation loss ΔE for a mean temperature of $T_m = 22\ °C$ needs to be determined. The interception capacity S equals 1.5 mm, and the interception volume stored in the canopy ICS when the evaporation process takes place equals 0.8 mm, thus the difference $S - ICS = 0.7$ mm mimics the amount water loss due to a virtual evaporation till E_0 (Figure 6). The corresponding initial condition t_0, calculated by Equation (20) provides 12 h. The pair (t_0, E_0) is illustrated in Figure 6 (point A).

Assuming that ΔE needs to be calculated after a $DS = 3$ h, Equation (19) yields $\Delta E = 0.111$ mm, which corresponds to the segment B–C in Figure 6, also indicating an evaporation loss $E = E_0 + \Delta E = 0.816$ mm (point C).

In order to consider that ΔE computation is limited by the antecedent volume stored water on the canopy, ICS, actually available to evaporate, the following condition was imposed, and a corrected ΔE, denoted as ΔE_* evaluated:

$$\begin{aligned} \Delta E_* &= ICS \quad if\ \Delta E > ICS \\ \Delta E_* &= \Delta E \quad\ \ otherwise \end{aligned} \tag{21}$$

In both interception and evaporation models previously introduced, neither the canopy actually covering the field (cover fraction), nor the proportion of rain which falls through the canopy without striking a surface (throughfall), were taken into account. In the next section, the latter issues are considered and then the water mass balance is analyzed.

5. Water Mass Balance

The fraction of ground covered by the canopy [40] plays a fundamental role in estimating interception losses, as well as the proportion of rain that falls through the canopy without striking any surface. In the following, in order to analyze the water mass balance, C_F denotes the fraction of ground covered by the canopy and p is the free throughfall coefficient [19,20]. Assuming $p = 0$, meaning that the canopy fully covers the ground at the plant scale, Figure 7 shows two very different conditions in terms of LAI and C_F, which could provide similar evaporation losses, since high C_F (Figure 7a) could be counterbalanced by low LAI (Figure 7b), and vice versa.

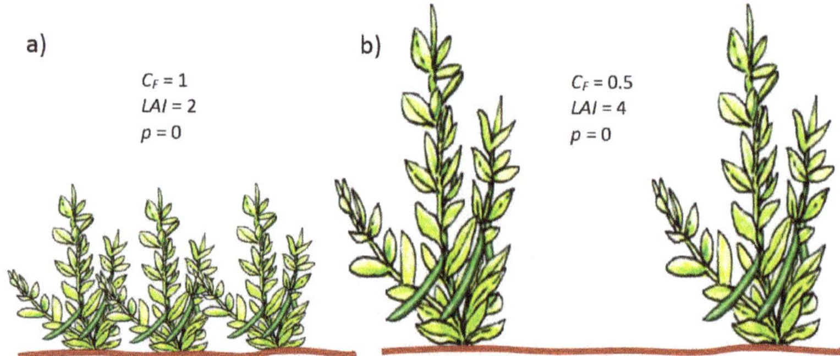

Figure 7. For faba bean, two different attributions of the parameters LAI, throughfall index, p, and fraction of ground covered by the canopy, C_F.

By considering both C_F and p, as weighting factors, for a gross rainfall depth R (mm), the water mass balance is described by the following relationship:

$$R = R_f + R_c + R_t = R\,(1 - C_F) + R\,C_F\,(1 - p) + R\,C_F\,p \tag{22}$$

where R_f (mm) is the portion of R that freely achieves the ground in between the plants, R_c (mm) is the fraction of R that achieves the canopy, and R_t (mm) is the throughfall.

Thus, depending on the plant species, both p and C_F affect R_c and could be simply determined by using RGB images [41,42]. According to Equation (22), to evaluate interception losses, in Equation (13) R has to be replaced by R_c.

By assuming $C_F = 1$ and $p = 0$ ($R = R_c$), for dry initial condition, $ICS_0 = 0$, for fixed parameters $S = 0.8$ mm, $LAI = 4$, $m = 0.047$, $n = 0.657$, $T_{ex} = 12\ °C$, and for three sequences of WS and DS, a simple application of the procedure is illustrated in Figure 8, where the denoted variables are also indicated. For simplicity, in the figure linear ICS and ΔE variations were assumed.

Figure 8 shows that for the third DS a high air temperature ($T_m = 20\ °C$) provides the condition $\Delta E > ICS$, and only the available water stored on the canopy may evaporate (ΔE_*). Moreover, for this step, according to the initial condition (t_0, E_0), the evaporation flux (i.e., the slope) is higher than those corresponding to the lower air temperatures.

For fixed $T_m = 12\ °C$, the evaporation flux is greater for the second DS step that starts from a higher ICS_0, which is near to the saturation compared to that of the first DS step. Similarly, for the second WS step, although the higher rainfall depth ($R = 2$ mm) with respect to the first WS step, ICS increases as in the first WS step, because of a higher initial condition ICS_0, agreeing with the dynamic flux of the both considered interception/evaporation models.

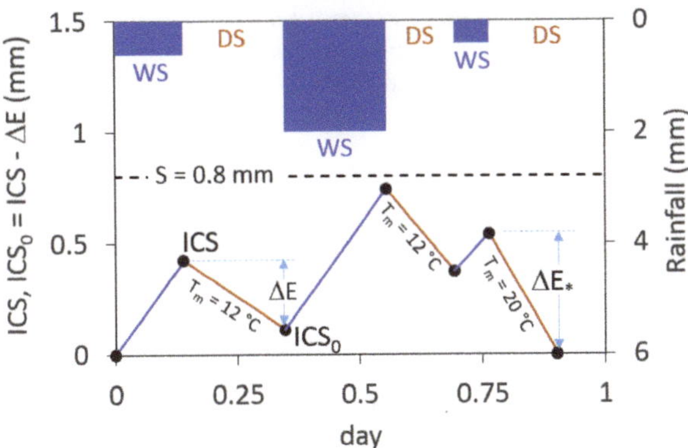

Figure 8. For a simple sequence of three *WSs* and *DSs*, interception loss during *WS*, and evaporation loss, ΔE, during *DS*, with $E_0 = S - ICS$ as initial condition. For the third *DS*, the condition described by Equation (21) occurs ($\Delta E > ICS$) and $\Delta E_* = ICS$.

The flow chart displayed in Figure 9 describes the suggested procedure to calculate the interception/evaporation water losses, which will be used in the following application, where the condition to be imposed (Equation (21)) is also indicated.

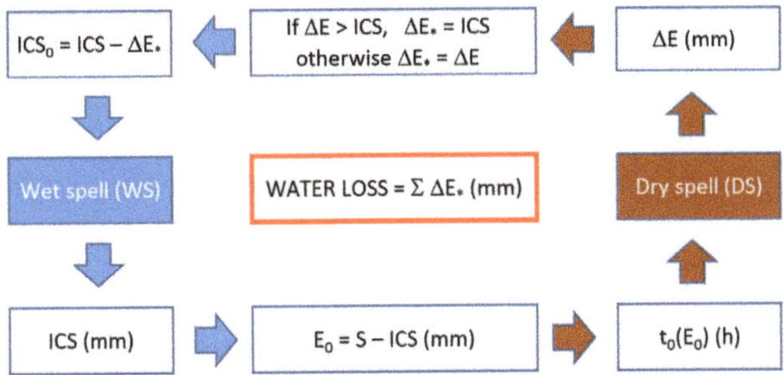

Figure 9. Schematic flowchart of the proposed methodology to calculate *ICS* during *WS*, and ΔE (or ΔE_*) during *DS*.

The water mass balance applied to the fraction of rainfall intercepted by the canopy, can be tested by checking the balance of the inflowing volumes as interception during the i *WSs* and the outflowing volumes, as evaporation from the canopy, during the j *DSs*, where of course the antecedent stored volumes ICS_0 are involved:

$$\sum_{WS=1}^{i}(ICS - ICS_0) = \sum_{DS=1}^{j}\Delta E_* \tag{23}$$

Once evaporation losses are calculated, the net rainfall R_n (mm) can be evaluated:

$$R_n = R - \sum_{DS=1}^{j}\Delta E_* \tag{24}$$

6. Interception Capacity LAI and Evaporation Measurements for Faba Bean

In order to show the results that can be obtained by the application of the simplified procedure, the 2019 rainfall (Figure 2) and temperature data series of Fontanasalsa station were considered, and the required empirical parameters of Equation (19) were obtained by experimental measurements carried out for faba bean. Faba bean is a species of flowering plant in the pea and bean family Fabaceae, originated in the continent of Africa, which is widely cultivated as a crop for human consumption and also used as a cover crop.

For a single plant, planted in mid-November, and sampled after 95 days, the temporal variation of water loss by evaporation starting from the interception capacity, the number of leaves, # Leaves, and the corresponding surface area were measured. Saturation of the canopy was achieved by using sprinkler irrigation, for a fixed outdoor temperature, $T_{ex} = 12\,^{\circ}\text{C}$.

Starting from the interception capacity condition, after dripping has ceased, water loss by evaporation was measured by weighing during the time. The # Leaves (45), and the corresponding surface area (0.043 m^2), made it possible to calculate the cumulative evaporation volume, per unit leaf surface area, E (mm).

Figure 10a plots the cumulate experimental E values during the time. This made it possible to calculate by a simple linear regression of the corresponding logarithmic values, the m and n parameters that are required to apply Equation (16). According to Babu et al. [36], Figure 10a shows that in a first stage evaporation rate is high, and then the evaporation rate decreases remaining almost constant in the time, for approximately $t = 1$ h (at least for this case), the behavior of which can be described by the power-law well.

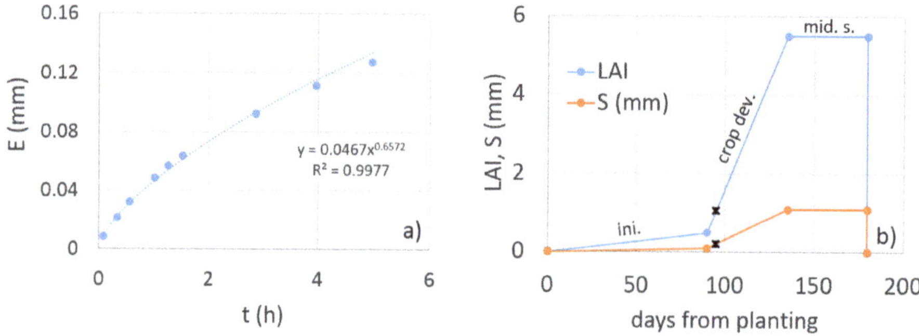

Figure 10. For Field bean, (**a**) relationship between experimental evaporation losses per unit *LAI*, *E/LAI*, during the time *t* (h), starting from the interception capacity, *S* (mm), and (**b**) assumed temporal variation of *LAI* and *S* (mm), from the day of planting (mid-November), considered for the application (Figure 11).

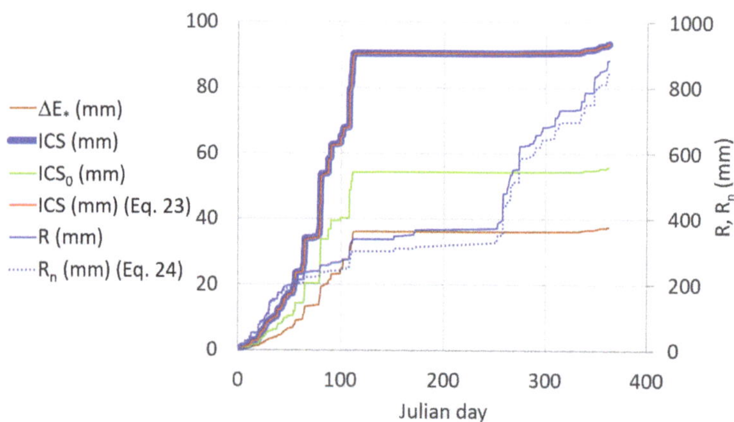

Figure 11. For the year 2009, evaporation loss ΔE_*, interception loss, *ICS*, antecedent volume stored on the canopy, ICS_0. The figure also illustrate the water mass balance (red line) that was checked by using Equation (23), and the gross and net rainfall depth, R and R_n (Equation (24)), respectively (secondary axis).

Of course, it should be noted that there are other secondary factors not considered in the present approach, such as the plant structure, the leaf angle and the light interception, although not easy to consider, which may affect the considered parameters [43], especially when upscaling is needed.

For the purpose of the present investigation, the relationship between the interception capacity and *LAI* was considered in order to extend to 2009, the corresponding relationships displayed in Figure 10a, which are related to the limited experimental *LAI* value. To consider the *LAI* temporal variations, some data for faba bean available in the literature were considered.

For faba bean planted in mid-November, as in this study, in El-Bosaily farm in the Northern coast of Egypt, Hegab et al. [44] measured a maximum *LAI* value equal to 5.5. This value was assumed for the mid-season stage according to FAO56 [35], which provides general lengths for four distinct growth stages, initial (ini.), crop development (crop dev.), mid-season (mid. s.) and late season (late), for various, crops, types of climates and locations.

For faba bean, FAO56 [35] reports ini. = 90, crop dev. = 45, mid.s. = 40, late s. = 0 days, thus no late season was considered since it was assumed that after the mid-season the faba bean was fully removed from the field. For the four growth stages, Figure 10b shows the *LAI* variations versus the days from planting that made it possible the applications described in the next section to be performed. In Figure 10b can be observed that it was assumed that the crop development stage, after 90 days from planting, starts with *LAI* = 0.5.

According to Brisson et al. [45], the interception capacity was assumed $S = 0.2\ LAI$. Thus, with increasing *LAI* during crop growth and decreasing *LAI* during senescence, S will increase and decrease, respectively. More recently, this relationship between S and *LAI* was also considered by Kozak et al. [46]. Figure 10b also graphs the S variation during the days from planting. The interception capacity in the day of sampling (cross symbol), i.e., after 95 days from the day of planting, was equal to 0.202, not too far from that experimentally measured (0.156).

It should be noted that other S vs. *LAI* relationships could be used in the presented methodology, and that specific methods, as that proposed by Aston [47] who also found a linear S vs. *LAI* relation, could be applied in order to experimentally determine such relationships that depend on the considered crops.

7. Example of Application

For the considered crop described in the previous section (Figure 10a,b), for $C_F = 1$ and $p = 0$, and for the 2019 rainfall data series, interception and evaporation losses were calculated according to the described procedure. Figure 11 plots the detailed results of the involved cumulative water balance components, i.e., the rainfall depth, R (mm), the interception depth at the end of each wet spell, ICS, and the evaporation losses, ΔE_*. The latter made it possible to evaluate the antecedent stored volume before each WS, $ICS_0 = ICS - \Delta E_*$.

The figure also illustrate the volume water mass balance (red line) that was checked by using Equation (23). Of course, the interception process mostly occurs during the growing season when it rains. For the considered year, 2009, the evaporation loss by canopy achieved 37.6 mm. Figure 11 also plots the net rainfall (dashed line), calculated according to Equation (24).

It is interesting to observe that the rainiest periods (1–110 and 250–365 Julian day) are not associated with the periods with the highest evaporation losses (1–110 Julian day). Of course, this is explained by the rainfall distribution, which has to be analyzed with respect to the growing period of the considered crop, playing a fundamental role in detecting evaporation losses by the rainfall intercepted by the canopy. Evaporation losses for the considered cover crop are 4.5% with respect to the yearly rainfall, and are thus lower than the minimum found in other studies [28], despite different cover crops being considered, whereas interception losses are of course higher and equal to 10.58%.

However, this author is of the opinion that comparison of these values with those obtained in other studies is not very meaningful, due to the high interactions between climate variability, rainfall distribution, and cover crop management (growing season, date of planting, etc.) that should be analyzed to explain the impact of inter-annual variability of interception/evaporation losses [48]. In this sense, the procedure described in this work could be useful to study in deep these interactions, also accounting for the other parameters as the cover fraction, the date of planting [42], the throughfall index and different rainfall regimes and their changes [33,49] that were not considered in this work.

8. Conclusions

In order to derive the evaporation losses by wet canopy, the suggested procedure combines a modified interception model proposed by Merrian, which is applied during rainfall wet spells, and a simple power-law equation to model evaporation during the dry spells. This simple approach makes it possible to estimate the evaporation losses during continuous simulations, and requires few parameters that consolidate climate and crop conditions. Moreover, it is shown that the Merrian model can be derived by considering a simple linear storage model that makes it possible to account of the antecedent intercepted volume, which is useful for applications. The crop considered in the application is the faba bean that was described according to the general lengths of four distinct growth stages considered in FAO56, whereas *LAI* and interception capacity were obtained from literature. Since the required parameters are few, this simple approach could be applied when a rough estimate of evaporation loss by wet canopy is necessary, in the absence of a detailed characterization of canopy and climate. Due to simplified schemes, the procedure can be easily applied at large scale, in order to study the important role of rainfall regime and crop growing stages, on this important component of the hydrologic cycle, and it could be suitable to be implemented according to a probabilistic line of approach.

Funding: This research received no external funding.

Data Availability Statement: All data, models, and code, generated or used during the study appear in the published article.

Acknowledgments: The author wish to thank Agata Novara for providing evaporation measurements, and the three anonymous reviewers for the helpful comments and suggestions during the revision stage.

Conflicts of Interest: The author declares no conflict of interest.

References

1. Brutsaert, W. *Hydrology: An Introduction*, 5th ed.; Cambridge University Press: Cambridge, UK, 2010.
2. Zhang, Y.; Li, X.Y.; Li, W.; Wu, X.C.; Shi, F.Z.; Fang, W.W.; Pei, T.T. Modeling rainfall interception loss by two xerophytic shrubs in the Loess Plateau. *Hydrol. Process.* **2017**, *31*, 1926–1937. [CrossRef]
3. Dunkerley, D. Measuring interception loss and canopy storage in dryland vegetation: A brief review and evaluation of available research strategies. *Hydrol. Process.* **2000**, *14*, 669–678. [CrossRef]
4. Savenije, H.H.G. The importance of interception and why we should delete the term evapotranspiration from our vocabulary. *Hydrol. Process.* **2004**, *18*, 1507–1511. [CrossRef]
5. Zhang, Y.F.; Wang, X.P.; Hu, R.; Pan, Y.X. Throughfall and its spatial variability beneath xerophytic shrub canopies within water-limited arid desert ecosystems. *J. Hydrol.* **2016**, *539*, 406–416. [CrossRef]
6. Baiamonte, G.; Singh, V.P. Overland Flow Times of Concentration for Hillslopes of Complex Topography. *J. Irrig. Drain. E-ASCE* **2016**, *142*, 04015059. [CrossRef]
7. Baiamonte, G. Simplified model to predict runoff generation time for well-drained and vegetated soils. *J. Irrig. Drain. E-ASCE* **2016**, *142*, 04016047. [CrossRef]
8. Cullotta, S.; Bagarello, V.; Baiamonte, G.; Gugliuzza, G.; Iovino, M.; La Mela Veca, D.S.; Maetzke, F.; Palmeri, V.; Sferlazza, S. Comparing different methods to determine soil physical quality in a Mediterranean forest and pasture land. *Soil Sci. Soc. Am. J.* **2016**, *80*, 1038–1056. [CrossRef]
9. Bagarello, V.; Baiamonte, G.; Caia, C. Variability of near-surface saturated hydraulic conductivity for the clay soils of a small Sicilian basin. *Geoderma* **2019**, *340*, 133–145. [CrossRef]
10. Baiamonte, G.; Bagarello, V.; D'Asaro, F.; Palmeri, V. Factors influencing point measurement of near-surface saturated soil hydraulic conductivity in a small Sicilian basin. *Land Degrad. Dev.* **2017**, *28*, 970–982. [CrossRef]
11. Gash, J.H.C.; Wright, I.R.; Lloyd, C.R. Comparative estimates of interception loss from three coniferous forests in Great Britain. *J. Hydrol.* **1980**, *48*, 89–105. [CrossRef]

12. Ubarana, V.N. Observations and Modelling of Rainfall Interception at Two Experimental Sites in Amazonia. In *Amazonian Deforestation and Climate*; Gash, J.H.C., Nobre, C.A., Roberts, J.M., Victoria, R.L., Eds.; John Wiley & Sons: Hoboken, NJ, USA, 1996; 611p.
13. Calder, I. *Evaporation in the Uplands*; Wiley: Chichester, UK; New York, NY, USA, 1990.
14. Muzylo, A.; Llorens, P.; Valente, F.; Keizer, J.J.; Domingo, F.; Gash, J.H.C. A review of rainfall interception modelling. *J. Hydrol.* **2009**, *370*, 191–206. [CrossRef]
15. Linsley, R.K., Jr.; Kohler, M.A.; Paulhus, J.L. *Applied Hydrology*; McGraw-Hill Book Co.: New York, NY, USA, 1988.
16. Horton, R.E. Rainfall interception. *Mon. Weather Rev.* **1919**, *47*, 603–623. [CrossRef]
17. Merriam, R.A. A note on the interception loss equation. *J. Geophys. Res.* **1960**, *5*, 3850–3851. [CrossRef]
18. Merriam, R.A. Fog drip from artificial leaves in a fog wind tunnel. *Water Resour. Res.* **1973**, *9*, 1591–1598. [CrossRef]
19. Rutter, A.J.; Kershaw, K.A.; Robins, P.C.; Morton, A.J. A predictive model of rainfall interception in forests, 1. Derivation of the model from observations in a plantation of Corsican pine. *Agric. Meteorol.* **1971**, *9*, 367–384. [CrossRef]
20. Rutter, A.J.; Morton, A.J.; Robins, P.C. A predictive model of rainfall interception in forests. II. Generalization of the model and comparison with observations in some coniferous and hardwood stands. *J. Appl. Ecol.* **1975**, *12*, 367–380. [CrossRef]
21. Calder, I. A model of transpiration and interception loss from a spruce forest in Plynlimon, Central Wales. *J. Hydrol.* **1977**, *33*, 247–265. [CrossRef]
22. Gash, J.; Morton, A. Application of the Rutter model to the estimation of the interception loss from Thetford forest. *J. Hydrol.* **1978**, *38*, 49–58. [CrossRef]
23. Monteith, J.L. Evaporation and environment. *Syrup. Soc. Exp. Biol.* **1965**, *19*, 205–234.
24. Gash, J.H.C. An analytical model of rainfall interception by forests. *Q. J. R. Meteorol. Soc.* **1979**, *105*, 43–55. [CrossRef]
25. Pearce, A.J.; Rowe, L.K.; Stewart, J.B. Nightime, wet canopy evaporation rates and the water balance of an evergreen mixed forest. *Water Resour. Res.* **1980**, *16*, 955–959. [CrossRef]
26. Dolman, A.J. Summer and winter rainfall interception in an oak forest. Predictions with an analytical and a numerical simulation model. *J. Hydrol.* **1987**, *90*, 1–9. [CrossRef]
27. Lloyd, C.R.; Gash, J.H.C.; Shuttleworth, W.J.; Marques, A.O. The measurement and modelling of rainfall interception by amazonian rain forest. *Agric. For. Meteorol.* **1988**, *43*, 277–294. [CrossRef]
28. Coutuntrn, D.E.; Ripley, E.A. Rainfall interception in mixed grass prairie. *Can. J. Plant Sci.* **1973**, *53*, 659–663.
29. Von Hoyningen-Hüne, J. Die Interception des Niederschlags in landwirtschaftlichen Beständen. *Schr. Des. DVWK* **1983**, *57*, 1–53.
30. Braden, H. Ein Energiehaushalts- und Verdunstungsmodell for Wasser und Stoffhaushaltsuntersuchungen landwirtschaftlich genutzer Einzugsgebiete. *Mittelungen Dtsch. Bodenkd. Geselschaft* **1985**, *42*, 294–299.
31. van Dam, J.C.; Huygen, J.; Wesseling, J.G.; Feddes, R.A.; Kabat, P.; van Walsum, P.E.V.; Groenendijk, P.; van Diepen, C.A. *Theory of SWAP Version 2.0. Simulation of Water Flow, Solute Transport and Plant Growth in the Soil-Water-Atmosphere-Plant Environment*; Technical Document 45; Wageningen Agricultural University; DLOW Winand Staring Centre: Wageningen, The Netherlands, 1997.
32. Agnese, C.; Baiamonte, G.; Cammalleri, C. Modelling the occurrence of rainy days under a typical Mediterranean climate. *Adv. Water Res.* **2014**, *64*, 62–76. [CrossRef]
33. Baiamonte, G.; Mercalli, L.; Cat Berro, D.; Agnese, C.; Ferraris, S. Modelling the frequency distribution of interarrival times from daily precipitation time-series in North-West Italy. *Hydrol. Res.* **2019**, *50*, 339–357. [CrossRef]
34. Baiamonte, G.; Agnese, C. Quick and Slow Components of the Hydrologic Response at the Hillslope Scale. *J. Irrig. Drain. E-ASCE* **2016**, *142*, 04016038. [CrossRef]
35. Allen, R.G.; Pereira, L.S.; Raes, D.; Smith, M. *Crop Evapotranspiration. Guidelines for Computing Crop Water Requirements*; FAO Irrigation and Drainage Paper 56; FAO: Rome, Italy, 1998; 300p.
36. Babu, A.K.; Kumaresan, G.; Raj, V.A.A.; Velraj, R. Review of leaf drying: Mechanism and influencing parameters, drying methods, nutrient preservation, and mathematical models. *Renew. Sustain. Energy Rev.* **2018**, *90*, 536–556. [CrossRef]
37. Black, T.A.; Gardner, W.R.; Thurtell, G.W. The prediction of evaporation, drainage, and soil water storage for a bare soil. *Soil Sci. Soc. Am. Proc.* **1969**, *33*, 655–660. [CrossRef]
38. Ritchie, J.T. Model for predicting evaporation from a row crop with incomplete cover. *Water Resour. Res.* **1972**, *8*, 1204–1213. [CrossRef]
39. Pumo, D. *L'Approvvigionamento Idrico per l'Agricoltura*; Aracne Editrice SRL: Rome, Italy, 2008; ISBN 978-88-548-1708-1. (In Italian).
40. Pereira, L.S.; Paredes, P.; Melton, F.; Johnson, L.; Wang, T.; López-Urrea, R.; Cancela, J.J.; Allen, R.G. Prediction of crop coefficients from fraction of ground cover and height. Background and validation using ground and remote sensing data. *Agric. Water Manag.* **2020**, *241*, 106197. [CrossRef]
41. Lee, K.J.; Lee, B.W. Estimating canopy cover from color digital camera image of rice field. *J. Crop Sci. Biotechnol.* **2011**, *14*, 151–155. [CrossRef]
42. Fernandez-Gallego, J.A.; Kefauver, S.C.; Kerfal, S.; Araus, J.L. Comparative canopy cover estimation using RGB images from UAV and ground. In Proceedings of the Remote Sensing for Agriculture, Ecosystems, and Hydrology XX, Berlin, Germany, 10–13 September 2018; p. 107830J. [CrossRef]
43. Knapp, A.K.; Smith, D.L. Leaf Angle, Light Interceptoon & Water Relations, Demonstrating How Plants Cope with Multiple Resource Limitations in the Field. *Am. Biol. Teach.* **1997**, *59*, 365–368.

44. Hegab, A.S.A.; Fayed, M.T.B.; Hamada, M.M.A.; Abdrabbo, M.A.A. Productivity and irrigation requirements of faba-bean in North Delta of Egypt in relation to planting dates. *Ann. Agric. Sci.* **2014**, *59*, 185–193. [CrossRef]
45. Brisson, N.; Mary, B.; Riposche, D.; Jeuffroy, M.; Ruget, F.; Gate, P.; Devienne-Barret, F.; Antonioletti, R.; Durr, C.; Nicoullaud, B.; et al. STICS: A generic model for the simulation of crops and their water and nitrogen balance. I. Theory and parameterization applied to wheat and corn. *Agronomie* **1998**, *18*, 311–316. [CrossRef]
46. Kozak, J.A.; Ahuja, L.R.; Green, T.R.; Ma, L. Modelling crop canopy and residue rainfall interception effects on soil hydrological components for semi-arid agriculture. *Hydrol. Process.* **2007**, *21*, 229–241. [CrossRef]
47. Aston, A.R. Rainfall interception by eight small trees. *J. Hydrol.* **1979**, *42*, 383–396. [CrossRef]
48. Meyer, N.; Bergez, J.E.; Constantin, J.; Belleville, P.; Justes, E. Cover crops reduce drainage but not always soil water content due to interactions between rainfall distribution and management. *Agric. Water Manag.* **2020**, *231*, 105998. [CrossRef]
49. Tripathi, S.; Govindaraju, R.S. Change detection in rainfall and temperature patterns over India. In *Proceedings of the Third International Workshop on Knowledge Discovery from Sensor Data (SensorKDD 2009), Paris, France, 28 June 2009*; Association for Computing Machinery: New York, NY, USA, 2009; pp. 133–141. [CrossRef]

Article

Estimation of Reference Evapotranspiration Using Spatial and Temporal Machine Learning Approaches

Ali Rashid Niaghi [1,*], Oveis Hassanijalilian [2] and Jalal Shiri [3]

[1] Benson Hill, Saint Louis, MO 63132, USA
[2] Agricultural and Biosystems Engineering Department, North Dakota State University, Fargo, ND 58102, USA; oveis.jalilian@ndsu.edu
[3] Water Engineering Department, Faculty of Agriculture, University of Tabriz, Tabriz 5166616471, Iran; j_shiri2005@yahoo.com
* Correspondence: arniaghi@gmail.com

Abstract: Evapotranspiration (ET) is widely employed to measure amounts of total water loss between land and atmosphere due to its major contribution to water balance on both regional and global scales. Considering challenges to quantifying nonlinear ET processes, machine learning (ML) techniques have been increasingly utilized to estimate ET due to their powerful advantage of capturing complex nonlinear structures and characteristics. However, limited studies have been conducted in subhumid climates to simulate local and spatial ET_o using common ML methods. The current study aims to present a methodology that exempts local data in ET_o simulation. The present study, therefore, seeks to estimate and compare reference ET (ET_o) using four common ML methods with local and spatial approaches based on continuous 17-year daily climate data from six weather stations across the Red River Valley with subhumid climate. The four ML models have included Gene Expression Programming (GEP), Support Vector Machine (SVM), Multiple Linear Regression (LR), and Random Forest (RF) with three input combinations of maximum and minimum air temperature-based (Tmax, Tmin), mass transfer-based (Tmax, Tmin, U: wind speed), and radiation-based (Rs: solar radiation, Tmax, Tmin) measurements. The estimates yielded by the four ML models were compared against each other by considering spatial and local approaches and four statistical indicators; namely, the root means square error (RMSE), the mean absolute error (MAE), correlation coefficient (r^2), and scatter index (SI), which were used to assess the ML model's performance. The comparison between combinations showed the lowest SI and RMSE values for the RF model with the radiation-based combination. Furthermore, the RF model showed the best performance for all combinations among the four defined models either spatially or locally. In general, the LR, GEP, and SVM models were improved when a local approach was used. The results showed the best performance for the radiation-based combination and the RF model with higher accuracy for all stations either locally or spatially, and the spatial SVM and GEP illustrated the lowest performance among the models and approaches.

Keywords: evapotranspiration; machine learning; local; spatial; subhumid climate

Citation: Rashid Niaghi, A.; Hassanijalilian, O.; Shiri, J. Estimation of Reference Evapotranspiration Using Spatial and Temporal Machine Learning Approaches. *Hydrology* **2021**, *8*, 25. https://doi.org/10.3390/hydrology8010025

Received: 24 December 2020
Accepted: 26 January 2021
Published: 2 February 2021

Publisher's Note: MDPI stays neutral with regard to jurisdictional claims in published maps and institutional affiliations.

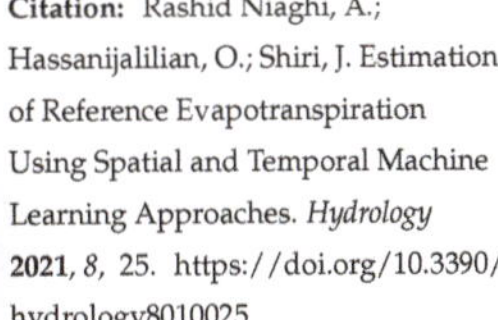

1. Introduction

Evapotranspiration (ET) is a combination of two separate processes that transfer huge volumes of water and energy from the soil (evaporation) and vegetation (transpiration) to the atmosphere [1,2]. ET is the second greatest component of hydrological cycle and a major component of the Earth's surface energy balance. ET closely relates to plant growth, droughts, gas efflux, and production. Since ET plays a crucial role in watershed and agricultural water management, accurate spatial and local estimation of crop water requirements (ET_a) at the scale of human influence is a critical need for a wide range of applications [3]. Quantifying ET_a from agricultural lands is vital to the management of water resources. Measurement methods of ET_a are available through water vapor transfer methods (e.g., Bowen ratio, eddy covariance) [4–6], water budget measurements (e.g., soil water balance, weighting lysimeters) [7–9], or remote sensing techniques [10–12]. However, the application of field scale methods is limited due to cost, complexity, and maintenance. Estimating ET_a using remote sensing models has developed in recent years, but cloud existence in many areas during the satellite passing dates limits the imagery's usefulness. Therefore, due to difficulties in acquiring direct ET_a measurements, ET_a can be estimated using by multiplying the calculated reference ET (ET_o), using different calculation methods and meteorological data, with the crop coefficient (K_c). However, the required meteorological data for ET_o calculation methods are not readily available for many locations, and methods with fewer variables must be considered when basic meteorological data are available [13]. However, the simplified and basic models are suited to estimating ET_o on a weekly or monthly basis instead of a daily basis [14].

The ET_o calculation is a complex process due to a large number of associated meteorological variables, and it is hard to develop an accurate empirical model to overcome all the complexities of the process [15]. Over the last few decades, machine learning (ML) techniques have attracted the interest of streams of researchers around the globe to overcome the ET_a estimation complexity. These methods are evolutionary computation techniques that can achieve the best relationship in a system with a data driving tool. Due to the capability of the ML methods in tackling the nonlinear relationship between dependent and independent variables [16], numerous ML techniques have been applied and proposed to predict ET_o for agricultural purposes including genetic programming (GP) [17,18]; kernel-based algorithms, e.g., support vector machine (SVM) [19,20]; artificial neural network [21–23]; wavelet neural network [14,24]; random forest (RF) [24,25]; and multiple linear regression (LR) [26,27].

Several authors have applied ML methods to detect ET_o values with minimum variation from the observed values [16,21,28]. Among these models, Gene Expression Programming (GEP) and SVM illustrated better estimation than other models [29,30]. Gene Expression Programming (GEP) is comparable to GP and both involve different sizes and shapes of computer programs encoded in linear chromosomes of fixed length [16]. The SVM method is a regression procedure that has been used successfully in the hydrology context [30–32] and agro-hydrology for ET_o modeling [19,33]. GEP has several advantages compared to GP such as generating valid structures, its multigenic nature, and the ability to surpasses the old GP system. Shiri et al. [17] evaluated GEP to estimate daily evaporation through spatial and local data scanning Kişi and Çimen [34] studied the potential of the SVM model in ET_o prediction and observed that the SVM model could be useful for ET_o estimation and hydrological modeling studies. Shiri and Kişi [35] compared the GEP and Adaptive neuro-fuzzy inference system (ANFIS) techniques for predicting short and long-term river flows. They also used a similar comparison to predict groundwater table fluctuations.

The LR model is one of the commonly used ML methods in hydrology [28,36]. It has been used to cover the study of the relationship between two or more hydrological variables and investigate the dependence and relationship between the successive values of hydrologic data. Some researchers have used the LR method to estimate the ET_o rate for different regions due to the simplicity of the method compared to other numerical methods [37,38]. Tabari et al. [27] compared the LR and SVM models' performance for the semi-arid region and found that the SVM model was superior to empirical and LR models.

Due to the high computational costs and complexity of the ML models, the tree-based ensemble models attract people by their simplicity and estimation power. The RF model as an ML tree-based model can produce a great result compared to the other ML models [25,39]. This model is known for its simplicity and the ability to perform both classification and regression tasks. The RF method is also widely used to predict the ET rate of different climate regions [40]. Feng et al. [26] applied the RF model for ET_o estimation daily for southwest China and indicated that the RF model performed slightly better than the general regression neural network model. Shiri [41] evaluated the coupled RF with wavelet algorithm to estimate ET_o for the southern part of Iran and obtained results stating that the coupled RF model showed a great improvement compared to the conventional RF and empirical models. To our knowledge, this model has not been applied in the Northern US for ET_o studies.

According to the literature, GEP and SVM models have been frequently applied across the world in various climate conditions for ET_o estimation, while the LR and RF models' applications were minimal. Besides, these models have not been compared with commonly used SVM and GEP models for subhumid climate conditions. Since the limited studies have been conducted to evaluate ML models for the Northern part of the US (which experiences a high variability of weather conditions and a huge amount of agricultural production), the objectives of this study were to (1) investigate the effect of different input combinations of meteorological data on the accuracy of daily ET_o estimation in subhumid climate using the GEP, SVM, LR, and RF methods, (2) compare the spatial and local prediction capabilities of the different ML models in ET_o estimation, and (3) evaluate the performance of the models based on the various study years and meteorological stations.

2. Materials and Methods

2.1. Study Area Climate and Reference Evapotranspiration (ET$_o$)

The weather data for the current study were obtained from the North Dakota (ND) Agricultural Weather Stations [42] located at Prosper (ND), Galesburg (ND), Leonard (ND), Sabin (MN: Minnesota), Perley (MN), and Fargo (ND) for 17 study years (January 2003–December 2016). Geographical location of the study stations is shown in Figure 1.

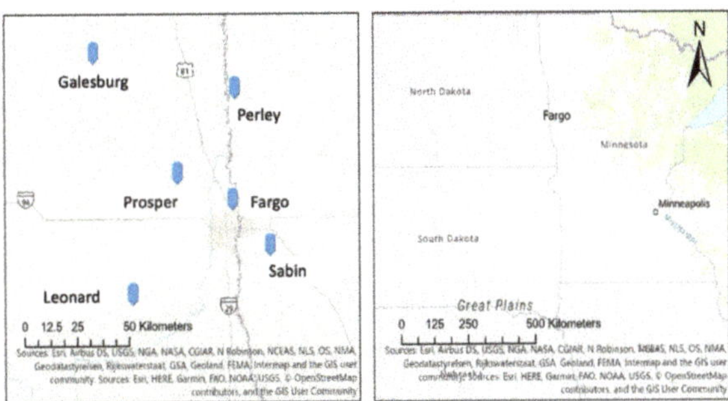

Figure 1. Location of weather stations used for the current study.

To calculate the daily reference evapotranspiration (ET_o) for each study station, the ASCE standardized reference evapotranspiration equation (ASCE-EWRI) was used for the alfalfa reference crop [43]. This equation provides a standardized calculation of ET_o demand that can be used in developing K_c and comparing it with other methods. Equation (1) presents the form of the standardized ET_o equation for all hourly and daily calculation time steps.

$$ET_o = \frac{0.408\,\Delta\,(R_n - G) + \gamma\frac{C_n}{T+273}u_2(e_s - e_a)}{\Delta + \gamma(1 + C_d u_2)} \tag{1}$$

where ET_o is reference evapotranspiration rate (mm d^{-1}), R_n is net solar radiation (MJ m^{-2} d^{-1}), G is soil heat flux (MJ m^{-2} d^{-1}), γ is psychrometric constant (KPa °C^{-1}), T is the mean daily air temperature, U_2 is the average wind speed at 2 m height (m s^{-1}), es is saturation vapor pressure (KPa), e_a is actual vapor pressure (KPa), Δ is the slope of the saturation vapor pressure–temperature relationship (KPa °C^{-1}), C_n and C_d are coefficients which are related to the crop and time step. The value for the constants C_n and C_d are 1600 and 0.38 for the alfalfa reference crop. Table 1 summarized weather parameters of the study locations for the study period.

Table 1. Statistical summary of the weather data for the study locations for the study period.

Station	Parameter	Unit	X_{max}	X_{min}	X_{mean}	S_X	C_V	C_{SX}
Prosper, ND	T_{max}	°C	37.9	24.3	11.3	14.3	1.27	−0.37
	T_{min}	°C	−29.8	−38.1	−0.8	13.0	−16.73	−0.28
	W_S	m s^{-1}	14.2	0.9	4.2	1.8	0.43	0.55
	R_h	%	100	13.8	68.6	15.6	0.23	−0.14
	R_S	MJ m^{-2}	31.1	0.3	13.2	7.9	0.60	0.51
	ET_o	mm	11.4	0	2.4	2.03	0.84	0.92
Galesburg, ND	T_{max}	°C	36.8	23.6	10.9	14.2	1.30	−0.33
	T_{min}	°C	−28.9	−37.3	−1.0	12.7	−12.41	−0.28
	W_S	m s^{-1}	12.8	0.7	3.9	1.6	0.41	0.45
	R_h	%	100	18.8	68.1	15.2	0.22	−0.09
	R_S	MJ m^{-2}	30.7	0.2	12.8	7.9	0.61	0.51
	ET_o	mm	10.6	0	2.3	1.97	0.85	1.03
Leonard, ND	T_{max}	°C	38.3	23.6	11.5	14.2	1.23	−0.39
	T_{min}	°C	−28.6	−37.7	−0.6	12.9	−21.05	−0.28
	W_S	m s^{-1}	13.2	0.9	4.2	1.7	0.42	0.50
	R_h	%	100	17.85	67.40	15.3	0.23	−0.02
	R_S	MJ m^{-2}	31.6	8.1	13.6	8.1	0.60	0.51
	ET_o	mm	10.6	0	2.5	2.09	0.85	0.77
Sabin, MN	T_{max}	°C	37.8	24.3	11.2	14.1	1.26	−0.33
	T_{min}	°C	−30.2	−38.5	−0.2	13.0	−73.34	−0.24
	W_S	m s^{-1}	12.7	0.5	4.0	1.7	0.42	0.46
	R_h	%	100	18.70	68.80	14.9	0.22	−0.08
	R_S	MJ m^{-2}	31.6	0.4	13.0	7.9	0.61	0.51
	ET_o	mm	10.1	0	2.4	2.02	0.86	0.85
Perley, MN	T_{max}	°C	37.3	24.1	10.9	14.3	1.31	−0.36
	T_{min}	°C	−30.5	−40.7	−0.7	13.1	−18.07	−0.30
	W_S	m s^{-1}	11.8	0.8	4.1	1.7	0.41	0.48
	R_h	%	100	17.22	69.10	14.9	0.22	−0.08
	R_S	MJ m^{-2}	31.3	0.4	12.8	7.9	0.61	0.51
	ET_o	mm	10.9	0	2.3	2.02	0.84	1.12
Fargo, ND	T_{max}	°C	39.6	25.6	11.4	14.2	1.24	−0.36
	T_{min}	°C	−29.5	−36.8	0.6	13.0	21.89	−0.23
	W_S	m s^{-1}	11.3	0.8	3.8	1.5	0.39	0.40
	R_h	%	100	15.55	66.19	14.9	0.23	−0.05
	R_S	MJ m^{-2}	31.0	0.1	12.8	7.9	0.61	0.52
	ET_o	mm	10.5	0	2.5	2.07	0.84	0.92

2.2. Models Structure and Application

To process the GEP and SVM algorithm, we used the GeneXpro program in Matlab, and to process the LR and RF models, a scikit-learn module embedded in the Python 3.2 programming language was used. GEP is an extension of GP [44] developed by Ferreira [45] that creates a computer program to investigate a relationship between input and output variables. GEP was developed to find a better solution for solving a particular problem relating to the understudied phenomena [46].

The application of GEP requires several steps. GEP is, like GAs and GP, a genetic algorithm, as it uses populations of individuals, selects them according to fitness, and introduces genetic variation using one or more genetic operators. The procedure to model daily evapotranspiration (as a dependent variable) by using weather variables (as independent variables) is as follows: 1. Selection of fitness function; 2. Choosing the set of terminals T and the set of functions F to create the chromosomes; 3. Choosing the chromosomal architecture; 4., Choosing the linking function; 5. Choosing the genetic operators.

The fitness function must be determined in the first step with a random generation of chromosomes of a certain program (initial population) and evaluated against a set of fitness cases [47]. Using weather station data as input variables (terminals) to model daily ET_0 involves the next general step. The selection of fitness functions (i.e., absolute error, relative error, and correlation coefficient) depends on the experience and intuition of the user. The GEP model in the current study was developed based on the recommended functions by Shiri et al. [17]. In the third step, the chromosomal architecture can be defined by having the weather variables as terminal and function set as chromosomes. The fourth step was to select a linking function that relates genes to each other in addition to linking the parse trees [17]. Finally, genetic operators' corresponding rates were chosen. Table 2 summarizes the commonly used parameters for each run.

Table 2. Parameters used per run of gene expression programming (GEP).

Number of Chromosomes	30	One-Point Recombination Rate	0.3
Head of the size	8	Two-point recombination rate	0.3
Number of genes	3	Gene recombination rate	0.1
Linking function	Addition	Gene transposition rate	0.1
Fitness function error type	RMSE	Insertion sequence transposition rate	0.1
Mutation rate	0.044	Root insertion sequence transposition	0.1
Inversion rate	0.1	Penalizing tool	parsimony pressure

The SVM was developed by Cortes and Vapnik [48] and is known as the classification and regression method [34] to solve problems by applying a flexible representation of the class boundaries and implementing an automatic complexity control to reduce overfitting. In SVM, the dependency of the dependent variable to a set of independent variables is evaluated. In regression estimation with Support Vector Regression (SVR), which is used to define SVMs in the literature, a functional dependency $f(x)$ between a set of sampled input points $X = \{x_1, x_2, x_3, \ldots, x_l\}$ (here, input sampled refer to meteorological variables) taken from R_n (input vector of n dimension) and target values $Y = \{y_1, y_2, y_3, \ldots, y_l\}$ (ET_0 as target values) with $y_i \in R_n$. More detail on SVM can be found in Vapnik [49].

The LR is a statistical method used to describe a quantitative relationship between a dependent variable and one or more independent variables [27,50]. In LR, the function is a linear equation and is expressed as:

$$Y_i = b_0 + b_1 \times_1 + \ldots + b_k x_k \tag{2}$$

where b_0–b_k are the fitting constants, y_i is the dependent variable, and x_1–x_k are the independent variables for this system.

The RF method combines a group of decision trees for either classification or regression purposes. Although each decision tree may not be capable of learning well, the combination of the decision trees results in a strong learner. Each decision tree predicts the outcome individually, and RF votes among the outcomes for classification or averages the outcomes for regression. Each decision tree is trained on a different subset of samples by a bagging extension of the RF model to reduce the risk of overfitting. Moreover, a different subset of input variables can be used in each tree to make it more useful in prediction for datasets with higher dimensions [51]. For this study, a small subset of data was used to find a good combination of parameters for the RF model. As a result, there were several trees in the forest and the minimum number of samples required for the leaf nodes were 50 and 35, respectively. The mean square error criteria are used as a procedure for estimation.

The calculated daily ET_0 was used to feed the GEP, SVM, LR, and RF models. Three treatments including temperature, radiation, and mass transfer-based combinations were used as input to feed the models, and each model of the combinations was assessed for spatial and local approaches. Different statistical analysis was performed to evaluate the accuracy and performance of the different combinations and approaches for each studied station. The combinations were as follows:

(i) T_{min}, T_{max}: temperature based (GEP1, SVM1, LR1, RF1)
(ii) T_{min}, T_{max}, R_s: radiation-based (GEP2, SVM2, LR2, RF2)
(iii) T_{min}, T_{max}, W: mass transfer based (GEP3, SVM3, LR3, RF3).

2.3. K-Fold Cross-Validation

Splitting the data into the sets of data for testing and training is a usual procedure for assessing the ML techniques. Using 10–30% of the complete dataset as a single test set is a common method for GEP evaluation. Therefore, the K-fold cross-validation technique was used to increase the evaluation performance and set of data for either training or testing purposes. Using K-fold cross-validation, the dataset was divided into K subsets, and the training process was repeated K times leaving each time a distinct set of patterns for testing until a complete testing scan for the dataset was fulfilled. Computation cost defines the minimum assembly size of the test set. Here, the minimum test size was fixed as one year for local modeling and one station for spatial modeling. Consequently, at a local scale, one year was held out each time for testing while the models were trained using the remaining 16 years; hence, a total of 612 models (17 years × 6 stations × 3 input combinations × 2 models) were established for the local k-fold testing. At the spatial scale, one station was considered as a test block each time and the models were trained using the patterns from the remaining stations; hence, a total number of 36 models (6 stations × 3 input combinations × 2 models) were constructed. The local and spatial approaches were noted with T and S in the figures.

2.4. Evaluation Criteria

To investigate the performance of the models for each combination and approaches, four statistical indicators were used, namely, the root mean square error (RMSE), the mean absolute error (MAE), correlation coefficient (r^2), and scatter index (SI), defined as follows:

$$RMSE = \sqrt{\frac{1}{n}\sum_{i=1}^{n}(ET_e - ET_o)^2} \tag{3}$$

$$MAE = \frac{\sum_{i=1}^{n}|ET_e - ET_o|}{n} \tag{4}$$

$$r^2 = \left(\frac{\sum_{i=1}^{n}(ET_o - \overline{ET_o})(ET_e - \overline{ET_e})}{\sqrt{\sum_{i=1}^{n}(ET_o - \overline{ET_o})^2 \sum_{i=1}^{n}(ET_e - \overline{ET_e})^2}}\right)^2 \tag{5}$$

$$SI = \frac{RMSE}{\overline{E_o}} \tag{6}$$

where ET_e and ET_o are simulated and calculated reference evapotranspiration at the *i*-th time step, respectively, n is a number of time steps, $\overline{ET_e}$ and $\overline{ET_o}$ are mean values of simulated and calculated ET_o, respectively.

The RMSE describes the average magnitude of errors and can take on values from 0 to ∞ indicate perfect and worst fit, respectively, and the MAE scores the error magnitudes without any specific weight to larger/smaller errors. Therefore, the lower value of the RMSE and MAE is desirable. The r^2 values around 1 indicate a perfect linear relationship between estimated and calculated values, where the closer a value is to zero, the more it demonstrates the poor relationship between simulation and calculation. Finally, SI is a dimensionless index of RMSE that gives a good insight to compare the performance of different models.

3. Results and Discussions

The local and spatial analysis of four models for six studied stations is shown in Table 3. According to the three combinations' performance, the radiation-based method illustrated the highest accuracy for either local or spatial approaches compared to the other combinations. The mass-transfer-based combination was the next most accurate combination. The results showed that the local trained models surpassed the spatially trained models because of using the same patterns for both training and testing models. However, the spatial models gave comparable results compared to the local model in some cases, especially for radiation-based combinations. Differences in temperature among the stations have dramatically affected the performance of both the temperature-based and the mass transfer-based models. In all cases, the minimum differences between the performance accuracy of the local and spatial models belonged to the LR model. This can be inferred to the mathematical structure of this technique, where only linear relationships can be supposed between the input and target parameters with a lower degree of flexibility compared to heuristic data driven models.

Table 3. Global average performance indicators of the gene expression programming (GEP), support vector machine (SVM), multiple linear regression (LR), and random forest (RF) methods for three input combinations of local (T) and spatial (S) approaches.

Evaluation Criteria		Input Combination											
		1 (Temperature-Based)				2 (Radiation-Based)				3 (Mass-transfer-based)			
	Approach	GEP	SVM	LR	RF	GEP	SVM	LR	RF	GEP	SVM	LR	RF
R^2	T	0.75	0.80	0.77	0.85	0.85	0.91	0.88	0.92	0.77	0.86	0.78	0.86
	S	0.78	0.75	0.77	0.84	0.87	0.85	0.88	0.93	0.77	0.77	0.78	0.88
RMSE (mm/day)	T	0.90	0.97	0.97	0.82	0.71	0.72	0.68	0.57	0.72	0.73	0.94	0.73
	S	1.07	1.13	0.98	0.80	0.76	0.77	0.69	0.55	0.91	0.93	0.95	0.69
MAE (mm/day)	T	0.64	0.71	0.76	0.58	0.50	0.54	0.51	0.38	0.64	0.62	0.75	0.53
	S	0.84	0.82	0.77	0.57	0.57	0.61	0.52	0.36	0.69	0.67	0.76	0.49
SI	T	0.38	0.41	0.40	0.34	0.29	0.30	0.28	0.24	0.35	0.33	0.39	0.30
	S	0.44	0.46	0.40	0.33	0.32	0.33	0.29	0.23	0.38	0.36	0.39	0.28

Among four models with three input combinations, the models relying on radiation, mass-transfer, and temperature-based combinations showed the lowest RMSE and MAE, respectively (Table 3). Comparing the GEP, SVM, LR, and RF models, the RF model illustrated the lowest rate of RMSE and MAE with the best performance for radiation-based approaches. However, the RF model improved 4.37, 5.76, and 1.49 percent from local to spatial approaches for temperature, radiation, and mass-transfer-based combinations, respectively, which was in contrast with the improvement's direction in the other models. Considering the models based on radiation combination, the spatial RF model exhibited the highest linear relationship ($r^2 = 0.927$) between calculated and estimated ET_0 in comparison with the other models. The local RF method was the next accurate approach to estimate ET_0 based on radiation-based data. This observation illustrated the ability of the RF algorithms to estimate ET_0 using data from local stations for training. Furthermore, the LR model had significant improvement for RMSE and MAE from spatial to local approaches for all three types of input combinations. For the LR model, the r^2 value was not changed for radiation-based and temperature-based combination from spatial to local approaches and the change was 0.13 percent for the mass-transfer-based model. Therefore, the LR model illustrated almost similar results for both spatial and local approaches among all models.

The GEP and SVM models illustrated the great improvement rate for all three input combinations from spatial to local condition with the highest improvement of 21 percent for mass-transfer-based combination. Specifically, the GEP model showed an improvement from spatial to local approach, however, the percentage of improvement was 2.3, 3.9, and 0.13 for radiation, temperature, and mass-transfer-based combinations, respectively. In term of obtained improvement for RMSE and MAE from spatial to local approaches, both of the GEP and SVM models gained similar results. The correlation coefficient of the SVM model decreased from spatial to local approaches for about 6.3, 6.5, and 10.9 percent for radiation, temperature, and mass-transfer-based combinations, respectively. By using local radiation data for training the models, the SI indicator for the GEP and SVM models showed an improvement of 8.2 and 10 percent from spatial to local approach, respectively. This improvement was about 6.6 and 8.7 percent for mass-transfer-based and 15 and 10.9 percent for temperature-based combinations, respectively.

Statistical analysis revealed the similar performance of the local GEP and SVM models. For RMSE and MAE statistical variables, GEP and SVM models showed a greater improvement in performance for mass-transfer and temperature-based combinations, respectively. By considering correlation coefficient values, it can be concluded that the improvement in accuracy of either GEP or SVM approaches was not significant and all illustrated the ability to estimate with acceptable accuracy. Therefore, if temperature data are not available at a particular station, but they are for other stations, the GEP and SVM approaches can be useful to estimate ET_0. However, due to the higher mapping ability of the GEP models, using either local or spatial GEP are preferable.

The models relying on the mass-transfer combination had slightly higher accuracy than the temperature-based approach, but lower accuracy compared to the radiation-based combination. All of the local and spatial GEP and SVM methods illustrated lower improvement compared to that for the temperature-based approach. This showed that wind speed can have a significant effect on the accurate estimation of spatial and local ET_0. Due to the flat topography of the study area and being faced with lots of high-speed winds during the growing season and almost all other seasons, including the wind as a parameter to build the model architecture and estimating the ET_0 can increase the accuracy of the approach.

Overall, the RF and the LR models illustrated the best performance among the four models, and comparing the GEP and SVM models, the GEP model showed better performance than the SVM model for all three input combinations.

A breakdown of the models' performance accuracy at each station is shown in Figures 2–4 for all of the three input combinations, respectively. In the case of the temperature-based combination (Figure 2), the local GEP and SVM models (shown as TGEP and TSVM) gave more accurate results than the spatial (shown as SGEP and SSVM) models. For the LR and RF models, the difference in accuracy between local (TLR and TRF) and spatial approaches (SLR and SRF) was not significant and both showed better performance than GEP and SVM models since they relied on the patterns of the same location used for training and testing the models. According to Table 2, station 6 (Fargo) had the highest, and station 2 (Galesburg) had the lowest range of recorded temperature among the study stations. This range may be caused to have the lowest performance for station 2; however, it was difficult to evaluate the model's performance in the climate context of each station due to the few number of study stations. The RF model showed the best performance with higher accuracy for all stations either locally or spatially, and the spatial SVM and GEP illustrated the lowest performance among other models and approaches.

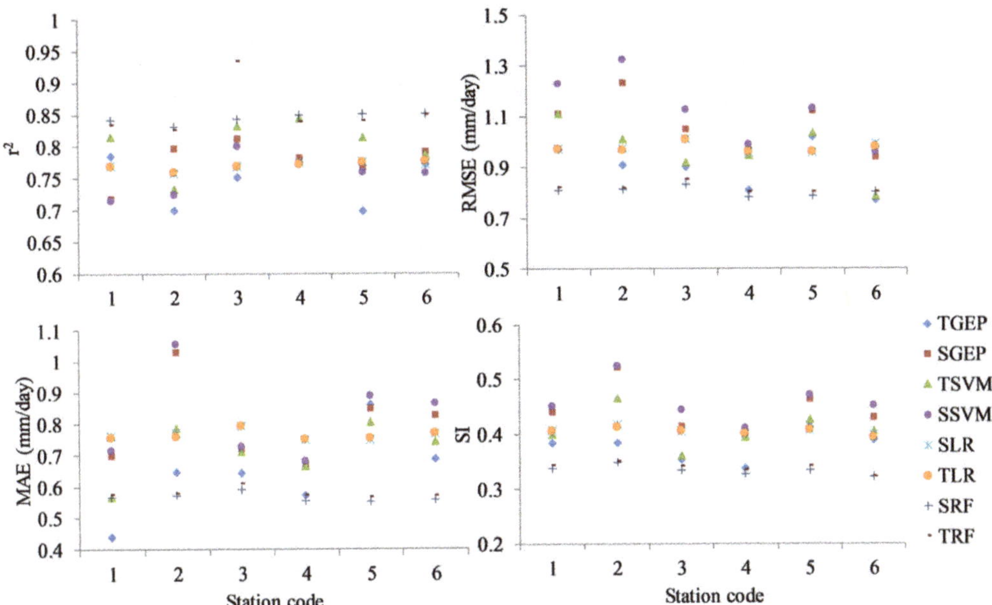

Figure 2. Statistical indices of the temperature-based combination for the four applied models (GEP, VM, LR, and RF) with local (T) and spatial (S) approaches.

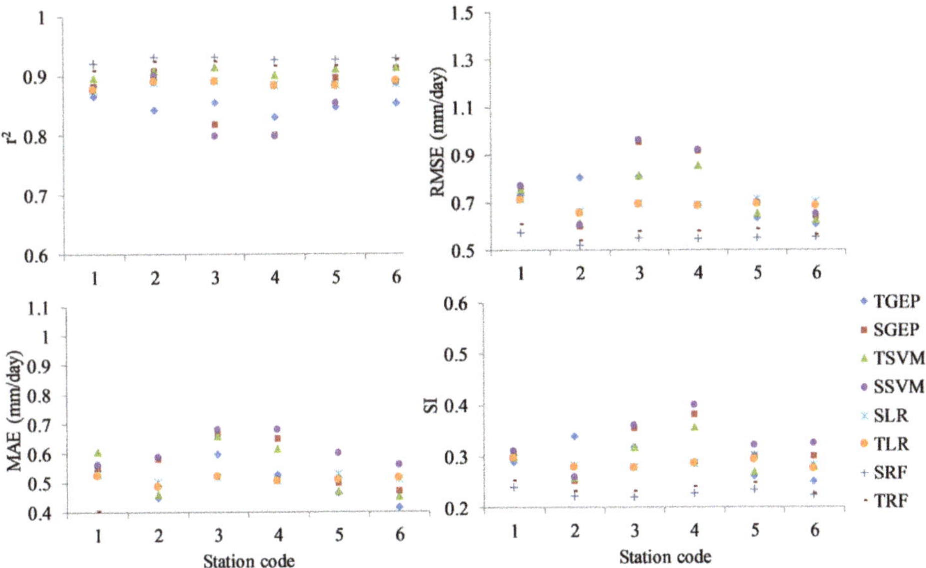

Figure 3. Statistical indices of radiation-based combination for the four applied models (GEP, SVM, LR, and RF) with local (T) and spatial (S) approaches.

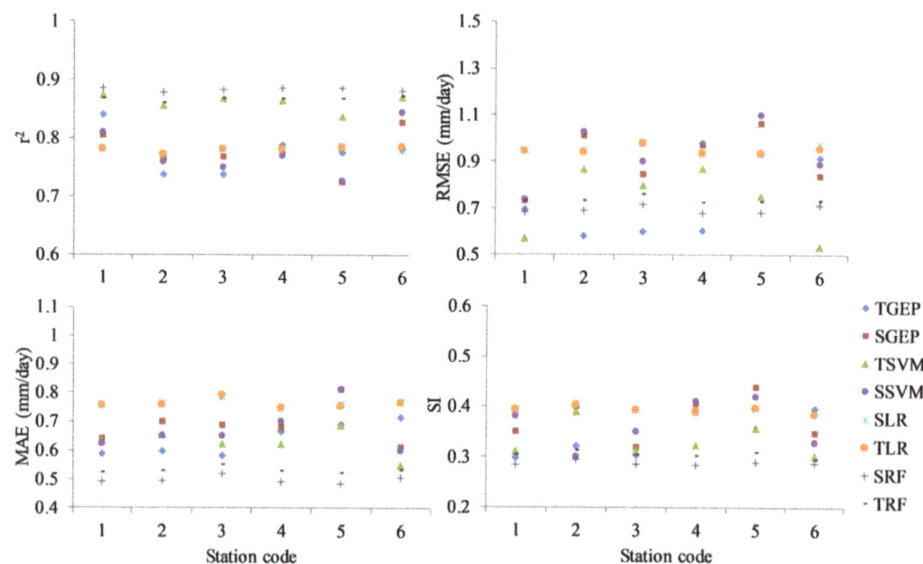

Figure 4. Statistical indices of mass transfer-based combination for the four applied models (GEP, SVM, LR, and RF) with local (T) and spatial (S) approaches.

The RF and LR methods showed the lowest range of SI compared to the spatial and local GEP and SVM methods. For temperature-based combinations, the spatial and local LR approaches had minimum SI ranges of 0.018 and 0.020, respectively, and the spatial SVM and GEP methods illustrated the highest SI ranges of 0.113 and 0.119, respectively. The spatial RF approaches with an average of 0.333, and spatial SVM, with an average of 0.457, showed the lowest and highest SI rate, respectively. Therefore, spatial RF approaches may be the most practical way to estimate the missing meteorological data of the study stations.

Figure 3 shows the evaluation result of the radiation-based combination for the four models with spatial and local approaches. The amount of received radiation for all study stations was similar. According to the global performance of the defined models, the radiation-based combination gained the best performance among the three input combinations. Besides, the radiation-based combination had the lowest rate of RMSE, MAE, and SI, and the highest rate of r^2 for each of the study stations. Among the spatial and local scenarios, the local approach had a better performance than the spatial approach. For the radiation-based combination, the spatial RF and local RF models had an accurate estimation of ET_o, respectively. For stations 3 and 4 (Leonard and Sabin) either spatial or local approaches of GEP and SVM models gained lower performance than the other stations. This could be due to the slightly higher magnitude and variations of solar radiation (Table 2) among the other stations during the study period.

Among the study stations comparison, the SI range of the spatial RF was 0.018, which showed the best performance compared to the other applied methods. As obtained from the temperature-based combination, the LR method performed well in the radiation-based combination too, with an SI indicator range of 0.021 and 0.024 for local and spatial LR approaches, respectively. The worst performance was observed for spatial GEP and SVM approaches, with SI indicator of 0.128 and 0.140, respectively. According to the GEP and SVM models, the local GEP performed well compared to other approaches of the SVM and GEP models. The statistical indicators were in agreement with the spatial RF performance in which they showed the lowest rate of RMSE and MAE and the highest value of r^2. However, comparing the MAE might not be a valid indicator due to taking into account the local order of magnitude of the target variable. The ranking of the SI indicators showed that spatial RF and LR could overcome the lack of meteorological data for the station. On the other hand, the averages of the SI values for all six study stations showed that the spatial RF and local RF had the lowest and the spatial GEP and spatial SVM had the highest rate of SI indicators. Therefore, either spatial or local RF methods could be useful to estimate the missing values for any of the stations.

Figure 4 shows the statistical indices of the mass-transfer-based combination. Similar to the previous combinations, the spatial and local RF gave a more accurate estimation than other methods. On the other hand, the local SVM approach showed better estimation than the spatial SVM and GEP methods for all stations except station 2, which had the lowest range of temperature variation. The fluctuations of the indices among the stations were higher than the radiation-based combination and lower than that for temperature-based combination, which showed mediocre accuracy compared to the other combinations.

By having wind speed and temperature data as the input variable for the mass transfer-based combination, the spatial RF approach gained the lowest SI and highest r^2 values for ET_o estimation compared to all other methods. The minimum and maximum SI values for mass transfer-based combination were obtained for the spatial RF and spatial SVM approaches, which were 0.011 and 0.120, respectively. According to the performance ranking of the models based on the SI indicator, spatial LR, local LR, and local RF showed better performance after spatial RF with SI values of 0.015, 0.018, and 0.018, respectively. The local SVM, local GEP, and spatial GEP had the SI values of 0.087, 0.10, and 0.119, respectively. The average of SI for all study stations showed that the spatial and local LR had the highest and spatial and local RF had the lowest SI values, respectively. Therefore, by having the lowest range of SI and lowest value of SI for the spatial RF approach, it might be more practical to apply the spatial RF for other stations without training a specific model for each station. Accordingly, no local dataset would be needed to train the local models. This could be helpful to estimate the ET_o for stations with partial or missing meteorological data.

To understand the yearly performance of the applied models based at each of the study stations, the models were assessed per test year. Figure 5 illustrates the SI values obtained from the three input combinations for each study year of the study stations. The SI values of the models fluctuated considerably for almost all stations during the test years.

As shown in Figure 5, the SI values fluctuated considerably within test years for all input combinations and approaches. Among the study stations, Prosper and Sabin stations showed the average maximum range for the SI values. The minimum average of the SI value 0.223 was observed for the RF radiation-based combination for the Fargo station, and the maximum average of the SI was obtained for the SVM temperature-based combination (SVM1) for the Galesburg station. The Galesburg station had the lowest temperature range among the study stations. According to Figure 4, the RF2 (radiation-based combination) model showed the lowest fluctuation compared to the other approaches with a similar trend among the study stations. For all of the study stations, test years, and input combinations, the RF models gave the best performance with the lowest SI values compared to the other models and combinations. The SVM and GEP models showed the worst SI averages for the temperature-based combinations. In this case, the order of the accuracy of the models was similar to that obtained from the station-based analysis. The radiation-based combination gave the most accurate results and estimation in comparison with the temperature or mass-transfer-based combination models.

Comparing the performance of the models relying on each of the combination methods, it can be observed that the performance of the approaches is similar. However, the range of the SI indicator for different approaches was different depending on the test years. For example, 2011, 2012, and 2013 were the dry, normal, and wet years among the study years, respectively. The result of SI per test years showed that 2012 had lower SI than 2011 and 2013 for all of the three input combinations, and the SI values of the various methods and approaches were close together. On the other hand, for the 2013 test year, some of the approaches illustrated a huge jump for the obtained SI from 2012 to 2013 due to the weakness of the model performance.

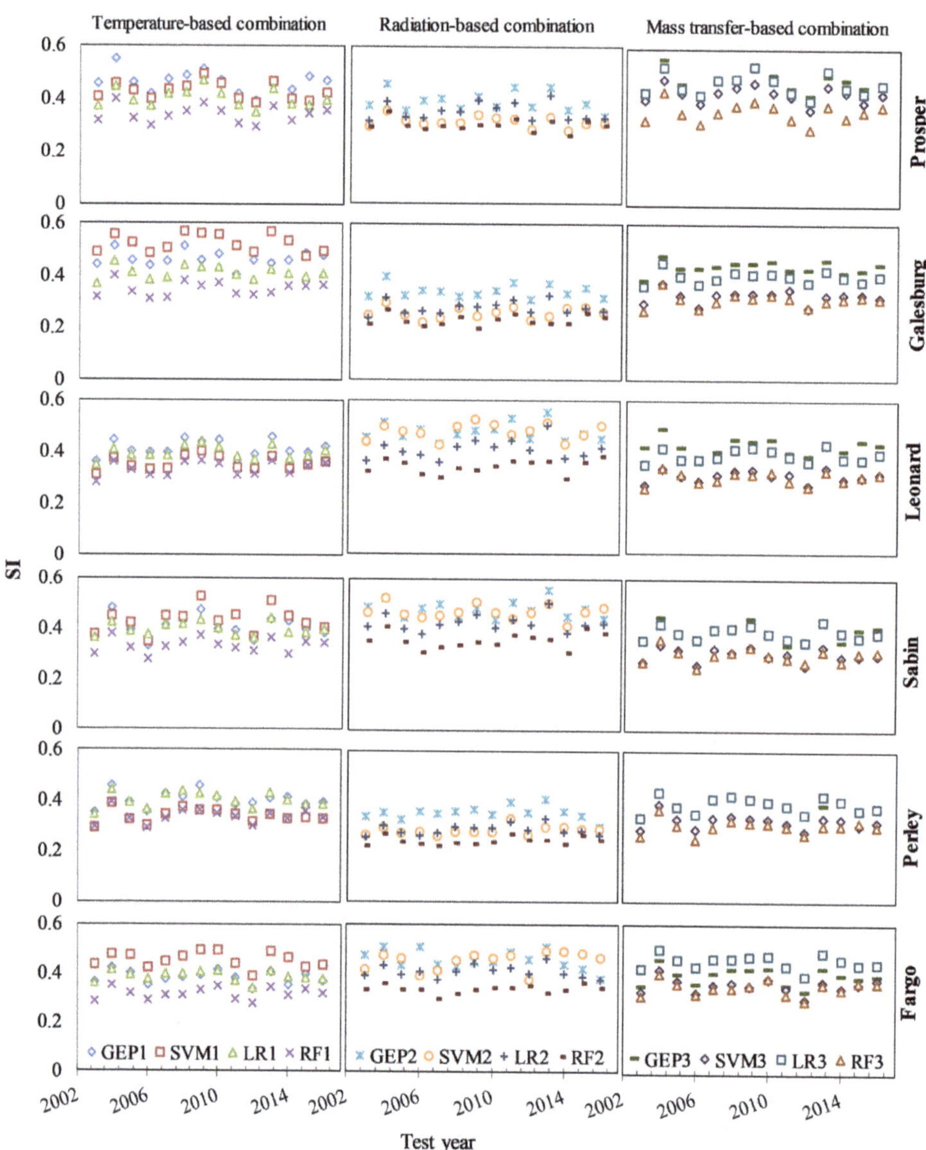

Figure 5. Scatter index (SI) values of GEP, SVM, LR, and RF models per test year and study station for three input combinations.

Due to the variability in the meteorological data and the climate pattern, the variability in performance accuracy at each station was expected. Similar variability in performance of the ML approaches was observed by Shiri et al. [17]. Selection of the training set and testing set plays an important role in model performance. The existence of any abnormal year in the test years in comparison with training datasets causes it to have an inaccurate estimation [19]. By lowering the number of the input values, the validity of the training set for estimation of test years decreases. Because of the lower input values, the variable rate would be low, and this type of input would only be valid for periods with very specific trends. This explanation may clarify the performance of spatial approaches, where the relationship encountered might be site-specific. Other researchers illustrated the site-specific performance of spatial approaches for the different study regions and climate conditions [34,35].

The comparison of the three input combinations showed that the performance of some approaches was similar for some years while the performance of methods for some years were far from each other. For example, the SVM model showed the most improvement from temperature to mass-transfer-based combination, which became like the RF method. However, depending on the station and test year, this similarity becomes even closer. All the test years and stations showed an improvement from temperature to radiation or mass-transfer-based combination except for the Prosper station, which is in agreement with the findings of Adnan et al. [15]. On the other hand, the Prosper station showed the best improvement for the SVM model for radiation-based approaches. Considering that all of the input combinations rely on the temperature and another variable (solar radiation or wind speed), it might be thought that the performance differences could be due to the effect of the second variable in the estimation of the output. Besides, when the performance of the models is similar, the impact of the secondary variable might be less than the primary variable (temperature). However, when the gap between the performance of the SI indicator increases, it proves the crucial impact of the second variable on the model performance for the specific test year or station. A similar conclusion could be drawn for the comparison between the input combinations. If each of the combinations shows a better performance than the other combination method for a specific year and station, the second variable effect should be important and might have a significant influence in the explanation of the output for that test year.

4. Conclusions

This paper aimed to evaluate the different ML methods including GEP, SVM, LR, and RF to estimate ET_0 in the Red River Valley. The external approach exempts using local data like spatial approaches in the current paper for simulating ET_0 values in a decisive way when local data is not available, reliable, or sufficient. Global comparison of the performance accuracy of the applied models revealed that the RF model was the best for all combinations among the four defined models. Furthermore, the RF model illustrated the best performance for spatial and local conditions for all input combinations. In general, the LR, GEP, and SVM models were improved when a local approach was used, except for the RF model, which was less accurate with a local approach. The radiation-based combination was the most accurate predictor among all models tested. As a result, this combination showed the lowest rate of improvement due to better performance in the first step.

The results showed that due to the flat topography of the study area with high wind speeds during the growing season, including the wind as a parameter to build the model architecture and estimate the ET_0 can increase the accuracy of the prediction. Besides, it might be more practical to apply the spatial RF model for stations with missing meteorological data without the need for local training. The recommended application of spatial RF using radiation combination allows for a more reliable estimate of ET_0 to fill the missing values for more precision water management purposes. Further research should confirm the current results in other geographical locations and for the various input combination methods.

Author Contributions: Conceptualization, A.R.N. & J.S.; methodology, A.R.N. & J.S.; software, A.R.N., J.S., and O.H.; validation, A.R.N. & J.S.; formal analysis, A.R.N., J.S., and O.H.; investigation A.R.N.; resources, A.R.N.; data curation, A.R.N.; writing—original draft preparation, A.R.N.; writing—review and editing, A.R.N.; visualization, A.R.N. & J.S.; supervision, J.S.; All authors have read and agreed to the published version of the manuscript.

Funding: This research received no external funding.

Institutional Review Board Statement: Not applicable.

Informed Consent Statement: Not applicable.

Conflicts of Interest: The authors declare no conflict of interest.

References

1. Allen, R.G.; Pereira, L.S.; Howell, T.A.; Jensen, M.E. Evapotranspiration information reporting: I. Factors governing measurement accuracy. *Agric. Water Manag.* **2011**, *98*, 899–920. [CrossRef]
2. Niaghi, A.R.; Majnooni-Heris, A.; Haghi, D.Z.; Mahtabi, G. Evaluate several potential evapotranspiration methods for regional use in Tabriz, Iran. *J. Appl. Environ. Biol. Sci.* **2013**, *3*, 31–41.
3. Anderson, M.C.; Allen, R.G.; Morse, A.; Kustas, W.P. Use of Landsat thermal imagery in monitoring evapotranspiration and managing water resources. *Remote Sens. Environ.* **2012**. [CrossRef]
4. Kabenge, I.; Irmak, S.; Meyer, G.E.; Gilley, J.E.; Knezevic, S.; Arkebauer, T.J.; Woodward, D.; Moravek, M. Evapotranspiration and surface energy balance of a common reed-dominated riparian system in the platte river Basin, Central Nebraska. *Trans. ASABE* **2013**, *56*, 135–153. [CrossRef]
5. O'Brien, P.L.; Acharya, U.; Alghamdi, R.; Niaghi, A.R.; Sanyal, D.; Wirtz, J.; Daigh, A.L.M.; DeSutter, T.M. Hydromulch Application to Bare Soil: Soil Temperature Dynamics and Evaporative Fluxes. *Agric. Environ. Lett.* **2018**, *3*, 180014. [CrossRef]
6. Niaghi, A.R.; Jia, X.; Scherer, T.F.; Steele, D.D. Measurement of Non-Irrigated Turfgrass Evapotranspiration Rate in the Red River Valley. *Vadose Zone J.* **2019**, *18*, 180202. [CrossRef]
7. Chávez, J.L.; Howell, T.A.; Copeland, K.S. Evaluating eddy covariance cotton ET measurements in an advective environment with large weighing lysimeters. *Irrig. Sci.* **2009**, *28*, 35–50. [CrossRef]
8. Valayamkunnath, P.; Sridhar, V.; Zhao, W.; Allen, R.G. Intercomparison of surface energy fluxes, soil moisture, and evapotranspiration from eddy covariance, large-aperture scintillometer, and modeling across three ecosystems in a semiarid climate. *Agric. For. Meteorol.* **2018**, *248*, 22–47. [CrossRef]
9. Niaghi, A.R.; Haji Vand, R.; Asadi, E.; Majnooni-Heris, A. Evaluation of Single and Dual Crop Coefficient Methods for Estimation of Wheat. *Adv. Environ. Biol.* **2015**, *9*, 963–971.
10. Elnmer, A.; Khadr, M.; Kanae, S.; Tawfik, A. Mapping daily and seasonally evapotranspiration using remote sensing techniques over the Nile delta. *Agric. Water Manag.* **2019**, *213*, 682–692. [CrossRef]
11. Nagler, P.L.; Scott, R.L.; Westenburg, C.; Cleverly, J.R.; Glenn, E.P.; Huete, A.R. Evapotranspiration on western U.S. rivers estimated using the Enhanced Vegetation Index from MODIS and data from eddy covariance and Bowen ratio flux towers. *Remote Sens. Environ.* **2005**, *97*, 337–351. [CrossRef]
12. Wagle, P.; Bhattarai, N.; Gowda, P.H.; Kakani, V.G. Performance of five surface energy balance models for estimating daily evapotranspiration in high biomass sorghum. *ISPRS J. Photogramm. Remote Sens.* **2017**, *128*, 192–203. [CrossRef]
13. Partal, T.; Cigizoglu, H.K. Prediction of daily precipitation using wavelet-neural networks. *Hydrol. Sci. J.* **2009**, *54*, 234–246. [CrossRef]
14. Torres, A.F.; Walker, W.R.; McKee, M. Forecasting daily potential evapotranspiration using machine learning and limited climatic data. *Agric. Water Manag.* **2011**, *98*, 553–562. [CrossRef]
15. Adnan, R.M.; Chen, Z.; Yuan, X.; Kisi, O.; El-Shafie, A.; Kuriqi, A.; Ikram, M. Reference evapotranspiration modeling using new heuristic methods. *Entropy* **2020**, *22*, 547. [CrossRef]
16. Fan, J.; Yue, W.; Wu, L.; Zhang, F.; Cai, H.; Wang, X.; Lu, X.; Xiang, Y. Evaluation of SVM, ELM and four tree-based ensemble models for predicting daily reference evapotranspiration using limited meteorological data in different climates of China. *Agric. For. Meteorol.* **2018**, *263*, 225–241. [CrossRef]
17. Shiri, J.; Kişi, Ö.; Landeras, G.; López, J.J.; Nazemi, A.H.; Stuyt, L.C.P.M. Daily reference evapotranspiration modeling by using genetic programming approach in the Basque Country (Northern Spain). *J. Hydrol.* **2012**, *414–415*, 302–316. [CrossRef]
18. Kisi, O.; Guven, A. Evapotranspiration Modeling Using Linear Genetic Programming Technique. *J. Irrig. Drain. Eng.* **2010**, *136*, 715–723.
19. Shiri, J.; Nazemi, A.H.; Sadraddini, A.A.; Landeras, G.; Kisi, O.; Fakheri Fard, A.; Marti, P. Comparison of heuristic and empirical approaches for estimating reference evapotranspiration from limited inputs in Iran. *Comput. Electron. Agric.* **2014**, *108*, 230–241. [CrossRef]
20. Wen, X.; Si, J.; He, Z.; Wu, J.; Shao, H.; Yu, H. Support-Vector-Machine-Based Models for Modeling Daily Reference Evapotranspiration With Limited Climatic Data in Extreme Arid Regions. *Water Resour. Manag.* **2015**, *29*, 3195–3209. [CrossRef]
21. Kisi, O. Evapotranspiration modelling from climatic data using a neural computing technique. *Hydrol. Process.* **2007**, *21*, 1925–1934. [CrossRef]
22. Odhiambo, L.O.; Yoder, R.E.; Yoder, D.C.; Hines, J.W. Optimization of Fuzzy Evapotranspiration Model Through Neural Training with Input–Output Examples. *Trans. ASAE* **2013**, *44*, 1625–1633. [CrossRef]
23. Trajkovic, S.; Todorovic, B.; Stankovic, M. Closure to "Forecasting of Reference Evapotranspiration by Artificial Neural Networks" by Slavisa Trajkovic, Branimir Todorovic, and Miomir Stankovic. *J. Irrig. Drain. Eng.* **2005**, *131*, 391–392. [CrossRef]
24. Falamarzi, Y.; Palizdan, N.; Huang, Y.F.; Lee, T.S. Estimating evapotranspiration from temperature and wind speed data using artificial and wavelet neural networks (WNNs). *Agric. Water Manag.* **2014**, *140*, 26–36. [CrossRef]

25. Hassan, M.A.; Khalil, A.; Kaseb, S.; Kassem, M.A. Exploring the potential of tree-based ensemble methods in solar radiation modeling. *Appl. Energy* **2017**, *203*, 897–916. [CrossRef]
26. Feng, Y.; Cui, N.; Gong, D.; Zhang, Q.; Zhao, L. Evaluation of random forests and generalized regression neural networks for daily reference evapotranspiration modelling. *Agric. Water Manag.* **2017**, *193*, 163–173. [CrossRef]
27. Tabari, H.; Kisi, O.; Ezani, A.; Hosseinzadeh Talaee, P. SVM, ANFIS, regression and climate based models for reference evapotranspiration modeling using limited climatic data in a semi-arid highland environment. *J. Hydrol.* **2012**, *444–445*, 78–89.
28. Shirsath, P.B.; Singh, A.K. A Comparative Study of Daily Pan Evaporation Estimation Using ANN, Regression and Climate Based Models. *Water Resour. Manag.* **2010**, *24*, 1571–1581. [CrossRef]
29. Pour-Ali Baba, A.; Shiri, J.; Kisi, O.; Fard, A.F.; Kim, S.; Amini, R. Estimating daily reference evapotranspiration using available and estimated climatic data by adaptive neuro-fuzzy inference system (ANFIS) and artificial neural network (ANN). *Hydrol. Res.* **2013**, *44*, 131. [CrossRef]
30. Feng, Y.; Peng, Y.; Cui, N.; Gong, D.; Zhang, K. Modeling reference evapotranspiration using extreme learning machine and generalized regression neural network only with temperature data. *Comput. Electron. Agric.* **2017**, *136*, 71–78. [CrossRef]
31. Vogt, M. Support Vector Machines for Identification and Classification Problems in Control Engineering. *Autom* **2008**, *56*, 391. [CrossRef]
32. Singh, R.; Helmers, M.J. Improving crop growth simulation in the hydrologic model drain mod to simulate corn yields in subsurface drained landscapes. *Am. Soc. Agric. Biol. Eng. Annu. Int. Meet. 2008* **2008**, *2*, 856–873.
33. Wu, W.; Wang, X.; Xie, D.; Liu, H. Soil water content forecasting by support vector machine in purple hilly region. In *IFIP International Federation for Information Processing*; Springer: Boston, MA, USA, 2008; Volume 258, pp. 223–230. ISBN 9780387772509.
34. Kisi, O.; Cimen, M. Reply to the Discussion of "Evapotranspiration modelling using support vector machines" by R. J. Abrahart et al. *Hydrol. Sci. J.* **2010**, *55*, 1451–1452. [CrossRef]
35. Shiri, J.; Kişi, Ö. Comparison of genetic programming with neuro-fuzzy systems for predicting short-term water table depth fluctuations. *Comput. Geosci.* **2011**, *37*, 1692–1701. [CrossRef]
36. Sanford, W.E.; Selnick, D.L. Estimation of Evapotranspiration across the Conterminous United States Using a Regression with Climate and Land-Cover Data. *J. Am. Water Resour. Assoc.* **2013**, *49*, 217–230. [CrossRef]
37. Perugu, M.; Singam, A.J.; Kamasani, C.S.R. Multiple Linear Correlation Analysis of Daily Reference Evapotranspiration. *Water Resour. Manag.* **2013**, *27*, 1489–1500. [CrossRef]
38. Khoshravesh, M.; Sefidkouhi, M.A.G.; Valipour, M. Estimation of reference evapotranspiration using multivariate fractional polynomial, Bayesian regression, and robust regression models in three arid environments. *Appl. Water Sci.* **2015**, *7*, 1911–1922. [CrossRef]
39. Alipour, A.; Yarahmadi, J.; Mahdavi, M. Comparative Study of M5 Model Tree and Artificial Neural Network in Estimating Reference Evapotranspiration Using MODIS Products. *J. Climatol.* **2014**, *2014*, 1–11. [CrossRef]
40. Pérez, J.Á.M.; García-Galiano, S.G.; Martin-Gorriz, B.; Baille, A. Satellite-based method for estimating the spatial distribution of crop evapotranspiration: Sensitivity to the priestley-taylor coefficient. *Remote Sens.* **2017**, *9*, 611. [CrossRef]
41. Shiri, J. Improving the performance of the mass transfer-based reference evapotranspiration estimation approaches through a coupled wavelet-random forest methodology. *J. Hydrol.* **2018**, *561*, 737–750. [CrossRef]
42. NDAWN North Dakota Agricultural Weather Network. Available online: https://ndawn.ndsu.nodak.edu/ (accessed on 21 December 2020).
43. ASCE-EWRI. The ASCE standardized reference evapotranspiration equation: ASCE-EWRI Standardization of Reference Evapotranspiration Task Committe Report. *Am. Soc. Civ. Eng.* **2005**. [CrossRef]
44. Angeline, P.J. Genetic programming: On the programming of computers by means of natural selection. *Biosystems* **2003**, *33*, 69–73. [CrossRef]
45. Ferreira, C. Gene Expression Programming: A New Adaptive Algorithm for Solving Problems. *arXiv* **2001**, arXiv:cs/0102027.
46. Sharifi, S.S.; Rezaverdinejad, V.; Nourani, V. Estimation of daily global solar radiation using wavelet regression, ANN, GEP and empirical models: A comparative study of selected temperature-based approaches. *J. Atmos. Solar-Terrestrial Phys.* **2016**, *149*, 131–145. [CrossRef]
47. Shiri, J.; Sadraddini, A.A.; Nazemi, A.H.; Kisi, O.; Landeras, G.; Fakheri Fard, A.; Marti, P. Generalizability of Gene Expression Programming-based approaches for estimating daily reference evapotranspiration in coastal stations of Iran. *J. Hydrol.* **2014**, *508*, 1–11. [CrossRef]
48. Cortes, C.; Vapnik, V. Support-Vector Networks. *Mach. Learn.* **1995**, *20*, 273–297. [CrossRef]
49. Vapnik, V. The Support Vector Method of Function Estimation. In *Nonlinear Modeling*; Springer US: Boston, MA, USA, 2011; pp. 55–85.
50. Azimi, V.; Salmasi, F.; Entekhabi, N.; Tabari, H.; Niaghi, A. Optimization of Deficit Irrigation Using Non-Linear Programming (Case Study: Mianeh Region, Iran). *Ijagcs.com* **2013**, 252–260.
51. Géron, A. *Hands-On Machine Learning with Scikit-Learn & TensorFlow*; O'Reilly Media, Inc.: Sebastopol, CA, USA, 2017; ISBN 9781491962299.

Article

Sensitivity of the Evapotranspiration Deficit Index to Its Parameters and Different Temporal Scales

Frank Joseph Wambura

Department of Urban and Regional Planning, Ardhi University, P.O. Box 35176, Dar es Salaam, Tanzania;
frank.wambura@aru.ac.tz or wamburafj@gmail.com

Abstract: Sound estimates of drought characteristics are very important for planning intervention measures in drought-prone areas. Due to data scarcity, many studies are increasingly using less data-intensive approaches, such as the evapotranspiration deficit index (ETDI), in estimations of agricultural droughts. However, little is known about the sensitivity of this specific ETDI formula to its parameters, and to data at different temporal scales. In this study, a general ETDI formula, homologous to the specific ETDI formula, was introduced and used to test the sensitivity of the ETDI to its parameters and to data at different temporal scales. The tests used time series of remotely sensed evapotranspiration data in the Ruvu River basin in Tanzania. The parameter sensitivity tests revealed that ETDI is sensitive to its parameters, and different parameter combinations resulted in different drought characteristics. The temporal scale sensitivity test showed that drought characteristics, such as the number of drought events and the total drought durations, decreased as the temporal scale increased. Thus, an inappropriate temporal scale may lead to the misrepresentation of drought characteristics. To reduce uncertainty and increase the accuracy of ETDI-based agricultural drought characteristics, ETDI requires parameter calibration and the use of data with small temporal scales, respectively. These findings are useful for improving estimations of ETDI-based agricultural droughts.

Keywords: agricultural drought; drought characteristics; evapotranspiration deficit index; parameter sensitivity; temporal scale sensitivity; water stress anomaly

Citation: Wambura, F.J. Sensitivity of the Evapotranspiration Deficit Index to Its Parameters and Different Temporal Scales. *Hydrology* **2021**, *8*, 26. https://doi.org/10.3390/hydrology8010026

Received: 6 January 2021
Accepted: 26 January 2021
Published: 2 February 2021

Publisher's Note: MDPI stays neutral with regard to jurisdictional claims in published maps and institutional affiliations.

1. Introduction

Drought is an environmental disaster that brings severe social, economic, and environmental impacts around the world. Thus, drought is usually categorized into four main operation-based types, namely, meteorological drought, hydrological drought, agricultural drought, and socio-economic drought [1–5]. Since drought is often caused by a decrease of precipitation below the normal amount, agricultural productivity is usually the most affected due to its direct dependence on water resources, especially soil moisture. Drought begins when the soil moisture available to plants drops to a level that adversely affects the crop yield and, consequently, agricultural production [6,7]. The decline of agricultural production indirectly causes critical issues such as food insecurity, which may eventually lead to socio-economic consequences. For this reason, understanding agricultural drought is vital for planning mitigation and adaptation measures in areas susceptible to drought.

Several indices have been developed to estimate agricultural drought using various water balance parameters. Most of these indices use precipitation, temperature, actual evapotranspiration (ET), and potential evapotranspiration (PET) data, and crop characteristics, crop management practices, etc. [8–11]. One of the prominent drought indices is the evapotranspiration deficit index (ETDI) [12]. The ETDI uses ET and PET data for estimating short-term agricultural drought [12]. ETDI can be scaled between -2 and $+2$ to compare with the standardized precipitation index [13–17], or between -4 and $+4$ to compare with the Palmer drought severity index [18]. Details about other drought indices are found in the studies by Sivakumar, et al. [19] and Zargar, et al. [20].

ETDI has been widely used to estimate drought in many parts of the world. Narasimhan and Srinivasan [12] used ETDI for monitoring the agricultural drought of six watersheds located in major river basins across Texas, United States. Trambauer, et al. [17] used ETDI to analyze hydrological drought in the Limpopo River basin, southern Africa. Esfahanian, et al. [21] used ETDI and other drought indices to develop a comprehensive drought index in the Saginaw watershed in Michigan, United States. Bayissa, et al. [2] used ETDI in comparisons of drought indices in the Upper Blue Nile Basin, Ethiopia. Wambura and Dietrich [22] used ETDI to analyze spatio-temporal drought in the Kilombero catchment, Tanzania. In all these studies, ETDI was computed using the specific ETDI formula. Thus, the sensitivity of ETDI to its parameters and to data at different temporal scales is hardly known.

Therefore, the objective of this study was to investigate the sensitivity of ETDI (1) to its parameters, and (2) to data at different temporal scales. First, a general ETDI formula homologous to the specific ETDI formula was introduced. Then the general ETDI formula was used to test the sensitivity of ETDI to its different parameter combinations. Finally, the sensitivity of ETDI to remotely sensed ET and PET data at different temporal scales (i.e., 8-day, 16-day, and 1-month) was also tested under a constant parameter combination.

2. Materials and Methods

2.1. Case Study

The study area was the Ruvu River basin located between 6°18′ S–7°46′ S and 37°15′ E–38°58′ E in eastern Tanzania (Figure 1). Its headwaters originate on the eastern slopes of the Uluguru Mountains and descend northeast towards the coast in a swampy estuary at the Indian Ocean. The basin area is approximately 17,693 km², and its elevation ranges between 4 and 2636 m above sea level (Figure 1) [23].

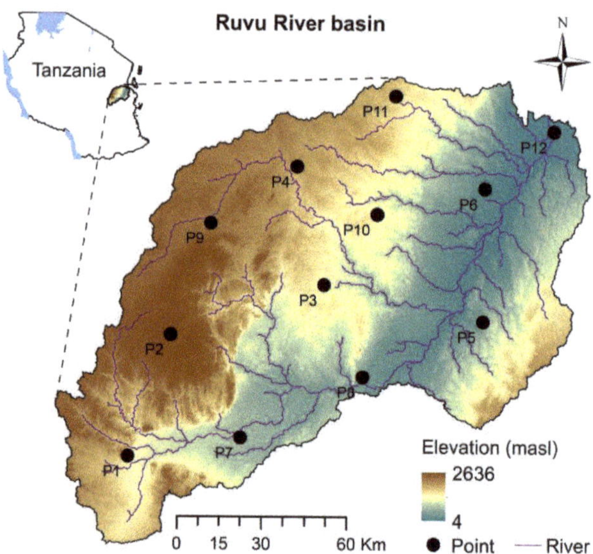

Figure 1. The Ruvu River basin showing elevation [23] and the points (P1 to P12) used to extract the time series of evapotranspiration and potential evapotranspiration from remote sensing images.

The average air temperature in the basin is between 18 °C in August and 32 °C in February, whereas the mean annual rainfall ranges from 800 mm to 2000 mm [24]. This region of coastal Tanzania is also known to have frequent and intense drought episodes [25]. Thus, the river basin has a very dynamic weather system. The Ruvu River basin was selected because of these dynamic weather systems, which are often very sensitive to even small changes in the western Indian Ocean sea surface temperature.

2.2. Main Datasets Used

Due to data scarcity in this region, ET and PET data used in this study were obtained from the Moderate Resolution Imaging Spectroradiometer (MODIS) imagery program [26]. The remotely sensed ET and PET products from the MODIS program were MOD16A2-v5 (from now on MODIS ET), and were available at a spatial resolution of 1 km and temporal resolution of 8 days and 1 month. The first dataset consisting of 690 images of 8-day MODIS ET covering the Ruvu River basin was downloaded from the MODIS repository (http://files.ntsg.umt.edu/data/NTSG_Products/, accessed on 15 October 2017). Another dataset consisting of 180 images of 1 month MODIS ET covering the river basin was also downloaded from the same repository (accessed on 10 July 2019). The two MODIS ET datasets spanned between the years 2000 and 2014.

Each of the twelve points (P1 to P12) spatially distributed in the Ruvu River basin (Figure 1) was used to extract two pairs of time series from the MODIS ET datasets. First, the twelve points extracted ET and PET time series from the 8-day MODIS ET dataset. Then the 8-day time series of ET and PET were aggregated to form a 16-day time series. The conversion to a 16-day timestep was necessary because MODIS ET products are only available at 8-day and 1-month timesteps. Finally, the twelve points were also used to extract monthly ET and PET time series from the monthly MODIS ET dataset. For illustration purposes, Figure 2a–c shows the 8-day, 16-day, and monthly ET and PET at point P1.

2.3. Evapotranspiration Deficit Index Approach

The ETDI approach involves three steps, first, the estimation of water stress (WS), then, the estimation of the water stress anomaly (WSA), and finally, the estimation of ETDI. The estimation of WS at a point uses Equation (1) [2,12].

$$WS_{i,j} = \frac{PET_{i,j} - ET_{i,j}}{PET_{i,j}} \tag{1}$$

where i represents a period (e.g., month i) in a given year, j. The years range between 2000 and 2014 with a timestep of one year. WS ranges from 0 (ET is the same as PET) to 1 (no ET).

The estimation of WSA at a point uses Equation (2) [12]. Equation (2) removes the seasonality inherent in the time series of WS.

$$WSA_{i,j} = \begin{cases} \frac{med\ WS_i - WS_{i,j}}{med\ WS_i - min\ WS_i} & if\ WS_{i,j} \leq med\ WS_i \\ \frac{med\ WS_i - WS_{i,j}}{max\ WS_i - med\ WS_i} & if\ WS_{i,j} > med\ WS_i \end{cases} \tag{2}$$

where $min\ WS$, $med\ WS$, and $max\ WS$ are long-term minimum, median, and maximum WS values at time i from all years in the time series. WSA ranges from -1 to $+1$ indicating extremely dry to extremely wet conditions, respectively.

The estimation of ETDI at a point uses a cumulating procedure similar to that of the soil moisture deficit index [12]. In analogy to the original formulation of the soil moisture deficit index, the change in ETDI is equal to the difference between two consecutive ETDIs (Equation (3)) [12].

$$\Delta ETDI_t = ETDI_t - ETDI_{t-1} \tag{3}$$

where $\Delta ETDI$ represents a change in ETDI. The subscripts t and $t - 1$ represent consecutive periods (e.g., month t and month $t - 1$, respectively) continuously ranging from the beginning to the end of the record.

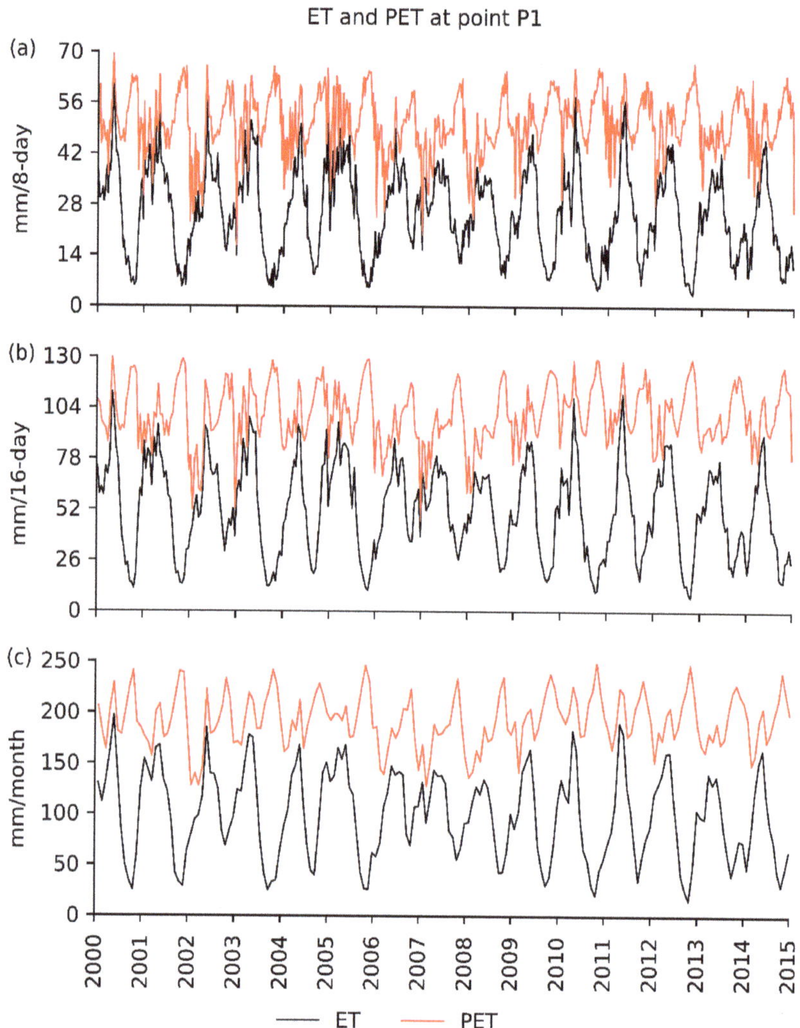

Figure 2. Typical Moderate Resolution Imaging Spectroradiometer (MODIS) evapotranspiration (ET) and potential evapotranspiration (PET) time series at (**a**) 8-day, (**b**) 16-day, and (**c**) 1-month temporal scales [26] at point P1 in the Ruvu River basin.

On the basis of the contribution of the previous drought severity, the change of the current ETDI depends on a weighted contribution of the previous ETDI, and the full contribution of the current WSA (Equation (4)) [12].

$$\Delta ETDI_t = c\, ETDI_{t-1} + \frac{WSA_t}{50} \tag{4}$$

where c controls the contribution of the previous ETDI. In Equation (4), Narasimhan and Srinivasan [12] scaled WSA between -100 and $+100$ (percentages). Thus, the value of 50 in this equation reduces WSA from ±100 to ±2, so that the ETDI of consecutive extreme drought events lies between -4 and $+4$.

By combining Equations (3) and (4), Narasimhan and Srinivasan [12] created the specific ETDI formula which states that the current ETDI is the sum of half of the previous ETDI and the current WSA (Equations (A1) and (A2) in Appendix A). The specific ETDI formula is a linear equation, and the coefficient of WSA was assumed to be one. Moreover, a residual term was also not addressed by the specific ETDI formula (Equations (A1) and (A2) in Appendix A).

From Equation (4) it is apparent that the importance of the previous ETDI or drought memory with respect to WSA cannot always be the same at different land cover types or climatic regions [27–30]. Therefore, this study modified Equation (4) to include a coefficient β to the WSA term, in order to facilitate the calibration of both drought memory and WSA at different places (Equation (5)). With regard to Equation (2), here WSA was considered to range between -1 and $+1$, so that the ETDI of consecutive extreme drought events does not exceed -2 and $+2$ [2,17]. In addition, a residual term, γ, was introduced because Equation (4) resembles a linear equation (Equation (5)).

$$\Delta ETDI_t = c\,ETDI_{t-1} + \beta\,WSA_t + \gamma \tag{5}$$

By combining Equations (3) and (5), this study obtained the ETDI formula that incorporates weighted contributions of both the previous ETDI and the current WSA (Equation (6)).

$$ETDI_t = (1+c)\,ETDI_{t-1} + \beta\,WSA_t + \gamma \tag{6}$$

The general ETDI formula (Equation (7)) was obtained by replacing $(1+c)$ in Equation (6) with an α. The general ETDI formula has three variables and three unknown coefficients, including the constant term. The general ETDI formula is homologous to the specific ETDI formula. Therefore, the specific ETDI formula (Equation (A2) in Appendix A) is a special case of the general ETDI formula (Equation (7)).

$$ETDI_t = \alpha\,ETDI_{t-1} + \beta\,WSA_t + \gamma \tag{7}$$

where α modulates the long-term memory of ETDI.

At an extremely dry boundary condition, consecutive dry periods have WSA_t equal to -1, and $ETDI_t$ and $ETDI_{t-1}$ equal to -2. Likewise, at an extremely wet boundary condition, consecutive wet periods have WSA_t equal to $+1$, and $ETDI_t$ and $ETDI_{t-1}$ equal to $+2$. By substituting these two boundary conditions in Equation (7), the γ-parameter becomes 0. Therefore, the general ETDI formula (Equation (7)) becomes Equation (8).

$$ETDI_t = \alpha\,ETDI_{t-1} + \beta\,WSA_t \tag{8}$$

Again, by substituting either of the two boundary conditions (i.e., extremely dry or extremely wet), Equation (8) turns into a parameter equation that governs the relationship between α and β parameters (Equation (9)). Figure 3 shows the straight line of Equation (9).

$$\beta = -2\alpha + 2 \tag{9}$$

Equation (9) indicates the presence of a large number of parameter combinations along the straight line. Table 1 shows the ranges of those parameter combinations at consecutive extremely dry and wet boundary conditions. Thus, for values of ETDI in Equation (8) to span between -2 and $+2$, values of α should range between 0 and 1, and values of β should range between 0 and 2 (Equation (9), Figure 3, Table 1). Therefore, the estimation of the ETDI time series at a point should use Equation (8), where parameters are governed by Equation (9), and at an initial condition, ETDI equals zero. In this study, an ETDI time series derived using (α, β)-parameters is hereafter referred to as an $ETDI_{(\alpha,\ \beta)}$ time series or curve.

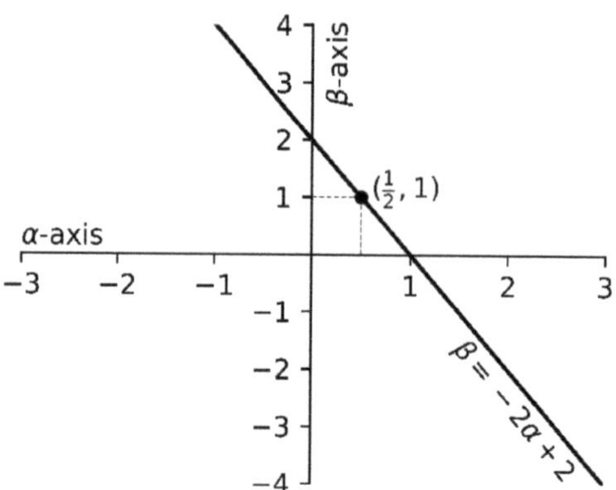

Figure 3. Straight line representing extremely dry and wet conditions using α and β parameters as coefficients of the previous evapotranspiration deficit index and the current water stress anomaly, respectively.

Table 1. Evapotranspiration deficit index (ETDI) of a point in time (t) at consecutive extremely dry and wet boundary conditions for three different ranges of (α,β)-parameter combinations. WSA represents water stress anomaly.

Extreme	ETDI_{t-1}	WSA_t	ETDI_t at $(\alpha < 0, \beta > 2)$	ETDI_t at $(0 \leq \alpha \leq 1, 2 \geq \beta \geq 0)$	ETDI_t at $(\alpha > 1, \beta < 0)$
Dry–Dry	−2	−1	−2	−2	−2
Wet–Wet	+2	+1	+2	+2	+2
Dry–Wet	−2	+1	>+2	−2 to +2	<−2
Wet–Dry	+2	−1	<−2	−2 to +2	>+2

2.3.1. Parameter Sensitivity Test

The parameter sensitivity test used 8-day ET and PET data at point P1 in the Ruvu River basin (Figures 1 and 2). Prior to the parameter sensitivity test, Equations (1) and (2) were used to estimate WS and WSA, respectively. Since the parameter sensitivity test intended to investigate how ETDI values from Equation (8) change relative to various α and β parameter combinations, a sample of eleven α-parameters from 0.0 to 1.0 at an interval of 0.1 was selected and used to obtain corresponding β values using Equation (9). Then by using Equation (8), WSA values at point P1 were used to generate an ETDI curve for each parameter combination.

ETDI curves for all parameter combinations at point P1 were finally used in a correlation analysis in order to investigate parameter combinations that have more or less similar ETDI curves. Estimations of drought events and total drought durations from ETDI curves at point P1 were also conducted in order to compare ETDI curves of different parameter combinations in terms of drought characteristics. A drought event was identified by the start and the end of a drought. The start of a drought event was the time when the ETDI was less or equal to −1.00 for at least eight consecutive, 8-day periods (approx. 2 months) [31]. The end of a drought event was the time when the ETDI returns to zero [32]. Total drought durations were the sum of all periods from all drought events in a time series.

2.3.2. Temporal Scale Sensitivity Test

The sensitivity of the ETDI to data at different temporal scales used 8-day, 16-day, 1-month ET, and PET data at all twelve points in the Ruvu River basin (Figure 1). Equations (1) and (2) were used to estimate WSs and WSAs at the points, respectively. Prior to the temporal scale sensitivity test, values of α and β equal to 0.5 and 1, respectively, were selected as the appropriate constant parameter combination, because they are in the middle of both parameter ranges. Moreover, this parameter combination is also commonly used in estimations of ETDI [2,12]. By using the constant parameter combination in Equation (8), the sensitivity of the ETDI to the three different temporal scales was investigated by estimating ETDI curves of 8-day, 16-day, and 1-month timesteps at each of the twelve points (P1 to P12) in the river basin.

Then drought events and total drought durations at each point were computed in order to compare ETDI curves at different temporal scales in terms of drought characteristics. Here, drought events for 8-day, 16-day, and 1-month timesteps had at least eight consecutive 8-day periods, four consecutive 16-day periods, and two consecutive months, respectively.

3. Results and Discussion

3.1. Parameter Sensitivity

In the parameter sensitivity test, eleven parameter combinations resulted in eleven $ETDI_{(\alpha,\beta)}$ time series. For illustration purposes, Figure 4 shows only five of the eleven $ETDI_{(\alpha,\beta)}$ time series. The $ETDI_{(0.0,2.0)}$ curve was the widest in both dry (negative ETDI) and wet (positive ETDI) axes. The peaks of $ETDI_{(0.1,1.8)}$ and $ETDI_{(0.5,1.0)}$ curves were smaller than those of the $ETDI_{(0.0,2.0)}$ curve. However, these three curves had similar patterns. On the other hand, the $ETDI_{(0.9,0.2)}$ curve was very different from other curves due to its shorter and smoother peaks (Figure 4). This is because the β-parameter of the curve was very small ($\beta = 0.2$), therefore, it diminished the influence of WSA_t (Equation (8)).

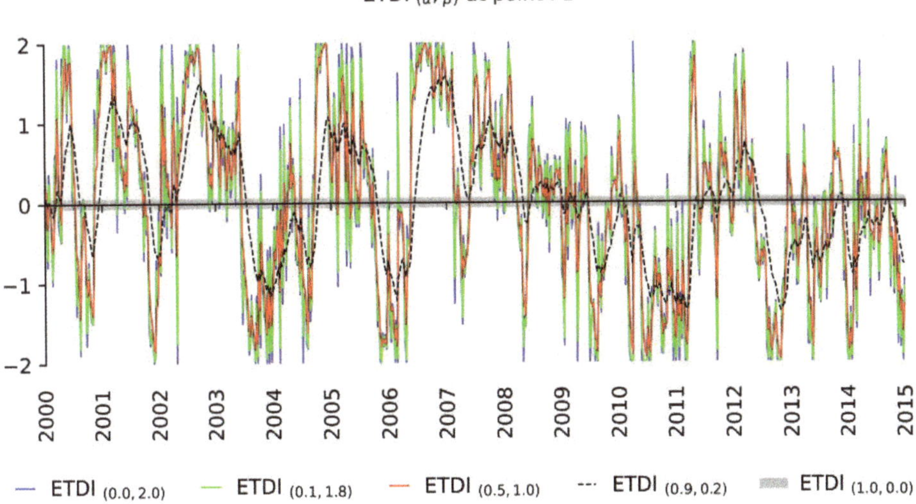

Figure 4. The 8-day evapotranspiration deficit index (ETDI) for five different (α,β)-parameter combinations at point P1 in the Ruvu River basin.

Unlike the curves of other parameter combinations, the $ETDI_{(1.0,0.0)}$ time series had zero values throughout the record length, thus coinciding with the time axis (Figure 4). Zero values occurred because WSA_t was nullified by the β-parameter, which was equal to 0.0, thus the $ETDI_{(1.0,0.0)}$ time series depended only on $ETDI_{t-1}$, which was initially assumed to be zero. In that case the $ETDI_{(1.0,0.0)}$ time series was excluded in both correlation analysis and drought characterization.

The $ETDI_{(0.0,2.0)}$ curve correlated highly with the $ETDI_{(0.1,1.8)}$ curve (Figure 5), they both show the highest number of drought events, and the lowest duration per event (4 months per event, Table 2). This means that the small α-parameters of these two curves reduced the influence of $ETDI_{t-1}$, while large β-parameters allowed the dominance of WSA_t (Equation (8)). This is inversely demonstrated by the $ETDI_{(0.9,0.2)}$ curve which had the lowest number of drought events and the highest duration per event (10 months per event, Table 2). Here, a large α-parameter allowed the dominance of $ETDI_{t-1}$, but the small β-parameter had already smoothened peaks of WSA_t (Equation (8)), thus causing wide, but few, peaks (cf. Figure 4).

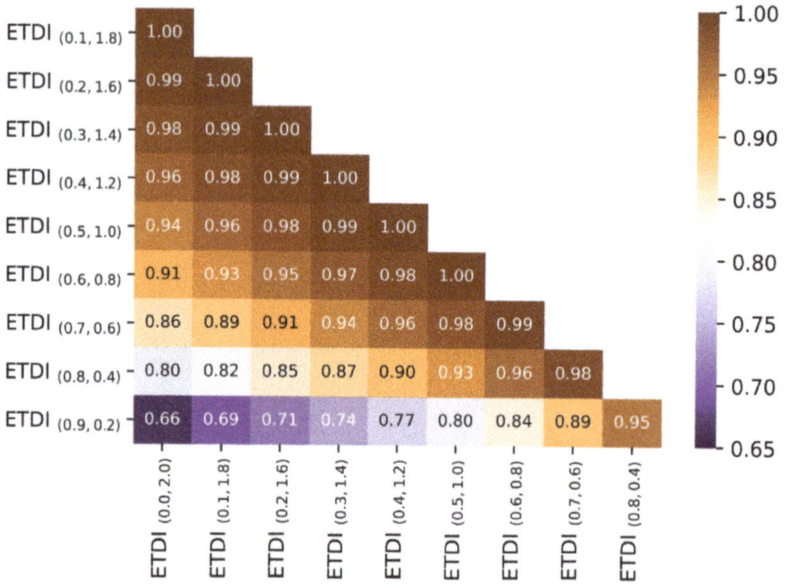

Figure 5. Pearson correlation coefficients between evapotranspiration deficit indices (ETDIs) at point P1 for various (α, β)-parameter combinations.

Table 2. Drought events, total drought durations, and duration per event at point P1 for various (α,β)-parameter combinations.

Parameter	Events	Total Duration (Month)	Duration per Event (Month)
$ETDI_{(0.0,2.0)}$	11	42	4
$ETDI_{(0.1,1.8)}$	10	38	4
$ETDI_{(0.2,1.6)}$	8	39	5
$ETDI_{(0.3,1.4)}$	8	41	5
$ETDI_{(0.4,1.2)}$	10	47	5
$ETDI_{(0.5,1.0)}$	10	51	5
$ETDI_{(0.6,0.8)}$	10	51	5
$ETDI_{(0.7,0.6)}$	9	50	6
$ETDI_{(0.8,0.4)}$	9	54	6
$ETDI_{(0.9,0.2)}$	4	40	10

In addition, the $ETDI_{(0.9,0.2)}$ and $ETDI_{(0.8,0.4)}$ curves highly correlated (Figure 5), but they had a substantially different number of drought events and total drought durations (Table 2). A high correlation between the two curves was due to the similarity of their patterns, which were not affected by minor parameter differences. However, the differences in drought characteristics were mainly due to the β-parameter, because it substantially reduced the WSA_t of the $ETDI_{(0.9,0.2)}$ curve more than that of the $ETDI_{(0.8,0.4)}$ curve. The $ETDI_{(0.4,1.2)}$, and $ETDI_{(0.6,0.8)}$ curves also highly correlated with the $ETDI_{(0.5,1.0)}$ curve, and had an equal number of drought events (Figure 5, Table 2); this is because the influence of their $ETDI_{t-1}$ and WSA_t were reduced to almost half by α-parameters, and were almost fully allowed by β-parameters (Equation (8)), respectively.

Generally, as the (α, β)-parameters deviated from the midpoint $(0.5, 1.0)$ towards the endpoint $(0.0, 2.0)$, $ETDI_t$ depended mostly on WSA_t, while $ETDI_{t-1}$ became substantially diminished (Equation (8)). However, when (α, β)-parameters equaled $(0.0, 2.0)$, the $ETDI_{(0.0,2.0)}$ curve did not substantially differ from the ETDI curve of the mid-point. That is why the correlation coefficient between the $ETDI_{(0.0,2.0)}$ curve and the ETDI curve of the mid-point was still very high (94%, Figure 5), and drought durations per event had minor differences (Table 2). As (α, β)-parameters approached $(0.9, 0.2)$, the $ETDI_{(0.9,\ 0.2)}$ curve deviated substantially from the ETDI curve of the mid-point. That is why their correlation coefficient was very small, (66%, Figure 5) and drought durations per event differed by 5 months (Table 2). This deviation was caused by diminishing WSA_t due to a declining β-parameter (Equation (8)). Thus the β-parameter is more influential than the α-parameter because it controls strong signals from WSA_t, whereas the latter modulates the long-term memory of $ETDI_{t-1}$.

Therefore, an arbitrary choice of parameter combination has drastic effects on drought characteristics. As a result, information about drought frequency, severity, and intensity can be misrepresented, leading to inappropriate intervention measures for mitigation or adaptation to drought. This uncertainty in the selection of an optimal parameter combination is enormous, because the range between the endpoints (see Figure 3) can be sub-divided into many parameter combinations depending on the required level of accuracy, i.e., decimal places. Despite its wide application, the mid-point is still not a universal parameter combination, because the contributions of $ETDI_{t-1}$ and WSA_t might vary from place to place. On the other hand, the endpoints, i.e., $(0.0, 2.0)$ and $(1.0, 0.0)$ are also not realistic because they neglect the contributions of $ETDI_{t-1}$ and WSA_t, respectively.

Like coefficients of the Palmer drought severity index, parameters of the ETDI might also be derived from local crop characteristics or land cover types in an area [19,33,34]. Apart from this, comparisons of ETDIs with other drought information could also be used to locally calibrate ETDI parameters [18]. This would involve testing of different parameter combinations to identify a pair that gives a satisfactory match between the time series of the ETDI and other drought indices, or between durations of the ETDI and historically severe drought events in an area. Locally calibrated ETDIs from different areas can be compared as long as they are scaled using the same range [35,36].

3.2. Temporal Scale Sensitivity

For illustration purposes, only ETDI curves of points P1 to P6 are graphically presented (Figures 6 and 7), the rest of the points are summarized in Table 3. The 8-day, 16-day, and 1-month temporal scales caused substantially different ETDI curves at the points in the Ruvu River basin.

Figures 6 and 7 show that 8-day ETDI curves were the widest in both dry (negative ETDI) and wet (positive ETDI) axes. Thus, 16-day ETDI curves were enclosed by 8-day ETDI curves throughout the time series. Similarly, monthly ETDI curves were also enclosed by both 8-day ETDI and 16-day ETDI curves. These ETDI curves showed that the effects of the aggregation of ET and PET from small to large temporal scales were propagated to ETDI values (cf. Figures 2, 6 and 7).

Table 3 shows that at all twelve points in the river basin, the number of drought events decreased as the size of the temporal scale increased. The difference in the number of drought events between consecutive temporal scales was mainly between 1 and 2, except at points P4 and P11, where the differences between 16-day and 1-month temporal scales were relatively large (about 5 drought events). The large differences in the number of drought events at these two points could be attributed to local effects, because they are both found in the northern part of the river basin (cf. Figure 1).

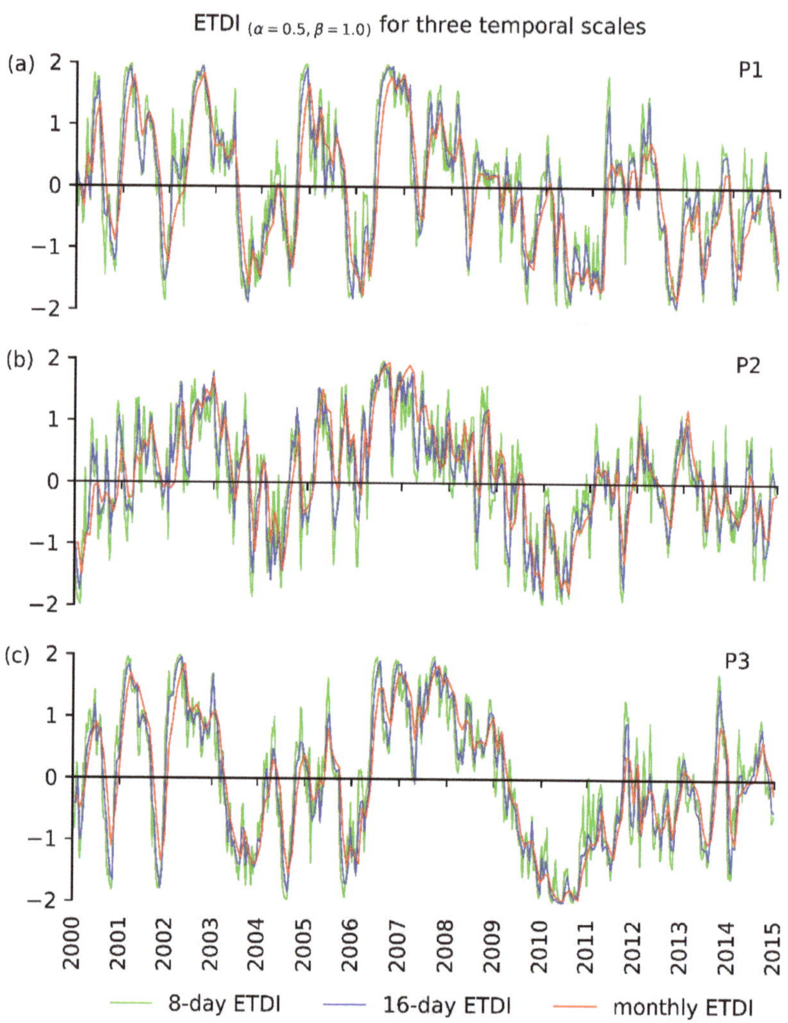

Figure 6. Evapotranspiration deficit index (ETDI) at 8-day, 16-day, and 1-month temporal scales at points P1 to P3 in the Ruvu River basin.

Although differences between the numbers of drought events at many points in the river basin were not large, their corresponding total drought durations differed by many months (Table 3). The total drought durations of 8-day ETDI curves were almost twice and thrice those of 16-day ETDI curves and monthly ETDI curves, respectively. Thus, total drought durations also decreased as the temporal scale increased. Moreover, almost all points in the river basin had durations per event ranging from 5 months for 8-day ETDI curves, to 2 months for monthly ETDI curves (Table 3).

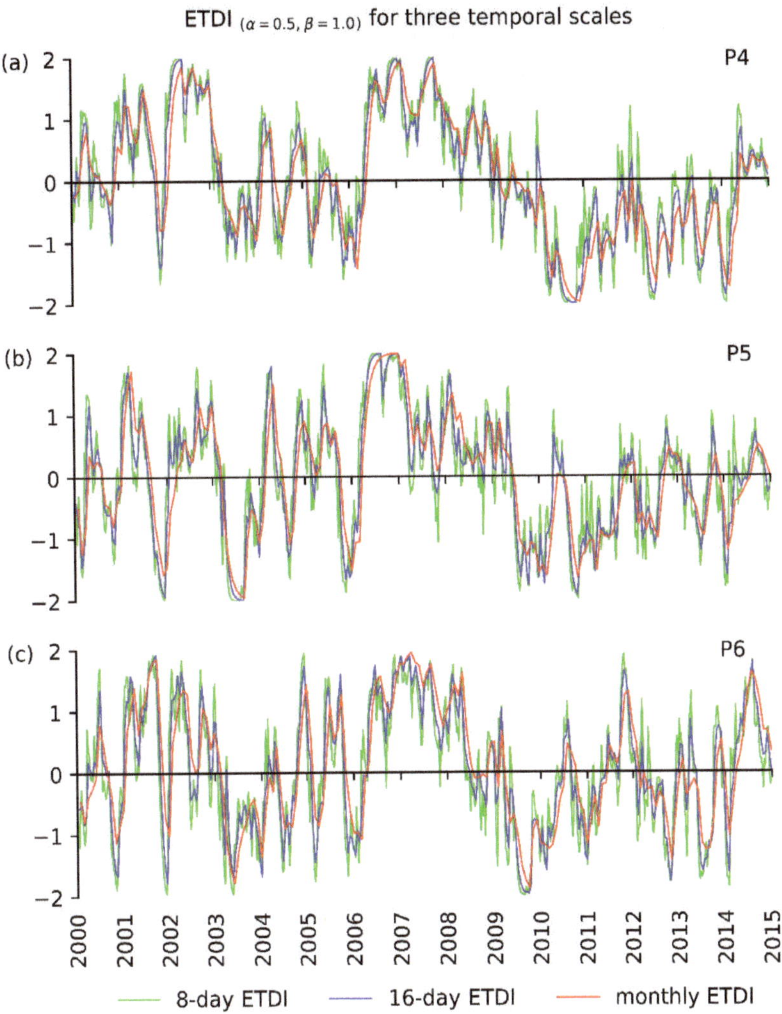

Figure 7. Evapotranspiration deficit index (ETDI) at 8-day, 16-day, and 1-month temporal scales at points P4 to P6 in the Ruvu River basin.

Since different numbers of drought events and drought durations usually lead to different drought severities and drought intensities [8,25,31], therefore, different temporal scales of ET and PET data lead to different ETDIs, and consequently different drought characteristics. By using the standardized precipitation index and effective drought index, Jain, et al. [37] also found that drought characteristics vary greatly with different temporal scales. Moreover, Ntale and Gan [27] argued that there are no objective rules for selecting an appropriate temporal scale. However, the largest number of drought events being captured by the 8-day temporal scale in this study (Table 3) indicates that small temporal scales can be useful because a region suffering from drought can return to a normal condition with only a few days' rainfall [27,38].

Table 3. Drought events, total drought durations, and duration per event at points P1 to P12 at 8-day, 16-day, and 1-month temporal scales in the Ruvu River basin.

Point	Time Series	Events	Total Duration (Months)	Duration per Event (Months)
P1	8-day	10	51	5
	16-day	9	29	3
	1-month	8	17	2
P2	8-day	7	33	5
	16-day	5	16	3
	1-month	5	9	2
P3	8-day	10	59	6
	16-day	9	31	3
	1-month	8	16	2
P4	8-day	7	51	7
	16-day	7	31	4
	1-month	2	15	7
P5	8-day	9	46	5
	16-day	10	29	3
	1-month	9	15	2
P6	8-day	11	54	5
	16-day	11	29	3
	1-month	8	12	2
P7	8-day	11	59	5
	16-day	9	30	3
	1-month	7	13	2
P8	8-day	9	59	7
	16-day	7	30	4
	1-month	6	15	3
P9	8-day	8	63	8
	16-day	8	30	4
	1-month	5	14	3
P10	8-day	9	54	6
	16-day	7	26	4
	1-month	8	14	2
P11	8-day	14	52	4
	16-day	12	30	3
	1-month	7	17	2
P12	8-day	15	54	4
	16-day	11	32	3
	1-month	9	17	2

4. Conclusions

This study used the general ETDI formula to test the sensitivity of the ETDI to its parameters and to data at different temporal scales. Data used were the MODIS ET time series for twelve points spatially distributed in the Ruvu River basin, Tanzania. The parameter sensitivity test revealed that ETDI is less sensitive when the (α, β)-parameters range from $(0.1, 1.8)$ to $(0.5, 1.0)$ inclusive, and more sensitive when they approach $(0.9, 0.2)$. Since the ETDI is sensitive to different parameter combinations, the selection of an optimal parameter combination might rely on information from specific locations. Moreover, an optimal parameter combination can also be obtained when ETDI is calibrated against other drought indices or durations of historically severe drought events. The temporal scale sensitivity test at the twelve points in the river basin showed that the number of drought events, the total drought durations, and durations per event decreases as the temporal scale increases. Therefore, small temporal scale ET data are highly recommended in order to increase the accuracy of ETDI-based drought characteristics.

Funding: This research received no external funding.

Acknowledgments: The author would like to thank the maintainer of the NTSG repository for freely providing MODIS ET datasets. Thanks to Festo Silungwe from the Sokoine University of Agriculture (Tanzania) for proofreading the manuscript.

Conflicts of Interest: The author declares no conflict of interest.

Appendix A

The specific evapotranspiration deficit index (ETDI) formula derived by Narasimhan and Srinivasan [12] is given by Equation (A1) below.

$$ETDI_t = (1 + c)\, ETDI_{t-1} + WSA_t \tag{A1}$$

where a subscript, t, represents a continuous timestep. c controls the amount of $ETDI_{t-1}$ that contributes to $ETDI_t$. WSA_t is scaled between -1 and $+1$.

At a boundary condition (i.e., extremely dry condition), WSA_t equals -1, and $ETDI_t$ and $ETDI_{t-1}$ equal -2. By substituting WSA and ETDI values in Equation (A1), c becomes equal to -0.5. Therefore, the final specific ETDI formula is shown in Equation (A2).

$$ETDI_t = 0.5\, ETDI_{t-1} + WSA_t \tag{A2}$$

The endpoints of the ETDI range, i.e., -2 and $+2$ indicate extremely dry and wet conditions, respectively.

References

1. Ziolkowska, J. Socio-Economic Implications of Drought in the Agricultural Sector and the State Economy. *Economies* **2016**, *4*, 19. [CrossRef]
2. Bayissa, Y.; Maskey, S.; Tadesse, T.; van Andel, S.; Moges, S.; van Griensven, A.; Solomatine, D. Comparison of the Performance of Six Drought Indices in Characterizing Historical Drought for the Upper Blue Nile Basin, Ethiopia. *Geosciences* **2018**, *8*, 81. [CrossRef]
3. Kim, T.-W.; Jehanzaib, M. Drought Risk Analysis, Forecasting and Assessment under Climate Change. *Water* **2020**, *12*, 1862. [CrossRef]
4. Liu, S.; Shi, H.; Niu, J.; Chen, J.; Kuang, X. Assessing Future Socioeconomic Drought Events under a Changing Climate over the Pearl River Basin in South China. *J. Hydrol. Reg. Stud.* **2020**, *30*, 100700. [CrossRef]
5. Peng, J.; Dadson, S.; Hirpa, F.; Dyer, E.; Lees, T.; Miralles, D.G.; Vicente-Serrano, S.M.; Funk, C. A Pan-African High-Resolution Drought Index Dataset. *Earth Syst. Sci. Data* **2020**, *12*, 753–769. [CrossRef]
6. Martínez-Fernández, J.; González-Zamora, A.; Sánchez, N.; Gumuzzio, A.; Herrero-Jiménez, C.M. Satellite Soil Moisture for Agricultural Drought Monitoring: Assessment of the SMOS Derived Soil Water Deficit Index. *Remote Sens. Environ.* **2016**, *177*, 277–286. [CrossRef]
7. Das, P.K.; Das, R.; Das, D.K.; Midya, S.K.; Bandyopadhyay, S.; Raj, U. Quantification of Agricultural Drought over Indian Region: A Multivariate Phenology-Based Approach. *Nat. Hazards* **2020**, *101*, 255–274. [CrossRef]
8. Hao, Z.; Singh, V.P. Drought Characterization from a Multivariate Perspective: A Review. *J. Hydrol.* **2015**, *527*, 668–678. [CrossRef]
9. Touma, D.; Ashfaq, M.; Nayak, M.A.; Kao, S.-C.; Diffenbaugh, N.S. A Multi-Model and Multi-Index Evaluation of Drought Characteristics in the 21st Century. *J. Hydrol.* **2015**, *526*, 196–207. [CrossRef]
10. Martínez-Fernández, J.; González-Zamora, A.; Sánchez, N.; Gumuzzio, A. A Soil Water Based Index as a Suitable Agricultural Drought Indicator. *J. Hydrol.* **2015**, *522*, 265–273. [CrossRef]
11. Yang, H.; Wang, H.; Fu, G.; Yan, H.; Zhao, P.; Ma, M. A Modified Soil Water Deficit Index (MSWDI) for Agricultural Drought Monitoring: Case Study of Songnen Plain, China. *Agric. Water Manag.* **2017**, *194*, 125–138. [CrossRef]
12. Narasimhan, B.; Srinivasan, R. Development and Evaluation of Soil Moisture Deficit Index (SMDI) and Evapotranspiration Deficit Index (ETDI) for Agricultural Drought Monitoring. *Agric. For. Meteorol.* **2005**, *133*, 69–88. [CrossRef]
13. Šebenik, U.; Brilly, M.; Šraj, M. Drought Analysis Using the Standardized Precipitation Index (SPI). *Acta Geogr. Slov.* **2017**, *57*. [CrossRef]
14. Shah, R.; Bharadiya, N.; Manekar, V. Drought Index Computation Using Standardized Precipitation Index (SPI) Method for Surat District, Gujarat. *Aquat. Procedia* **2015**, *4*, 1243–1249. [CrossRef]

15. Pramudya, Y.; Onishi, T. Assessment of the Standardized Precipitation Index (SPI) in Tegal City, Central Java, Indonesia. *IOP Conf. Ser. Earth Environ. Sci.* **2018**, *129*, 012019. [CrossRef]
16. Li, J.Z.; Wang, Y.X.; Li, S.F.; Hu, R. A Nonstationary Standardized Precipitation Index Incorporating Climate Indices as Covariates. *J. Geophys. Res. Atmos.* **2015**, *120*, 12082–12095. [CrossRef]
17. Trambauer, P.; Maskey, S.; Werner, M.; Pappenberger, F.; van Beek, L.P.H.; Uhlenbrook, S. Identification and Simulation of Space–Time Variability of Past Hydrological Drought Events in the Limpopo River Basin, Southern Africa. *Hydrol. Earth Syst. Sci.* **2014**, *18*, 2925–2942. [CrossRef]
18. Jacobi, J.; Perrone, D.; Duncan, L.L.; Hornberger, G. A Tool for Calculating the Palmer Drought Indices. *Water Resour. Res.* **2013**, *49*, 6086–6089. [CrossRef]
19. Sivakumar, M.V.K.; Motha, R.P.; Wilhite, D.A.; Wood, D.A. Agricultural Drought Indices. In Proceedings of the WMO/UNISDR Expert Group Meeting on Agricultural Drought Indices (AGM-11, WAOB-2011), Murcia, Spain, 2–4 June 2010; WMO/TD No. 1572. World Meteorological Organization: Geneva, Switzerland, 2011.
20. Zargar, A.; Sadiq, R.; Naser, B.; Khan, F.I. A Review of Drought Indices. *Environ. Rev.* **2011**, *19*, 333–349. [CrossRef]
21. Esfahanian, E.; Nejadhashemi, A.P.; Abouali, M.; Adhikari, U.; Zhang, Z.; Daneshvar, F.; Herman, M.R. Development and Evaluation of a Comprehensive Drought Index. *J. Environ. Manag.* **2017**, *185*, 31–43. [CrossRef]
22. Wambura, F.J.; Dietrich, O. Analysis of Agricultural Drought Using Remotely Sensed Evapotranspiration in a Data-Scarce Catchment. *Water* **2020**, *12*, 998. [CrossRef]
23. Jarvis, A.; Reuter, H.I.; Nelson, A.; Guevara, E. Hole-Filled SRTM for the Globe Version 4. 2008. Available online: http://srtm.csi.cgiar.org (accessed on 28 October 2017).
24. Kashaigili, J.J. Rapid Environmental Flow Assessment for the Ruvu River, A Consultancy Report submitted to iWASH. 2011. Available online: http://www.suaire.sua.ac.tz/bitstream/handle/123456789/1481/Kashaigili17.pdf (accessed on 28 October 2017).
25. Hassan, I.H.; Mdemu, M.V.; Shemdoe, R.S.; Stordal, F. Drought Pattern along the Coastal Forest Zone of Tanzania. *Atmos. Clim. Sci.* **2014**, *4*, 369–384. [CrossRef]
26. Mu, Q.; Zhao, M.; Running, S.W. Improvements to a MODIS Global Terrestrial Evapotranspiration Algorithm. *Remote Sens. Environ.* **2011**, *115*, 1781–1800. [CrossRef]
27. Ntale, H.K.; Gan, T.Y. Drought Indices and Their Application to East Africa. *Int. J. Climatol.* **2003**, *23*, 1335–1357. [CrossRef]
28. Sharafati, A.; Nabaei, S.; Shahid, S. Spatial Assessment of Meteorological Drought Features over Different Climate Regions in Iran. *Int. J. Climatol.* **2020**, *40*, 1864–1884. [CrossRef]
29. Pei, Z.; Fang, S.; Wang, L.; Yang, W. Comparative Analysis of Drought Indicated by the SPI and SPEI at Various Timescales in Inner Mongolia, China. *Water* **2020**, *12*, 1925. [CrossRef]
30. Raible, C.C.; Bärenbold, O.; Gómez-Navarro, J.J. Drought Indices Revisited—Improving and Testing of Drought Indices in a Simulation of the Last Two Millennia for Europe. *Tellus A Dyn. Meteorol. Oceanogr.* **2017**, *69*, 1287492. [CrossRef]
31. Brito, S.S.B.; Cunha, A.P.M.A.; Cunningham, C.C.; Alvalá, R.C.; Marengo, J.A.; Carvalho, M.A. Frequency, Duration and Severity of Drought in the Semiarid Northeast Brazil Region. *Int. J. Climatol.* **2018**, *38*, 517–529. [CrossRef]
32. Spinoni, J.; Naumann, G.; Vogt, J.; Barbosa, P. European Drought Climatologies and Trends Based on a Multi-Indicator Approach. *Glob. Planet. Chang.* **2015**, *127*, 50–57. [CrossRef]
33. Karl, T.R. The Sensitivity of the Palmer Drought Severity Index and Palmer's Z-Index to Their Calibration Coefficients Including Potential Evapotranspiration. *J. Clim. Appl. Meteorol.* **1986**, *25*, 77–86. [CrossRef]
34. Palmer, W.C. Meteorological Drought. In *Research Paper No. 45*; US Department of Commerce Weather Bureau: Washington, DC, USA, 1965; Volume 30.
35. Ogunrinde, A.T.; Oguntunde, P.G.; Olasehinde, D.A.; Fasinmirin, J.T.; Akinwumiju, A.S. Drought Spatiotemporal Characterization Using Self-Calibrating Palmer Drought Severity Index in the Northern Region of Nigeria. *Results Eng.* **2020**, *5*, 100088. [CrossRef]
36. Wells, N.; Goddard, S.; Hayes, M.J. A Self-Calibrating Palmer Drought Severity Index. *J. Clim.* **2004**, *17*, 2335–2351. [CrossRef]
37. Jain, V.K.; Pandey, R.P.; Jain, M.K.; Byun, H.-R. Comparison of Drought Indices for Appraisal of Drought Characteristics in the Ken River Basin. *Weather Clim. Extrem.* **2015**, *8*, 1–11. [CrossRef]
38. Byun, H.-R.; Wilhite, D.A. Objective Quantification of Drought Severity and Duration. *J. Clim.* **1999**, *12*, 2747–2756. [CrossRef]

MDPI
St. Alban-Anlage 66
4052 Basel
Switzerland
Tel. +41 61 683 77 34
Fax +41 61 302 89 18
www.mdpi.com

Hydrology Editorial Office
E-mail: hydrology@mdpi.com
www.mdpi.com/journal/hydrology